The
SCIENTIFIC AMERICAN
Book of
Astronomy

The
SCIENTIFIC AMERICAN
Book of
Astronomy

The Editors of
Scientific American

Introduction by Timothy Ferris

The Lyons Press

Designed by Compset, Inc.

Printed in the United States of America

10 9 8 7 6 5 4 3 2 1

The Library of Congress Cataloging-in-Publication Data is available on file.

CONTENTS

INTRODUCTION

For much of human history we have honored imagination as a divine gift, and rightly so, while regarding, with less justification, the material universe as gross and mundane. The findings of science have changed that perception, by revealing that the universe is richer, more varied, and more creative than we'd imagined—that, as Aldous Huxley put it, "Thought is crude, matter unimaginably subtle." Although deliberately provocative, Huxley's remark bears thinking upon. The world does look very different once we begin to think of the universe as in some sense subtler and more creative than we are. Consider some of the cosmic wonders explored in this book, and ask yourself what poet or artist ever imagined anything so strange.

Gamma-ray bursters flare up unexpectedly. Some bursters are estimated to generate, as Bradley E. Schaefer writes, "more power per unit of volume than any other object in the known universe." Then they disappear. Nobody yet knows what they are.

Cosmic rays are high-energy particles that come speeding through space and slam into the earth's atmosphere, splitting atoms in the upper atmosphere to produce showers of secondary particles that physicists on the ground have been observing for a half century. Nobody knows where they originate, either. Comets have hit the earth repeatedly over the eons, possibly triggering the mass-extinction events that wiped out most living creatures on the planet and cleared the way for the appearance of new species—including, as a result of the most recent such event, *Homo sapiens*. As they report in this book, Eugene and Carolyn Shoemaker and David Levy discovered a comet in 1993 that had been orbiting the giant planet

Jupiter for about 65 years. The world watched when, 16 months later, this fragmented iceberg smashed into Jupiter—an event that, to my knowledge, had never been imagined by anyone, not even the most admirably inventive authors of science-fiction tales.

Venus, which the science-fiction writers had envisioned to be a lush, tropical planet, turns out instead to be hotter and drier than the inside of a kitchen oven—a disappointment, perhaps, but also a cautionary tale. In terms of size, mass, and approximate distance from the sun, Venus is the earth's twin. Until we understand what went wrong on Venus, we are driving blind in our management of the earth's environment. As Janet Luhmann, James Pollack, and Lawrence Colin write, "A better understanding of the global environment on Venus could . . . provide a perspective on the nature of the environment on the earth and on the delicate balance of physical processes that keep our world habitable."

Farther out, things get stranger still. There are binary neutron stars—pairs of objects, each of which is, as Tsvi Piran reports, "a very dense star of 1.4 solar masses in a volume no larger than an asteroid"—that emit radio pulses so regular that they have been used to test Einstein's general theory of relativity. There are "cataclysmic" variable stars whose flare-ups may be produced from the interactions of binary stars orbiting each other so closely that, as John Cannizzo and Ronald Kaitchuck note, "the typical cataclysmic system could fit inside our sun."

The visible universe is comprised of galaxies, great aggregations of hundreds of billions of stars whose existence was not established until the twentieth century, although something like galaxies had been imagined earlier by the philosopher Immanuel Kant and a few other visionaries. By age-dating stars, as Sidney van den Bergh and James Hesser tell us, astronomers, like archaeologists, "look at small, disparate clues to determine how our galaxy and others like it were born about a billion years after the big bang and took on their current shapes." Studies of galactic evolution suggest that the shapes and composition of galaxies are to some extent determined by their environments—specifically, by their interactions with neighboring galaxies and gas clouds. In some instances the results of such interactions can be spectacular, as Sylvain Veilleux and his colleagues explain in their essay "Colossal Galactic Explosions." Quasars, the brilliant galactic nuclei that Michael Disney describes as "the most luminous objects in the universe," may flare up when giant black holes at the center of galaxies are fed by material falling in from the galaxy or from a gravitationally interacting neighbor. But as Vera Rubin reminds us, a proper understanding of the dynamics and evolution of galaxies will require that astronomers determine the nature of the "dark matter" nonluminous material that "must exceed the density of the luminous matter by a factor of roughly 70."

Meanwhile, it is possible to directly observe stages in cosmic history. Since light takes billions of years to reach us from distant galaxies, observers using powerful telescopes can see what the universe looked like billions of years ago. It's as if historians had a machine that enabled them to view ancient Athens from the air, or catch glimpses of the pyramids being built in Egypt. F. Duccio Macchetto and Mark Dickinson inform us that stars are forming today at a rate less than one tenth what was typical when the cosmos was one quarter to a half its present age, indicating that "the universe has apparently settled into a quiet maturity."

At the limits of space-time and human thought, physicists are investigating the thermonuclear chemistry of the big bang, contemplating the "singularity" from which the universe may have sprung—a mind-boggling state in which all places and times were one—and studying black holes, which provide small-scale models of singularity and raise thought-provoking paradoxes that may help revise our concepts of space, time, and information. It is even possible, as Andrei Linde tells us, to envision our observable universe as one among many space-time bubbles in a much vaster cosmos.

All such speculations must in the end report to, and be modified by, the observational data, and this book aptly concludes with a look at the technology of astronomy and space exploration. The scope of human vision greatly exceeds the reach of our travels, and always will, I suppose, but there's still no substitute for being there.

As reports come in from astronomers at their telescopes, physicists replicating bits of the big bang, and astronauts studying the sun and other stars from orbit, the question remains: To what extent will their findings remain exclusively within the province of science, and to what extent will they be woven into the fabric of our wider culture, continuing to redefine the relationship between imagination and reality? *Scientific American* has long played a leading role in this mission by popularizing science accurately and evocatively without talking down to its readers. If, as Emerson put it, "The very design of imagination is to domesticate us in another, a celestial nature," then the articles collected here are works of visionary poetry as well as fact, and redound not only to the mind but also to the human spirit.

—Timothy Ferris
Rocky Hill Observatory

I

Rays, Waves, and Particles

GAMMA-RAY BURSTERS

Intense flashes of high-energy radiation appear unpredictably
in the sky. Limited observational data have frustrated
attempts to determine their cause, but likely
mechanisms have been proposed

Bradley E. Schaefer

Approximately once per day a burst of very intense gamma radiation emanates from some completely unpredictable part of the sky. The duration of the burst is typically between 1 second and 10 seconds, although some bursts have been as short as .01 second and others have been as long as 80 seconds. During this time the burst brightens into visibility, varies randomly in intensity, and then fades back to invisibility. With few exceptions, no more than one burst has come from exactly the same direction and none has been positively identified with a previously known object.

At their peak these bursts are by far the brightest emissions of gamma radiation in the sky. In fact, if we make reasonable assumptions concerning the distance to the bursts' sources (which are called gamma-ray bursters), it seems that they generate more power per unit of volume than any other object in the known universe.

One estimate is that they generate as much energy in one second as the sun does in one week.

In spite of the tremendous amount of energy involved in each emission, however, little is known about gamma-ray bursters. The bursts are unpredictable, and they fall in a region of the spectrum that is extremely difficult to measure precisely. The paucity of observational data allows a great number of proposed models. To date nearly 40 models of bursters have been published, including such exotic ideas as exploding black holes, ultrarelativistic dust grains, and the fission of superheavy elements. It is embarrassing that most of these models are consistent with current observational data.

Even with the limited amount of available information, however, astronomers have recently begun to come to a consensus on what the general characteristics of a gamma-ray burster must be. In addition instruments are being built that may answer many of the outstanding questions about the properties of bursts. It may be possible in the near future to determine precisely what causes a gamma-ray burst, thus ending the speculation these mysterious phenomena have aroused since their discovery more than a decade ago.

Gamma-ray bursters were first observed by a set of Vela satellites operated by the U.S. Department of Defense. The satellites which were designed to monitor Soviet compliance with a treaty banning nuclear tests in space, carried detectors to record the sudden flash of gamma radiation that accompanies a nuclear explosion. After a number of flashes had been detected it was realized that they represented not atomic tests but an entirely new class of previously unknown astronomical objects. The discovery of gamma-ray bursters was soon confirmed by Thomas L. Cline and Upendra D. Desai of the National Aeronautics and Space Administration's Goddard Space Flight Center. Soon thereafter a number of other groups were able to find gamma-ray bursts in data from earlier experiments designed for other purposes.

In the late 1970s a second generation of experiments was begun. In order to pinpoint the direction of a gamma-ray burst as accurately as possible, investigators placed a network consisting of roughly a dozen detectors on various spacecraft that were to travel within the inner solar system.

In one of these experiments, called the Konus experiment ("Konus" is derived from the Russian for "cone" and refers to the spacecraft's configuration), four Soviet Venera space probes were equipped with extremely sensitive detectors. The experimenters, E. P. Mazets, S. V. Golenetskii, and their colleagues at the A. F. Joffe Institute of Physics and Technology in Leningrad, used these detectors to compile a catalog of the times, positions, spectra, and variations in brightness of hundreds of separate bursts.

Although these more recent observations have determined the position of some bursts, it is difficult to acquire more specific information about any single burst. Because they are so brief and because they occur in unpredictable locations, it is impossible to find them using a sensitive detector with a narrow field of view. Instead, to have a practical hope of detecting any bursts at all, an observer must use a relatively insensitive, wide-angle detector.

Further difficulties rise because the bursts consist almost entirely of gamma radiation, which is absorbed by the earth's atmosphere. Detectors must therefore be flown in space, an expensive and complicated process. In addition gamma-ray detectors are inherently less sensitive than detectors of radiation at lower energies.

In the face of these experimental difficulties, what kinds of data are there that could make it possible to distinguish among models? Fundamentally all gamma-ray observations consist solely of "light curves," records of the variation in brightness of a burst. Since each burst consists of radiation at several levels of energy, different light curves (one for each energy level) must be recorded for each burst. So far light curves have been collected for several hundred bursts, and curves for a single burst have often been recorded by more than one satellite. This small collection of light curves constitutes almost the entire body of our knowledge about gamma-ray bursters.

Meager as this may seem, there are ways of extracting useful information from light curves. For example, by comparing the intensities and times of curves recorded by detectors on different satellites, investigators can determine a burst's position in the sky with some accuracy. It is also possible, by comparing the light curve of different energy levels within the same burst, to determine its spectrum: the relative intensities of the different energies of radiation that make it up.

The shape of an object's spectrum can often reveal information about the nature of that object, such as its temperature and size. Many investigators have examined the observed spectra of the bursts in attempts to identify the emission process that causes them. Although several different radiation processes are known that would generate spectra like those observed for gamma-ray bursts, no one of them can satisfactorily explain the spectral shape of all the bursts.

Indeed, it would be surprising if any of these theoretical mechanisms was able to explain the burst spectra. Each mechanism is valid only under idealized conditions—a single temperature, magnetic field orientation and intensity, and instant of time—and the true situation probably bears a complicated relation to those depicted in the idealized models.

Even if spectral information is not particularly useful in determining the properties of the region that emits the radiation, it may still hold information about the volume of space surrounding the emitting region. For example, if that region were to absorb or emit radiation at one of the frequencies contained in the burst, then a line would appear in the burster's observed spectrum. In fact, the Soviet Konus experiment found just such lines, usually within the range of 40 to 70 KeV (thousand electron volts), in roughly 15 percent of the bursts it detected. Independent evidence for the existence of these lines was found by Geoffrey J. Hueter of the University of California at San Diego, in data recorded by the HEAO-1 satellite.

No detector other than those aboard the Konus and HEAO satellites has found such lines, and some investigators have speculated that the Konus and HEAO results may have derived from errors in calibration. Indeed, it is difficult to calibrate the efficiency of a detector in the 40-to-70-KeV range because of a complicated response due to iodine in the instrument. Detailed analysis of the Konus and HEAO-1 experiments indicates, however, that they are probably correctly calibrated. It seems safe to assume that the spectral lines are in fact contained within the bursts.

The most popular hypothesis concerning the origin of the lines is that they are due to "cyclotron emission," radiation produced at sharply defined frequencies when electrons perform rapid circular motions in an intense magnetic field.

Another explanation is that the burst radiation is due to two sources very close together, each producing a different spectrum. In this picture the angular separation of the two sources, as seen from the earth, would be so small that the radiation coming from them would appear to come from a single source. It is possible that two different spectra could, when added together, produce a spectrum having such lines.

Whereas spectral information can be used to determine the mechanism behind the bursts, information about the intensities of the observed bursts can be used to find whether the sources are distributed isotropically (evenly) throughout space. One method is to pilot a graph of the number of bursts that are more intense than any given level of brightness. When the points are plotted on a logarithmic scale, they should fall on a straight line with a particular slope if the bursts are distributed evenly. If many of the gamma-ray bursters lie at one particular distance from the earth, the graph will show an irregularity at the intensity corresponding to bursters observed at that particular distance.

At low intensities the graph does indeed curve away from the straight line, although there are too few data points to be sure of the statistical significance of this deviation. The easiest explanation for the observed shape of the curve is that nearby bursters are evenly distributed but distant bursters are clumped in space in some way, leaving large regions of space with few, if any, bursters. Such a situation could occur naturally if bursters are distributed in a disk-shaped region of the galaxy. Unfortunately many other types of distribution would also satisfy the data.

Furthermore, the curve is apparently inconsistent with our knowledge of the positions of observed bursts. The curve implies that distant bursters are not evenly distributed throughout space. Hence they should appear concentrated in some direction of the sky (such as that of the galactic plane or some nearby cluster of galaxies). As Mark Jennings of the University of California at Riverside observed in 1981, however, the faint bursts seem to be evenly distributed. To date five acceptable yet mutually exclusive explanations have been advanced as possible solutions to this problem.

So far the picture I have painted is a grim one. Much of the basic observational information is flawed by inadequacies inherent in the detection apparatus. Even if the data are accepted, they are invariably open to many different explanations. This raises the question of whether any nontrivial facts are known concerning the nature of gamma-ray bursters.

Fortunately the situation may not be as bad as I have made it appear. In recent years investigators have reached a consensus about certain properties of gamma-ray bursters. It is centered on three basic points: that a gamma-ray-burst system contains a neutron star, that this neutron star has an intense magnetic field, and that most of the observed bursters are situated within our own Milky Way galaxy. Although none of these points has been proved conclusively—and reasonable disagreement could be raised against each point—together they provide the simplest explanation for the available evidence.

The keystone of the recent consensus is the conviction that a neutron star is somehow involved in the burst of gamma rays. A neutron star is a very small and dense star formed during the later stages of stellar evolution. (The radius of a neutron star is roughly six miles, whereas its central density can be greater than 1,000 million tons per cubic inch.) The extremely large surface gravity and magnetic field of a neutron star certainly hold enough energy to generate a burst of gamma rays, and many mechanisms have been proposed that would convert energy from these sources into gamma radiation. Another reason theorists are attracted to models that include neutron stars is that neutron stars are known to exist and are relatively common entities in our galaxy.

These theoretical reasons show it is plausible that a gamma-ray burster might contain a neutron star; certain observational facts make it probable that it does. One such fact is the very short time within which bursts change their intensity. Some bursts have been as short as .01 second, whereas a burst that occurred on March 5, 1979, rose in intensity in .0002 second. Since a source cannot significantly change brightness in a time shorter than the time it takes light to travel across the source region, the size of the March 5 burster must be smaller than .0002 light-second, or about 40 miles. There are few astronomical objects that meet the size limitations or have enough available energy to power a burst. A neutron star satisfies both of these requirements.

Another observational argument for the involvement of neutron stars in gamma-ray bursts is based on a peculiarity found in approximately 7 percent of the burst spectra. These spectra have a line at roughly 420 KeV. There are few processes that could emit a prominent line at that energy; the most plausible hypothesis is that the lines are formed when electrons meet their antiparticles, positrons. Such a collision would result in the total annihilation of both particles and the conversion of their mass energy into two gamma rays. Each gamma ray would normally have an energy of 511 KeV; if this mutual annihilation took place near the surface of a neutron star, however, then before the gamma rays could reach the earth they would first have to emerge from the neutron star's gravitational well. In doing so they would lose energy. (This loss of energy is the so-called gravitational red shift predicted by the general theory of relativity.) In fact, the gamma rays would lose just enough energy to make up the difference between the 511 KeV at which they were emitted from the electron-positron collision and the 420 KeV at which some gamma-ray bursters show emission lines.

Further evidence for the idea that neutron stars are involved in gamma-ray bursters is provided by two unique characteristics of the March 5, 1979, burst. One of these characteristics is that the burst's position in the sky was very near that of a supernova remnant. If this association is taken at face value, the source of the burst is probably related to the remains of a supernova. Since neutron stars are often created by supernova explosions, it is reasonable to suppose a neutron star is at least partly responsible for the March 5 burst.

A second unique characteristic of this burst was that its brightness oscillated with a period of approximately eight seconds. Many periodic astronomical emissions can be explained by the rotation of a star: as the star rotates, a beam emitted by one area of the star's surface periodically strikes the earth. An eight-second rotation period is too fast for most types of star, but it is typical for neutron stars.

The second feature of the recent consensus on models for gamma-ray bursters is that an intense magnetic field is part of the phenomenon. Again, none of the evidence in support of this hypothesis is convincing in its own right; each item can be interpreted in a manner that does not require a magnetic field. In sum, however, they provide a reasonable basis for the inclusion of a strong magnetic field in the theory of gamma-ray bursters.

One of the strongest of these arguments relies on the observation of emission or absorption lines in the region of 40 to 70 KeV. If these lines are indeed due to the circular motion of electrons, the magnetic field forcing the electrons to follow circular paths must be on the order of 10^{12} gauss. (In comparison, the earth's magnetic field is roughly half a gauss.) If the field were much smaller, the electrons would emit cyclotron radiation at a lower energy.

A second observational argument is based on the eight-second periodicity seen in the burst of March 5, 1979. This modulation was probably caused by the rotation of a neutron star; some emitting region of the star rotated periodically into and out of the earth's field of view. Such a region could not lie on one of the poles of the star, where the axis of rotation meets the star's surface, or it would not rotate and hence would not pass through the earth's line of sight. There must therefore be some mechanism, not symmetric about the axis of rotation that determines the location of the emitting region on the star's surface. A likely candidate for such a "symmetry breaking" mechanism is a magnetic field whose axis is misaligned with the rotation axis of the neutron star.

The hypothesis that a magnetic field is present in a gamma-ray burster is supported by theoretical as well as observational arguments. One of the strongest of these is that any region that emits intense gamma radiation must be confined by some force; otherwise radiation pressure generated by the gamma rays would cause it to expand very rapidly. If the region expanded too far, its density would become so low that it would no longer emit gamma rays. Even the tremendous surface gravity of a neutron star is not adequate to confine an intense source of gamma rays, but a magnetic field of about 10^{12} gauss would probably be sufficient.

The third feature of the recent consensus is that most of the gamma-ray bursters are in our own galaxy. One of the strongest arguments for this hypothesis is based on the great amount of energy the bursters would have to produce if they were extragalactic. A burster in the Milky Way would have to generate roughly 10^{38} ergs of energy to achieve the brightness observed on the earth; an extragalactic burster would have to generate at least 10^{46} ergs to achieve the same level of brightness. Whereas it is relatively easy to devise models of processes that would generate 10^{38} ergs, there are many serious difficulties in producing a model that would generate as much as 10^{46} ergs.

An observational argument, based on the graph of the number of bursts brighter than certain intensities, also supports the hypothesis that bursters are within the Milky Way galaxy. If most of the observed bursters were outside the galaxy, they would be most likely to exist within other large accumulations of mass (for example, other galaxies or clusters of galaxies). The observed distribution of bursters in space would therefore be irregular. Large concentrations of them would fall at various distances from the earth. In that case the graph would show a series of irregularities, corresponding to the distances to various concentrations of bursters. The curve shows no such irregularity.

Although the current consensus is a good starting point, its theoretical usefulness is limited by the difficulty of obtaining good data from bursts of high-energy photons such as gamma rays. Astrophysicists have realized for a long time that many observational difficulties could be solved if there were some way to observe lower-energy photons coming from the bursters. To this end much effort has been spent in determining accurate positions for as many bursts as possible. These specific positions have then been examined with optical telescopes, to see if any optical phenomena are associated with the bursts.

With two exceptions the results of this search have been consistently negative. The first exception was the unusual burst of March 5, 1979, which has been tentatively identified with the remains of a supernova. The second exception consists of three optical flashes, which I found in 1981 while I was a graduate student at the Massachusetts Institute of Technology. In my search for optical flashes from gamma-ray bursters I examined approximately 30,000 photographs, part of the archival collection of 500,000 photographs owned by the Harvard College Observatory. I searched each of the archival photographs in Harvard's collection that recorded an area of the sky where a gamma-ray burst had later been observed. In three photographs (of three separate regions), taken in 1901, 1928, and 1944 respectively, I found starlike images that did not appear in other photographs taken of the same regions. These images, which were caused by optical flashes, are certainly related to the gamma-ray bursts that were observed later in the same locations.

Although it would be useful to have more data, much can be learned from these three flashes alone. For example, in all three cases the energy contained in the recent gamma-ray burst is almost precisely 1,000 times the energy contained in the optical burst. Taken together with various theoretical arguments, this ratio can be used to show that the neutron star causing the bursts must have a companion of some kind, perhaps a faint star or an accretion disk of colder matter.

Since the positions of the bursters can be determined more accurately from optical data than from gamma-ray data, the three optical positions have enabled us to make deep searches for quiescent bursters, using large optical telescopes. Although the results are not conclusive for these three positions, it is apparent that bursters must be extremely faint when they are not bursting.

Statistical analysis of the three optical flashes shows that gamma-ray bursters must emit such flashes at an average rate of once per year: I examined an average of three months' worth of archival photographs for each of 12 positions known to contain gamma-ray bursters—in effect, I examined a cumulative exposure of three years. In this three-year collection of photographs I found three flashes, which indicates that the average time between optical flashes of a gamma-ray burster is approximately one year. This result may preclude those models that do not allow the possibility of yearly optical flashes.

Perhaps the most important implication of these three flashes is that they have forcefully reminded astrophysicists that it is both practical and desirable to observe gamma-ray bursters at lower energies.

The next few years will see the initiation of several experiments that have great potential for acquiring new information on gamma-ray bursters. The most exciting of these experiments is the burst detector on the Gamma Ray Observatory satellite, scheduled for launching in 1988. This instrument will not be fundamentally different from earlier detectors, except that it will be many times more sensitive. The Gamma Ray Observatory instrument will detect bursts 10 to 100 times fainter than the Konus experiment could have, and it will position these bursts on the sky to within one degree. In addition a special modification of the detector will record spectra over a wide range of energies for each observed burst.

A second major experiment, now in the final stages of construction, is a search for optical flashes. It will consist of two parts. The first, constructed by Roland Vanderspek and George R. Ricker, Jr., of M.I.T., will monitor most of the visible sky for any sudden brightening. Within a second of the start of any flash its position will be relayed to the second part of the experiment, an optical telescope that can point to any region of the sky within one second. (This part of the experiment is being built by Bonnard J. Teegarden, Ravi Kaipa, Tycho T. von Rosenvinge, and Cline at the Goddard Space Flight Center.) The investigators expect to detect roughly two dozen bursts per year.

The results we expect from these and other experiments should finally reveal the underlying cause of gamma-ray bursts. In spite of the current consensus, the burst phenomenon remains one of the most mysterious of all classes of astrophysical events. I look forward to the day when data will be available that will enable us to understand it.

—February 1985

Cosmic Rays at the Energy Frontier

These particles carry more energy than any others in the universe. Their origin is unknown but may be relatively nearby

James W. Cronin, Thomas K. Gaisser, and Simon P. Swordy

Roughly once a second a subatomic particle enters the earth's atmosphere carrying as much energy as a well-thrown rock. Somewhere in the universe, that fact implies, there are forces that can impart to a single proton 100 million times the energy achievable by the most powerful earthbound accelerators. Where and how?

Those questions have occupied physicists since cosmic rays were first discovered in 1912 (although the entities in question are now known to be particles, the name "ray" persists). The interstellar medium contains atomic nuclei of every element in the periodic table, all moving under the influence of electrical and magnetic fields. Without the screening effect of the earth's atmosphere, cosmic rays would pose a significant health threat; indeed, people living in mountainous regions or making frequent airplane trips pick up a measurable extra radiation dose.

Perhaps the most remarkable feature of this radiation is that investigators have not yet found a natural end to the cosmic-ray spectrum. Most well-known sources of charged particles—such as the sun, with its solar wind—have a characteristic energy limit; they simply do not produce particles with energies above this limit. In contrast cosmic rays appear, albeit in decreasing numbers, at energies as high as astrophysicists can measure. The data run out at levels around 300 billion times the rest-mass energy of a proton because there is no detector large enough to sample the very low number of incoming particles predicted.

Nevertheless, evidence of ultrahigh-energy cosmic rays has been seen at intervals of several years as particles hitting the atmosphere create myriad secondary particles (which are easier to detect). On October 15, 1991, for example, a cosmic-ray observatory in the Utah desert registered a shower of secondary particles from a 50-joule (3×10^{20} electron volts) cosmic ray. Although the cosmic-ray flux decreases with higher energy, this decline levels off somewhat above about 10^{18} eV, suggesting that the mechanisms responsible for ultrahigh-energy cosmic rays are different from those for rays of more moderate energy.

In 1960 Bernard Peters of the Tata Institute in Bombay suggested that lower-energy cosmic rays are produced predominantly inside our own galaxy, whereas those of higher energy come from more distant sources. One reason to think so is that a cosmic-ray proton carrying more than 10^{19} eV, for example, would not be deflected significantly by any of the magnetic fields typically generated by a galaxy, so it would travel more or less straight. If such particles came from inside our galaxy, we might expect to see different numbers coming from various directions because the galaxy is not arranged symmetrically around us. Instead the distribution is essentially isotropic, as is that of the lower-energy rays, whose directions are scattered.

Such tenuous inferences reveal how little is known for certain about the origin of cosmic rays. Astrophysicists have plausible models for how they might be produced but have no definitive answers. This state of affairs may be the result of the almost unimaginable difference between conditions on the earth and in the regions where cosmic rays are born. The space between the stars contains only about one atom per cubic centimeter, a far lower density than the best artificial vacuums we can create. Furthermore, these volumes are filled with vast electrical and magnetic fields, intimately connected to a diffuse population of charged particles even less numerous than the neutral atoms.

■ SUPERNOVA PUMPS

This environment is far from the peaceful place one might expect: the low densities allow electrical and magnetic forces to operate over large distances and timescales in a manner that would be quickly damped out in material of terrestrial densities. Galactic space is therefore filled with an energetic and turbulent plasma of partially ionized gas in a state of violent activity. The motion is often hard to observe on human timescales because astronomical distances are so large; nevertheless, those same distances allow even moderate forces to achieve impressive results. A particle might zip through a terrestrial accelerator in a few microseconds, but it could spend years or even millennia in the accelerator's cosmic counterpart. (The timescales are further complicated by the strange, relativity-distorted framework that ultrahigh-energy cosmic rays inhabit. If we could observe such a particle for 10,000 years, that period would correspond to only a single second as far as the particle is concerned.)

Astronomers have long speculated that the bulk of galactic cosmic rays—those with energies below about 10^{16} eV—originate with supernovae. A compelling reason for this theory is that the power required to maintain the observed supply of cosmic-ray nuclei in our Milky Way galaxy is only slightly less than the average kinetic energy delivered to the galactic medium by the three supernova explosions that occur every century. There are few, if any, other sources of this amount of power in our galaxy.

When a massive star collapses, the outer parts of the star explode at speeds of up to 10,000 kilometers (6,000 miles) per second and more. A similar amount of energy is released when a white dwarf star undergoes complete disintegration in a thermonuclear detonation. In both types of supernovae the ejected matter expands at supersonic velocities, driving a strong shock into the surrounding medium. Such shocks are expected to accelerate nuclei from the material they pass through, turning them into cosmic rays. Because cosmic rays are charged, they follow complicated paths through interstellar magnetic fields. As a result, their directions as observed from the earth yield no information about the location of their original source.

By looking at the synchrotron radiation sometimes associated with supernova remnants, researchers have found more direct evidence that supernovae can act as accelerators. Synchrotron radiation is characteristic of high-energy electrons moving in an intense magnetic field of the kind that might act as a cosmic-ray accelerator, and the presence of synchrotron x-rays in some supernova remnants suggests particularly high energies. (In earthbound devices, synchrotron emission limits a particle's energy because the emission rate increases as a particle goes faster; at some point the radiation bleeds energy out of an accelerating particle as fast as it can be pumped in.) Recently the Japanese x-ray satellite Asca made images of the shell of Supernova 1006, which exploded nearly 1,000 years ago. Unlike the radiation from the interior of the remnant, the x-radiation from the shell has the

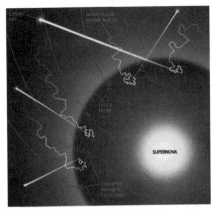

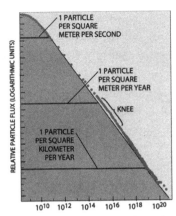

Cosmic-ray accelerator is believed to arise from a supernova explosion. Astrophysicists hypothesize that atomic nuclei crossing the supernova shock front will pick up energy from the turbulent magnetic field embedded in the shock. A particle may be deflected in such a way that it crosses the boundary of the shock hundreds or even thousands of times, picking up more energy on each passage until it escapes as a cosmic ray. Most of the particles travel on paths that result in relatively small accelerations, accounting for the general shape of the cosmic-ray energy spectrum (right), which falls off at higher energies. The "knee," or bend, in the curve suggests that most of the particles are accelerated by a mechanism incapable of imparting more than about 10^{15} electron volts. The relative excess of ultrahigh-energy particles indicates an additional source of acceleration whose nature is yet unknown. (George Kelvin, left. Jennifer C. Christiansen, right)

features characteristic of synchrotron radiation. Astrophysicists have deduced that electrons are being accelerated there at up to 10^{14} eV.

The EGRET detector on the Compton Gamma Ray Observatory has also been used to study point sources of gamma rays identified with supernova remnants. The observed intensities and spectra (up to a billion electron volts) are consistent with an origin from the decay of particles called neutral pions, which could be produced by cosmic rays from the exploding star's remnants colliding with nearby interstellar gas. Interestingly, however, searches made by the ground-based Whipple Observatory for gamma rays of much higher energies from some of the same remnants have not seen signals at the levels that would be expected if the supernovae were accelerating protons to 10^{14} eV or more.

A complementary method for testing the association of high-energy cosmic rays with supernovae involve the elemental composition of cosmic-ray nuclei. The size of the orbit of a charged particle in a magnetic field is proportional to its total momentum per unit charge, so heavier nuclei have greater total energy for a given orbit size. Any process that limits the particle acceleration on the basis of orbit size (such as an accelerating region of limited extent) will thus lead to an excess of heavier nuclei at high energies.

Eventually we would like to be able to go farther and look for elemental signatures of acceleration in specific types of supernovae. For example, the supernova of a white dwarf detonation would accelerate whatever nuclei populate the local interstellar medium. A supernova that followed the collapse of a massive star, in contrast, would accelerate the surrounding stellar wind, which is characteristic of the outer layers of the progenitor star at earlier stages of its evolution. In some cases, the wind could include an increased fraction of helium, carbon, or even heavier nuclei.

The identity of high-energy cosmic rays is all but lost when they interact with atoms in the earth's atmosphere and form a shower of secondary particles. Hence, to be absolutely sure of the nuclear composition, measurements must be made before the cosmic rays reach dense atmosphere. Unfortunately, to collect 100 cosmic rays of energies near 10^{15} eV, a one-square-meter detector would have to be in orbit for three years. Typical exposures at present are more like the equivalent of one square meter for three days.

Researchers are attacking this problem with some ingenious experiments. For example, the National Aeronautics and Space Administration has developed techniques to loft large payloads (about three metric tons) with high-altitude balloons for many days. These experiments cost a tiny fraction of what an equivalent satellite detector would. The most successful flights of this type have taken place in Antarctica, where the upper-atmosphere winds blow in an almost constant circle around the South Pole.

A payload launched at McMurdo Sound on the coast of Antarctica will travel at a nearly constant radius from the pole and return eventually to near the launch site. Some balloons have circled the continent for 10 days. One of us (Swordy) is collaborating with Dietrich Müller and Peter Meyer of the University of Chicago on a 10-square-meter detector that could measure heavy cosmic rays of up to 10^{15} eV on such a flight. There are efforts to extend the exposure times to roughly 100 days with similar flights nearer the equator.

■ ACROSS INTERGALACTIC SPACE

Studying even higher-energy cosmic rays—those produced by sources as yet unknown—requires large ground-based detectors, which overcome the problem of low flux by watching enormous areas for months or years. The information, however, must be extracted from cascades of secondary particles—electrons, muons, and gamma rays—initiated high in the atmosphere by an incoming cosmic-ray nucleus. Such indirect methods can only suggest general features of the composition of a cosmic ray on a statistical basis, rather than identifying the atomic number of each incoming nucleus.

At ground level, the millions of secondary particles unleashed by one cosmic ray are spread over a radius of hundreds of meters. Because it is impractical to blanket such a large area with detectors, the detectors typically sample these air showers at a few hundred or so discrete locations.

Technical improvements have enabled such devices to collect increasingly sophisticated data sets, thus refining the conclusions we can draw from each shower. For example, the CASA-MIA-DICE experiment in Utah, in which two of us (Cronin and Swordy) are involved, measures the distributions of electrons and muons at ground level. It also detects Cerenkov light (a type of optical shock wave produced by particles moving faster than the speed of light in their surrounding medium) generated by the shower particles at various levels in the atmosphere. These data enable us to reconstruct the shape of the shower more reliably and thus take a better guess at the energy and identity of the cosmic ray that initiated it.

The third one of us (Gaisser) is working with an array that measures showers reaching the surface at the South Pole. This experiment works in conjunction with AMANDA, which detects energetic muons produced in the same showers by observing Cerenkov radiation produced deep in the ice cap. The primary goal of AMANDA is to catch traces of neutrinos produced in cosmic accelerators, which may generate upward-streaming showers after passing through Earth.

Cosmic rays with energies above 10^{20} eV strike Earth's atmosphere at a rate of only about one per square kilometer a century. As a result, studying them requires an air-shower detector of truly gigantic proportions. In addition to the 1991 event in Utah, particles with energies above 10^{20} eV have been seen by groups elsewhere in the U.S., in Akeno, Japan, in Haverah Park, U.K., and in Yakutsk, Siberia.

Particles of such high energy pose a conundrum. On the one hand, they are likely to come from outside our galaxy because no known acceleration mechanism could produce them and because they approach from all directions even though a galactic magnetic field is insufficient to bend their path. On the other hand, their source cannot be more than about 30 million light-years away, because the particles would otherwise lose energy by interaction with the universal microwave background—radiation left over from the birth of the cosmos in the big bang. In the relativistic universe that the highest-energy cosmic rays inhabit, even a single radio-frequency photon packs enough punch to rob a particle of much of its energy.

If the sources of such high-energy particles were distributed uniformly throughout the cosmos, interaction with the microwave background would cause a sharp cutoff in the number of particles with energy above 5×10^{19} eV, but that is not the case. There are as yet too few events above this nominal threshold for us to know for certain what is going on, but even the few we have seen provide us with a unique opportunity for theorizing. Because these rays are essentially undeflected by the weak intergalactic magnetic fields, measuring the direction of travel of a large-enough sample should yield unambiguous clues to the locations of their sources.

It is interesting to speculate what the sources might be. Three recent hypotheses suggest the range of possibilities: galactic black-hole accretion disks, gamma-ray bursts, and topological defects in the fabric of the universe.

Astrophysicists have predicted that black holes of a billion solar masses or more, accreting matter in the nuclei of active galaxies, are needed to drive relativistic jets of matter far into intergalactic space at speeds approaching that of light; such jets have been mapped with radio telescopes. Peter L. Biermann of the Max Planck Institute for Radioastronomy in Bonn and his collaborators suggest that the hot spots seen in these radio lobes are shock fronts that accelerate cosmic rays to ultrahigh energy. There are some indications that the directions of the highest-energy cosmic rays to some extent follow the distribution of radio galaxies in the sky.

The speculation about gamma-ray bursts takes off from the theory that the bursts are created by relativistic explosions, perhaps resulting from the coalescence of neutron stars. Mario Vietri of the Astronomical Observatory of Rome and Eli Waxman of Princeton University independently noted a rough match between the energy available in such cataclysms and that needed to supply the observed flux of the highest-energy cosmic rays. They argue that the ultrahigh-speed shocks driven by these explosions act as cosmic accelerators.

■ RARE GIANTS

Perhaps most intriguing is the notion that ultrahigh-energy particles owe their existence to the decay of monopoles, strings, domain walls, and other topological defects that might have formed in the early universe. These hypothetical objects are believed to harbor remnants of an earlier, more symmetrical phase of the fundamental fields in nature, when gravity, electromagnetism, and the weak and strong nuclear forces were merged. They can be thought of, in a sense, as infinitesimal pockets preserving bits of the universe as it existed in the fractional instants after the big bang.

As these pockets collapse, and the symmetry of the forces within them breaks, the energy stored in them is released in the form of supermassive particles that immediately decay into jets of particles with energies up to 100,000 times greater than those of the known ultrahigh-energy cosmic rays. In this scenario the ultrahigh-energy cosmic rays we observe are the comparatively sluggish products of cosmological particle cascades.

Whatever the source of these cosmic rays, the challenge is to collect enough of them to search for detailed correlations with extragalactic objects. The AGASA array in Japan currently has an effective area of 100 square kilometers and can capture only a few ultrahigh-energy events a year. The new Fly's Eye High Resolution experiment in Utah can see out over a much larger area, but only on clear, moonless nights.

For the past few years Cronin and Alan A. Watson of the University of Leeds have spearheaded an initiative to gather an even larger sample of ultrahigh-energy cosmic rays. This development is named the Auger Project, after Pierre Auger, the French scientist who first investigated the phenomenon of correlated showers of particles from cosmic rays.

The plan is to provide a detection area of 6,000 square kilometers with a 100 percent duty cycle that is capable of measuring hundreds of high-energy events a year. A detector field would consist of many stations on a 1.5-kilometer grid; a single event might trigger dozens of stations. To cover the entire sky, two such detectors are planned, one each for the Northern and Southern Hemispheres.

An Auger Project design workshop held at the Fermi National Accelerator Laboratory in 1995 has shown how modern off-the-shelf technology such as solar cells, cellular telephones, and Global Positioning System receivers can make such a system far easier to construct. A detector the size of Rhode Island could be built for about $50 million.

Plans exist to cover even larger areas. Detectors in space could view millions of square kilometers of the atmosphere from above, looking for flashes of light signaling the passage of ultrahigh-energy particles. This idea, which goes by the name of OWL (Orbiting Wide-angle Light collectors) in the U.S. and by Airwatch in Europe, was first suggested by John Linsley of the University of New Mexico. To succeed, the project requires developing new technology for large, sensitive, finely segmented optics in space to provide the resolution needed. This development is under way by the U.S. National Aeronautics and Space Administration and in Italy.

As researchers confront the problem of building and operating such gigantic detector networks, the fundamental question remains: Can nature produce even more energetic particles than those we have seen? Could there be still higher-energy cosmic rays, or are we already beginning to detect the highest-energy particles our universe can create?

—*Scientific American Presents,* Spring 1998

Gamma-Ray Bursts

New observations illuminate the most powerful explosions in the universe

Gerald J. Fishman and Dieter H. Hartmann

About three times a day our sky flashes with a powerful pulse of gamma rays, invisible to human eyes but not to astronomers' instruments. The sources of this intense radiation are likely to be emitting, within the span of seconds or minutes, more energy than the sun will in its entire 10 billion years of life. Where these bursts originate, and how they come to have such incredible energies, is a mystery that scientists have been attacking for three decades. The phenomenon has resisted study—the flashes come from random directions in space and vanish without trace—until very recently.

On February 28, 1997, we were lucky. One such burst hit the Italian-Dutch Beppo-SAX satellite for about 80 seconds. Its gamma-ray monitor established the position of the burst—prosaically labeled GRB 970228—to within a few arc minutes in the Orion constellation, about halfway between the stars Alpha Tauri and Gamma Orionis. Within eight hours, operators in Rome had turned the spacecraft around to look in the same region with an x-ray telescope. They found a source of x-rays (radiation of somewhat lower frequency than gamma rays) that was fading fast, and they fixed its location to within an arc minute.

Never before has a burst been pinpointed so accurately and so quickly, allowing powerful optical telescopes, which have narrow fields of view of a few arc minutes, to look for it. Astronomers on the Canary Islands, part of an international team led by Jan van Paradijs of the University of Amsterdam and the University of Alabama in Huntsville, learned of the finding by electronic mail. They had some time available on the 4.2-meter William Herschel Telescope, which they had been using to study the locations of other bursts. They took a picture of the area 21 hours after GRB 970228. Eight days later they looked again and found that a spot of light seen in the earlier photograph had disappeared.

On March 13 the New Technology Telescope in La Silla, Chile, took a long, close look at those coordinates and discerned a diffuse, uneven glow. The Hubble Space Telescope later resolved it to be a bright point surrounded by a somewhat elongated background object. In a few days the Hubble reexamined the position and still found the point—now very faint—as well as the fuzzy glow, unaltered. Many of us believe the latter to be a galaxy, but its true identity remains unknown.

Even better, on the night of May 8, Beppo-SAX operators located a 15-second burst, designated GRB 970508. Soon after, Howard E. Bond of the Space Telescope Science Institute in Baltimore photographed the region with the .9-meter optical telescope on Kitt Peak in Arizona; the next night a point of light in the field had actually brightened. Other telescopes confirm that after becoming most brilliant on May 10, the source began to fade. This is the first time that a burst has been observed reaching its optical peak—which, astonishingly, lagged its gamma-ray peak by a few days.

Also for the first time, on May 13 Dale Frail, using the Very Large Array of radio telescopes in New Mexico, detected radio emissions from the burst remnant. Even more exciting, the primarily blue spectrum of this burst, taken on May 11 with the Keck II telescope on Hawaii, showed a few dark lines, apparently caused by iron and magnesium in an intervening cloud. Astronomers at the California Institute of Technology find that the displacement of these absorption lines indicates a distance of more than seven billion light-years. If this interpretation holds up, it will establish that bursts occur at cosmological distances. In that case gamma-ray bursts must represent the most powerful explosions in the universe.

■ CONFOUNDING EXPECTATIONS

For those of us studying gamma-ray bursts, this discovery salves two recent wounds. In November 1996 the Pegasus XL launch vehicle failed to release the

High Energy Transient Explorer (HETE) spacecraft equipped with very accurate instruments for locating gamma-ray bursts. And in December the Russian Mars '96 spacecraft, with several gamma-ray detectors, fell into the Pacific Ocean after a rocket malfunction. These payloads were part of a set designed to launch an attack on the origins of gamma-ray bursts. Of the newer satellites equipped with gamma-ray instruments, only Beppo-SAX—whose principal scientists include Luigi Piro, Enrico Costa, and John Heise—made it into space, on April 20, 1996.

Gamma-ray bursts were first discovered by accident, in the late 1960s, by the Vela series of spacecraft of the U.S. Department of Defense. These satellites were designed to ferret out the U.S.S.R.'s clandestine nuclear detonations in outer space—perhaps hidden behind the moon. Instead they came across spasms of radiation that did not originate from near the earth. In 1973 scientists concluded that a new astronomical phenomenon had been discovered.

These initial observations resulted in a flurry of speculation about the origins of gamma-ray bursts—involving black holes, supernovae, or the dense, dark star remnants called neutron stars. There were, and still are, some critical unknowns. No one knew whether the bursts were coming from a mere 100 light-years away or a few billion. As a result the energy of the original events could only be guessed at.

By the mid-1980s the consensus was that the bursts originated on nearby neutron stars in our galaxy. In particular theorists were intrigued by dark lines in the spectra (component wavelengths spread out, as light is by a prism) of some bursts, which suggested the presence of intense magnetic fields. The gamma rays, they postulated, are emitted by electrons accelerated to relativistic speeds when magnetic-field lines from a neutron star reconnect. A similar phenomenon on the sun—but at far lower energies—leads to flares.

In April 1991 the space shuttle Atlantis launched the Compton Gamma Ray Observatory, a satellite that carried the Burst And Transient Source Experiment (BATSE). Within a year BATSE had confounded all expectations. The distribution of gamma-ray bursts did not trace out the Milky Way, nor were the bursts associated with nearby galaxies or clusters of galaxies. Instead they were distributed isotropically, with any direction in the sky having roughly the same number. Theorists soon refined the galactic model: the bursts were now said to come from neutron stars in an extended spherical halo surrounding the galaxy.

One problem with this scenario is that the earth lies in the suburbs of the Milky Way, about 25,000 light-years from the core. For us to find ourselves near the center of a galactic halo, the latter must be truly enormous, almost 800,000 light-years in outer radius. If so, the halo of the neighboring Andromeda galaxy should be as extended and should start to appear in the distribution of gamma-ray bursts. But it does not.

This uniformity, combined with the data from GRB 970508, has convinced most astrophysicists that the bursts come from cosmological distances, on the order of 3 billion to 10 billion light-years away. At such a distance, though, the bursts should show the effects of the expansion of the universe. Galaxies that are very distant are moving away from the earth at great speeds; we know this because the light they emit shifts to lower, or redder, frequencies. Likewise, gamma-ray bursts should also show a "redshift," as well as an increase in duration.

Unfortunately BATSE does not see, in the spectrum of gamma rays, bright or dark lines characterizing specific elements whose displacements would betray a shift to the red. (Nor does it detect the dark lines found by earlier satellites.) In April astronomers using the Keck II telescope in Hawaii obtained an optical spectrum of the afterglow of GRB 970228—smooth and red, with no telltale lines. Still, Jay Norris of the National Aeronautics and Space Administration Goddard Space Flight Center and Robert Mallozzi of the University of Alabama in Huntsville have statistically analyzed the observed bursts and report that the weakest, and therefore the most distant, show both a time dilation and a red-shift. And the dark lines in the spectrum of GRB 970508 are substantially shifted to the red.

■ A COSMIC CATASTROPHE

One feature that makes it difficult to explain the bursts is their great variety. A burst may last from about 30 milliseconds to almost 1,000 seconds—and in one case, 1.6 hours. Some bursts show spasms of intense radiation, with no detectable emission in between, whereas others are smooth. Also complicated are the spectra—essentially, the colors of the radiation, invisible though they are. The bulk of a burst's energy is in radiation of between 100,000 and 1 million electron volts, implying an exceedingly hot source. (The photons of optical light, the primary radiation from the sun, have energies of a few electron volts.) Some bursts evolve smoothly to lower frequencies such as x-rays as time passes. Although this x-ray tail has less energy, it contains many photons.

If originating at cosmological distances, the bursts must have energies of perhaps 10^{52} ergs. (About 1,000 ergs can lift a gram by one centimeter.) This energy must be emitted within seconds or less from a tiny region of space, a few tens of kilometers across. It would seem we are dealing with a fireball.

The first challenge is to conceive of circumstances that would create a sufficiently energetic fireball. Most theorists favor a scenario in which a binary neutron-star system collapses. Such a pair gives off gravitational energy in the form of radiation. Consequently, the stars spiral in toward each other and may ultimately merge to form a black hole. Theoretical models estimate that one such event occurs every 10,000 to 1 million years in a galaxy. There are about 10 billion galaxies in the volume of space that BATSE observes; that yields up to 1,000 bursts a year in the sky, a number that fits the observations.

Variations on this scenario involve a neutron star, an ordinary star, or a white dwarf colliding with a black hole. The details of such mergers are a focus of intense study. Nevertheless, theorists agree that before two neutron stars, say, collapse into a black hole, their death throes release as much as 10^{53} ergs. This energy emerges in the form of neutrinos and antineutrinos, which must somehow be converted into gamma rays. That requires a chain of events: neutrinos collide with antineutrinos to yield electrons and positrons, which then annihilate one another to yield photons. Unfortunately, this process is very inefficient, and recent simulations by Max Ruffert and Hans-Thomas Janka of the Max Planck Institute in Munich, as well as by other groups, suggest it may not yield enough photons.

Worse, if too many heavy particles such as protons are in the fireball, they reduce the energy of the gamma rays. Such proton pollution is to be expected, because the collision of two neutron stars must yield a potpourri of particles. But then all the energy ends up in the kinetic energy of the protons, leaving none for radiation. As a way out of this dilemma, Peter Mészáros of Pennsylvania State University and Martin J. Rees of the University of Cambridge have suggested that when the expanding fireball—essentially hot protons—hits surrounding gases, it produces a shock wave. Electrons accelerated by the intense electromagnetic fields in this wave then emit gamma rays.

A variation of this scenario involves internal shocks, which occur when different parts of the fireball hit one another at relativistic speeds, also generating gamma rays. Both the shock models imply that gamma-ray bursts should be followed by long afterglows of x-rays and visible light. In particular Mario Vietri of the Astronomical Observatory of Rome has predicted detectable x-ray afterglows lasting for a month—and also noted that such afterglows do not occur in halo models. GRB 970228 provides the strongest evidence yet for such a tail.

There are other ways of generating the required gamma rays. Nir Shaviv and Arnon Dar of the Israel Institute of Technology in Haifa start with a fireball of unknown origin that is rich in heavy metals. Hot ions of iron or nickel could then interact with radiation from nearby stars to give off gamma rays. Simulations show that the time profiles of the resulting bursts are quite close to observations, but a fireball consisting entirely of heavy metals seems unrealistic.

Another popular mechanism invokes immensely powerful magnetic engines, similar to the dynamos that churn in the cores of galaxies. Theorists envision that instead of a fireball, a merger of two stars—of whatever kind—could yield a black hole surrounded by a thick, rotating disk of debris. Such a disk would be very short-lived, but the magnetic fields inside it would be astounding, some 10^{15} times those on the earth. Much as an ordinary dynamo does, the fields would extract rotational energy from the system, channeling it into two jets bursting out along the rotation axis.

The cores of these jets—the regions closest to the axis—would be free of proton pollution. Relativistic electrons inside them can then generate an intense, focused pulse of gamma rays. Although quite a few of the details remain to be worked out, many such scenarios ensure that mergers are the leading contenders for explaining bursts.

Still, gamma-ray bursts have been the subject of more than 3,000 papers—about one publication per recorded burst. Their transience has made them difficult to observe with a variety of instruments, and the resulting paucity of data has allowed for a proliferation of theories.

If one of the satellites detects a lensed burst, astronomers would have further confirmation that bursts occur at cosmological distances. Such an event might occur if an intervening galaxy or other massive object serves as a gravitational lens to bend the rays from a burst toward the earth. When optical light from a distant star is focused in this manner, it appears as multiple images of the original star, arranged in arcs around the lens. Gamma rays cannot be pinpointed with such accuracy; instead they are currently detected by instruments that have poor directional resolution.

Moreover, bursts are not steady sources like stars. A lensed gamma-ray burst would therefore show up as two bursts coming from roughly the same direction, having identical spectra and time profiles but different intensities and arrival times. The time difference would come from the rays' traversing curved paths of different lengths through the lens.

To further nail down the origins of the underlying explosion, we need data on other kinds of radiation that might accompany a burst. Even better would be to identify the source. Until the observation of GRB 970228 such "counterparts" had proved exceedingly elusive. To find others, we must locate the bursts very precisely.

Since the early 1970s Kevin Hurley of the University of California at Berkeley and Thomas Cline of the NASA Goddard Space Flight Center have worked to establish "interplanetary networks" of burst instruments. They try to put a gamma-ray detector on any spacecraft available or to send aloft dedicated devices. The motive

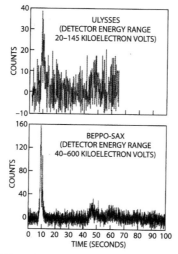

Time profile of GRB 970228 taken by the Ulysses spacecraft (top) and by Beppo-SAX (bottom) shows a brief, brilliant flash of gamma rays. (Courtesy of Kevin Hurley, University of California at Berkeley, Space Sciences Lab)

is to derive a location to within arc minutes, by comparing the times at which a burst arrives at spacecraft separated by large distances.

From year to year, the network varies greatly in efficacy, depending on the number of participating instruments and their separation. At present there are five components: BATSE, Beppo-SAX, and the military satellite DMSP, all near the earth; Ulysses, far above the plane of the solar system; and the spacecraft Wind, orbiting the sun. The data from Beppo-SAX, Ulysses, and Wind were used to triangulate GRB 970228 (BATSE was in the earth's shadow at the time). The process, unfortunately, is slow—eight hours at best.

■ WATCHING AND WAITING

Time is of the essence if we are to direct diverse detectors at a burst while it is glowing. Scott Barthelmy of the Universities Space Research Association at the NASA Goddard Space Flight Center has developed a system called GCN (Gamma-ray burst Coordinate Network) to transmit within seconds BATSE data on burst locations to ground-based telescopes.

BATSE consists of eight gamma-ray detectors pointing in different directions from eight corners of the Compton satellite; comparing the intensity of a burst at these detectors provides its location to roughly a few degrees but within several seconds. Often GCN can locate the burst even while it is in progress. The location is transmitted over the Internet to several dozen sites worldwide. In five more seconds, robotically controlled telescopes at Lawrence Livermore National Laboratory, among others, slew to the location for a look.

Unfortunately only the fast-moving smaller telescopes, which would miss a faint image, can contribute to the effort. The Livermore devices, for instance, could not have seen the afterglow of GRB 970228. Telescopes that are 100 times more sensitive are required. These midsize telescopes would also need to be robotically controlled so they can slew very fast, and they must be capable of searching reasonably large regions. If they do find a transient afterglow, they will determine its location rather well, allowing much larger telescopes such as Hubble and Keck to look for a counterpart.

The long-lasting afterglow following GRB 970228 gives new hope for this strategy. The HETE mission, directed by George Ricker of the Massachusetts Institute of Technology, is to be rebuilt and launched in about two years. It will survey the sky with x-ray detectors that can localize bursts to within several arc minutes. Ground-based optical telescopes will receive these locations instantly and start searching for transients.

Of course we do not know what fraction of bursts exhibit a detectable afterglow. Moreover, even a field as small as arc minutes contains too many faint objects to make a search for counterparts easy.

To further constrain the models, we will need to look at radiation of both higher and lower frequency than that currently observed. The Compton satellite has seen a handful of bursts that emit radiation of up to 10 billion electron volts. Better data in this regime from the Gamma-ray Large Area Space Telescope (GLAST), a satellite being developed by an international team of scientists, will greatly aid theorists. Photons of even higher energy—of about a trillion electron volts—might be captured by special ground-based gamma-ray telescopes. At the other end of the spectrum soft x-rays, which have energies of up to roughly one kilo electron volt (KeV), can help test models of bursts and obtain better fixes on position. In the range of .1 to 10 KeV, there is a good chance of discovering absorption or emission lines that would tell volumes about the underlying fireball.

When the Hubble telescope was pointed to the location of GRB 970508, it picked up the fading light from the optical transient. Much to our surprise, however, it saw no galaxy in the immediate vicinity—not even a hint of one. This absence emphasizes a potential problem noted by Bradley E. Schaefer of Yale University: bursts do not occur in the kind of bright galaxies within which one would expect an abundance of stars. So whereas astrophysicists now have strong evidence of the cosmological distances of bursts, we are still confounded as to their host environments and physical origins.

Just in time for New Year's 1998, nature provided a third afterglow from a gamma-ray burst. Again, Beppo-SAX discovered the initial event, following which Jules P. Halpern of Columbia University and John R. Thorstensen of Dartmouth College used the 2.4-meter telescope on Kitt Peak to find an optical transient. The glow dimmed in a manner similar to that of the previous two transients. As this article goes to press, we wait for Hubble to discern if this burst, GRB 971214, has a bright underlying galaxy or not.

—Scientific American Presents, Spring 1998

II

Comets, Asteroids, and Meteorites

COMET SHOEMAKER-LEVY 9 MEETS JUPITER

*Images of a comet that broke apart and plummeted into
Jupiter continue to dazzle astronomers a year afterward*

David H. Levy, Eugene M. Shoemaker,
and Carolyn S. Shoemaker

We were working underneath the dome of the small Schmidt telescope at Palomar Observatory in California, in a cramped room cluttered with papers, books, and a laptop computer. It was May 22, 1993. Carolyn sat hunched over her stereomicroscope, an instrument she has used to examine photographs for asteroids and comets for more than a decade, since she joined her husband, Gene, in his survey of these small wanderers in the sky.

Gene has spent a significant part of his career examining such objects. His studies in the 1950s demonstrated how the large pockmark in the desert east of Flagstaff,

Arizona, formed after a small asteroid struck the earth. He later investigated craters on our own moon and on the moons of the outer planets, as well as the remnants of ancient collisions in the Australian outback. More recently Gene, along with Carolyn, has been engaged in a systematic search for asteroids capable of striking the earth.

Peering at his computer that day, David checked his E-mail to see whether any newly detected comets or asteroids needed to be added to the observing schedule. Writer and lecturer by day, amateur astronomer by night, David has accumulated 21 comet-hunting trophies, eight of them for sightings he made using a 16-inch-diameter telescope in his backyard. Since joining forces in 1989, we three together have found 13 comets. Despite this level of combined experience, the revelation of May 1993 took us—and the rest of the scientific community—by complete surprise.

David's E-mail conveyed astonishing news from the International Astronomical Union's Central Bureau for Astronomical Telegrams—a kind of wire service for astronomers. A comet we had discovered two months earlier would strike Jupiter in July 1994. After a professional lifetime examining impact craters and the bodies that make them, Gene might actually get to see a collision.

■ THE IMPACT OF IMPACTS

As anyone who has gazed at the moon through even the smallest telescope knows, the lunar surface is studded with impact craters. The moon itself probably formed from the remains of a collision. During our planet's youth, a body the size of Mars may have struck the earth, melting it and sending into orbit a stream of debris that eventually congealed to form the moon. Tectonically static and without air or water, the moon can retain its crater-scarred face indefinitely. Erosion and the deposition of sediments constantly smooth the earth's surface, which consequently shows few craters, although our world has been hit far more often than the moon. For example, comets showered the earth during its formative period between 3.9 and 4.6 billion years ago, bringing carbon, hydrogen, nitrogen, and oxygen—critical elements that allowed life to evolve.

Such collisions have also taken life away. Sixty-five million years ago an object somewhat larger than Halley's comet slammed into what is now the coast of Mexico's Yucatán peninsula. The impact gouged a crater 170 kilometers across and launched debris worldwide. As the multitude of tiny ballistic missiles fell back toward the earth, meteors filled the sky, and the atmosphere became red hot. Fires erupted over the earth's surface, but the global inferno was soon followed by persistent darkness, as dust lifted into the atmosphere blocked the sun's rays. Months of planetwide cooling then gave way to centuries of greenhouse warming from the carbon dioxide released during the impact from the target rocks. Many species became extinct.

That ancient catastrophe demonstrates that projectiles from space can indeed affect this planet significantly. Our research program at Palomar was one of several designed to assess the rate at which interplanetary intruders of this kind collide with the planets and satellites. We did not, however, expect to witness such a colossal impact in the near future.

■ A LUCKY DISCOVERY

Our discovery of the comet began quietly enough. Little could we imagine then that we were about to make some of the most important observations of our lives. It was a dark and soon-to-be stormy night—March 23, 1993—that found us at our usual tasks, around the smallest of Palomar Observatory's four regularly used telescopes, an instrument with a 26-inch mirror and an 18-inch correcting lens designed to survey broad areas of the sky. We were accompanied on this observing run by Philippe Bendjoya, an astronomer visiting from the University of Nice.

Clouds slowly filled the sky, and although the haze did not completely hide the heavens, we knew it would obscure fainter stars, asteroids, and possible comets on our films. So we stopped our normal observing routine. Instead we decided just to expend some film we knew had been partially exposed to light. (The poor weather conditions seemed to merit using this problematic film.) One of our standard fields of view contained the planet Jupiter and was in the clear. We took three exposures—one with Jupiter and two of nearby parts of the sky—before clouds closed the gap. Later that night a short break in the overcast allowed us to take a second view of the Jupiter field.

Two day's later Carolyn began scanning the images taken on that cloudy night. Using her stereomicroscope, she was looking for the three-dimensional effect caused by the slight shift in position of an asteroid or comet relative to the background stars. Suddenly she sat up straight in her chair and announced, "I don't know what this is, but it looks like a squashed comet." Carolyn was not exaggerating. The object really did look like a comet that someone had stepped on. A typical comet has a nucleus several kilometers across composed of ices, rocky material, and organic compounds. When it nears the sun, the ices turn directly from solid to gas and release dust to form a light-scattering halo called a coma. The pressure of solar radiation then blows this material into an elongate tail. But instead of a single coma and tail, our new comet had a bar-shaped agglomeration of comae, with a composite tail stretching to the north. The strangest observation was that on either end of the bar was a pencil-thin line of light.

Our weird discovery needed confirmation with a better telescope. We contacted our colleague James V. Scotti of the University of Arizona, who was observing that night from the Spacewatch telescope atop Kitt Peak in Arizona. Jim agreed to take high-resolution television images of the comet. He was stunned. "There are at least five discrete comet nuclei side by side," Scotti explained to us over the telephone as he described the view, "but comet material exists between them. I suspect that there are more nuclei that I'll see when the sky clears" *[see color plate 1]*.

We immediately reported this bizarre comet to Brian G. Marsden, director of the Central Bureau for Astronomical Telegrams at the Harvard-Smithsonian Center for Astrophysics, and Scotti followed with his observations. The next day Marsden's office announced the discovery. The description of the object was so unusual that astronomers around the world began to examine it at once. Jane Luu of Stanford University and David Jewitt of the University of Hawaii obtained a magnificent image using the 88-inch reflector at Jewitt's institution. They later resolved 21 separate nuclei strung out, they wrote, "like pearls on a string" *[see color plate 1]*.

In accordance with a tradition that has gone on since the time of the French comet hunter Charles Messier more than two centuries ago, this comet was named after its discoverers. Because it was the ninth in a series we had found that traveled around the sun in short-period orbits, it took the formal title "Periodic Comet Shoemaker-Levy 9." We call it S-L 9, for short.

■ CLOSE ENCOUNTERS

By the middle of April 1993 Marsden, Syuichi Nakano in Japan, and Donald K. Yeomans of the Jet Propulsion Laboratory in Pasadena, California, had determined that the comet we had uncovered was actually in an orbit about Jupiter. They also ascertained that the comet had passed very close to the planet about eight months before we located it. Such proximity would explain why there were multiple fragments.

On July 7, 1992, S-L 9 had approached within about 20,000 kilometers of Jupiter's cloud tops. As it made a hairpin turn around the giant planet, it came apart because the pieces nearest Jupiter were deflected more sharply than those farther away. The difference in orbital paths resulted from the decrease in the strength of Jupiter's gravitational attraction between the near and far sides of the comet. The stress on S-L 9 was extremely weak, but it nonetheless broke up the comet easily. This behavior suggests that the original body was merely a pile of fragments held together by their weak gravitational attraction for one another.

Although astronomers had earlier established that comets have orbited Jupiter for brief periods in the past, S-L 9 is the first comet that anyone has seen in orbit about a planet. Jupiter indeed had not 1 but 21 tiny new moons. Yet these recently acquired satellites were not to last long. After further calculations, Marsden announced that the fractured comet would crash down on Jupiter in July of 1994.

Astronomers and planetary scientist immediately wondered what the impacts would entail. Would they see immense fireworks during the collisions, or would the event be a cosmic fizzle? H. Jay Melosh of the University of Arizona, for instance, suggested that the comets would penetrate so deeply into Jupiter's atmosphere before exploding that the planet would essentially swallow them with scarcely a trace. In contrast Thomas J. Ahrens and Toshiko Takata of the California Institute of Technology, Kevin Zahnle of the National Aeronautics and Space Administration Ames Research Center, and Mordecai-Mark Mac Low of the University of Chicago all proposed that each nucleus would dig a "tunnel of fire" in Jupiter's atmosphere, explode, and send a spectacular fireball back into space through the newly excavated cavity. David A. Crawford and Mark B. Boslough of Sandia National Laboratories believed a tremendous plume of hot gas would erupt chiefly from the upper part of the tunnel.

But even if most of those forecasts were right, would the astronomical community get to see any of this display? The answer hinged on just where on Jupiter the nuclei would hit. Early calculations were not encouraging: the comets were predicted to strike well over on Jupiter's nightside, where they would be hidden from the earth's view by the body of the planet. Jupiter would have to rotate eastward for at least an hour before any remains could be visible from the earth. Nature was to put on the biggest impact extravaganza in history, and it seemed that our seat was to be behind a post.

We accepted this assessment through the summer and fall of 1993; Jupiter and the sun were too close to each other in the sky for any further observations of S-L 9 to take place. But in early December Scotti obtained new positions for the comet fragments as Jupiter rose just before dawn. From these measurements came another revelation: the comets would strike Jupiter much closer to the side facing the earth.

■ A GLOBAL OBSERVING SESSION

As "impact week" approached in the summer of 1994, it became clear that this event was so extraordinary that it deserved observation time on as many telescopes as possible. Just as extraordinary was the good fortune that events had allowed astronomers a full 14 months to coordinate their programs. Heading the list of powerful telescopes to be aimed at Jupiter was the Hubble Space Telescope, whose newly corrected optics had already captured the comet nuclei with amazing clarity. For a team led by Harold A. Weaver of the Space Telescope Science Institute in Baltimore, Hubble's wide-field planetary camera would monitor the comet nuclei as they moved closer to Jupiter. A group led by Heidi B. Hammel of the Massachusetts Institute of Technology used the telescope to take detailed images of the entire planet on the day before the first collision, to compare with later views to come during the week. The telescope would also collect spectrographic signatures of elements and gases released during the explosions. That is, of course, if something could still be seen when the impact sites rotated into the earth's view *[see color plate 2]*.

But even if the nightside strikes were to be invisible from the earth, there was another means to examine them. On its way toward a rendezvous with Jupiter, the Galileo space probe was in a position that would give its cameras and other instruments a direct view of the impact sites. Controllers at the Jet Propulsion Laboratory instructed the spacecraft to collect and return data on several of the impacts.

Many of the world's great telescopes were destined to play a vital role in recording the strikes and the related phenomena. The collisions would occur over a period lasting almost six days; thus, telescopes spread over the globe were needed. Palomar's venerable five-meter telescope, other large telescopes in Spain, Chile, Hawaii, and Australia, and a host of smaller telescopes participated. The NASA Kuiper Airborne Observatory, flying out of Melbourne, Australia, captured key spectroscopic measurements. In addition teams of radio astronomers monitored Jupiter for the effects of the impacts on the Jovian magnetosphere.

Using the Keck Observatory's giant 10-meter telescope atop Mauna Kea in Hawaii, Imke de Pater of the University of California at Berkeley and her colleagues planned to record infrared images in the wavelengths of light absorbed by cold methane gas. Because Jupiter's methane-rich atmosphere absorbs these wavelengths, filters that pass light only in the "methane band" would darken the face of the planet and highlight anything happening very high in or above the planet's atmosphere. These measurements, Imke and others reasoned, should be sensitive enough to catch any spots left by the collisions and, possibly, the impact plumes themselves. The South Pole Infrared Explorer telescope (SPIREX) was primed to make similar observations.

■ JULY 16, 1994: SHOW TIME

After 14 months of waiting, the first word was electrifying: Calar Alto Observatory in Spain had recorded the infrared signature of a collapsing plume from the first impact (Nucleus A). The detection was confirmed by the European Southern Observatory in Chile. Not only was the impact detectable, it was spectacular. As we were soon to find out, the plume shot some 3,000 kilometers above the clouds of Jupiter. But even with this news, for astronomers awaiting the first data from Hubble, tension was high. The telescope employed different filters and detectors from those used at the observatory in Spain, and everyone wondered what the mighty eye in space would record.

The entire Hubble comet team huddled around a single video monitor at the Space Telescope Science Institute shortly after the first images had been returned. The first few had not shown any obvious disturbance, and anxiety mounted. But then a spot appeared above the edge of the planet, and everyone in the room began to breathe again. The next image showed the plume rising and brightening over Jupiter. The fireworks were in clear view, and the comet team erupted in celebration.

By that first magical day, it was clear that the meticulous planning had paid off handsomely. The international collection of observatories, on the earth and in space, was responding like a symphony orchestra, with Marsden's frequent electronic messages acting as conductors. They allowed observers to know what everyone else was doing so that their programs could be altered to keep up with the emerging picture.

■   A BATTERED PLANET

From the outset the performance of the comets was intriguing. As Jupiter rotated a large spot left by Nucleus A came into view. It was made up of three distinct parts: a central streak, an expanding ring, and a peculiar crescent-shaped outer cloud. In the visible part of the spectrum the markings looked extraordinarily dark, but in the infrared light of a methane absorption band the spot appeared bright against the dark planet. The entire spot was as large as the earth. Several hours later Nucleus B struck Jupiter with quite different effects. Even though B had been brighter than A, the plume that rose from its impact was so much smaller that only the largest telescope in the world, the 10-meter Keck, recorded it easily. Nucleus B may have consisted of a swarm of small house-size subnuclei that split off from Nucleus C sometime after the initial breakup. An observer on Jupiter would have seen a fabulous storm of meteors, but little was detected from the earth.

Nuclei C and E crashed with much the same effects as A. Two days later there was great anticipation as Nucleus G—which had a bright coma and presumably large mass—made its final descent. Hubble had a clear view of Jupiter, but that night all the big telescopes at Mauna Kea Observatories were closed because of fog and drizzle. Yet miraculously, only a minute before the impact, the clouds above Mauna Kea parted. The observatory domes raced open, and the telescopes captured images of the strike before more fog and rain forced them to close again only 10 minutes later. They were lucky to get a view: Nucleus G hit with such tremendous energy that the collapsing plume was much brighter than the entire planet in the infrared methane band. Nucleus G left the same imprint as the earlier major impacts of A, C, and E, but the scar was much bigger. The great flash of energy was well recorded in Australia and at the South Pole.

At this point Hubble had detected expanding rings from impacts A, E, and G in the clear regions between the inner dark core clouds and the outer dark crescents. It was found that these were expanding outward at about 450 meters per second. Interpretation of these features fell to Andrew P. Ingersoll of Caltech. Soon after impact week Ingersoll realized the rings were not moving out fast enough to be sound waves—they were not the "boom from the plume," as he had originally thought. But the speed of the waves was the same for all impacts. Ultimately Ingersoll and Hiroo Kanamori, also at Caltech, found that an "internal gravity" wave had been produced, somewhat like the waves formed by a stone thrown into a pond.

As impact week continued Nucleus L left the largest spot yet, once again complete with a central core and outer, crescent-shaped cloud. By this time amateur astronomers around the world had found that these dark features on Jupiter were so large and dense that they could be seen by using small telescopes. The nuclei of H, K, and L were all preceded by a long train of particles whose entry into the atmosphere produced a rising infrared glow before the arrival of the main part of the nucleus. The Galileo spacecraft took an engaging series of "snapshots" of the brilliant meteor and incandescent rising plume from the impact of W, the final nucleus, as it tore into Jupiter. The Hubble image sequence of the same fall ended with a view of the plume collapsing directly on top of the spot made earlier by Nucleus K.

■   REVIEWS STILL COMING IN

Despite the many observations of this dramatic episode, important questions are not yet fully answered. How large were the nuclei? Were they mostly swarms of small bodies, or were there large individual fragments? How much energy did they release when they hit? The diversity of effects and the sheer mass of data—more than for any other single event in the history of astronomy—preclude a simple analysis. Just as scientific discussions and meetings before the impact emphasized the need to coordinate observations, sessions convened afterward have concentrated on comparing the data to see which ideas fit best.

Comet S-L 9 probably began its wanderings in the outer solar system beyond the orbit of Neptune. A series of close encounters with Jupiter gradually altered its orbital period from one revolution about the sun every several thousand years to about once a decade. The latest orbital calculations, by Paul W. Chodas of the Jet Propulsion Laboratory, indicate that probably about 1929 (the year an unrelated crash hit the earth's stock market) the comet made a slow approach to Jupiter that allowed the planet to capture the comet as a moon. The resulting two-year-long orbit about the planet was, hoswever, unstable. Some revolutions followed narrow ellipses; others were roughly circular. In 1992, when the orbit was highly elliptical, the comet passed so close to Jupiter that it was broken apart.

The initial disintegration dispersed the cometary material into a long swarm of debris. Erik I. Asphaug of the NASA Ames Research Center and Willy Benz of the University of Arizona have shown that the loose string of rubble could have then coalesced into a set of distinct nuclei under the mutual gravitational attraction of the fragments. We suspect that large coherent pieces of fractured comet were present in some nuclei but not in others.

After the main disruption event, additional nuclei split from some of the earlier-formed nuclei. Just how this latter fracturing occurred is not understood. Possibly internal gas pressure ruptured large chunks, or perhaps the force of collisions between fragments traveling in the swarm knocked them apart. The largest individual nuclei in the entire train probably were no more than a kilometer or two across. These nuclei did not complete even one more orbit before striking Jupiter's flank. When they hit, the energy from each of the largest impacts probably equaled hundreds of thousands of large hydrogen bombs exploding simultaneously.

The great dark scars left on Jupiter gradually spread, merged, and slowly faded in the months after the impacts. Yet as this article goes to press, almost a year after the collision, a faint dark band along the line of impact sites is still visible through even small telescopes. Such dark clouds have never been seen on Jupiter before, and one wonders just how rare such a dramatic event must be.

The frequency of impacts depends on the scale of the body involved, and we are still uncertain about the size of this comet before it broke apart. But by making some reasonable assumptions, we can estimate that the crash of a string of nuclei such as S-L 9 probably occurs less than once every few thousand years. Thus, we feel fortunate to be living at this moment, to have found the comet on its way toward Jupiter, and to have witnessed its demise in a blaze of glory.

—August 1995

COLLISIONS WITH COMETS AND ASTEROIDS

The chances of a celestial body colliding with the earth are small, but the consequences would be catastrophic

Tom Gehrels

Are we going to be hit by an asteroid? Planetary scientists are divided on how worrisome the danger is. Some refuse to take it seriously; others believe the risk of dying from such an impact might even be greater than the risk of dying in an airplane crash. After years of studying the problem, I have become convinced that the danger is real. Although a major impact is unlikely, the energies released could be so horrendous that our fragile society would be obliterated.

Early in our planet's history asteroids and comets made life possible by accreting into the earth and then by bringing water to the newborn planet. And they have already destroyed, at least once, an advanced form of that life. The dinosaurs were killed by such an impact, making way for the age of the mammals. Now for the first time, creatures have evolved to a point where they can wrest control of their fate from the heavenly bodies, but humans must come to grips with the danger.

Some four and a half billion years ago the solar system formed out of a swirling cloud of gas and dust. Initially the planetesimals—coarse collections of rocky materials—coagulated, merging with one another to create planets. Because of the energy released by the colliding rocks, the earth began as a molten globe, so hot that the volatile substances—water, carbon dioxide, ammonia, methane, and other gases—boiled off. As the material of the inner solar nebula was mopped up by the growing planets, the bombardment of the earth slowed. The glowing planet cooled, and a crust solidified. Only then did water—the life-giving fluid that covers three quarters of the earth's surface—return, borne on cold comets arriving from the solar system's distant reaches. Fossil records show that simple life-forms started evolving almost right away.

Comets and asteroids are, in fact, leftover planetesimals. Most asteroids inhabit the vast belt between the orbits of Mars and Jupiter. Being quite close to the sun, they were formed hot; as on the early earth, the high temperatures vaporized the lighter substances, such as water, leaving mostly silica, carbon, and metals. (Only recently have astronomers found some rare asteroids that contain crystalline water embedded in rocks.)

Comets, on the other hand, hover at the outer edges of the solar system. As the solar system was formed, a good deal of matter was thrown outward, beyond the orbits of Uranus and Neptune. Coalescing far from the sun, the comets were born cold, at temperatures as low as -260 degrees Celsius. They retained their volatile materials, the gas, ice, and snow. Sometimes called dirty snowballs, these objects are usually tenuous aggregates of carbon and other light elements.

■ FIERY VISITORS

In 1950 Jan H. Oort, professor of astronomy at Leiden University in the Netherlands, was teaching a class that I was allowed to attend as an undergraduate. While reviewing astronomical calculations for his students, Oort noted that a number of known comets reach their farthest point from the sun—called the aphelion—at a great distance. He went on to formulate the idea that a cloud of comets exists as a diffuse spherical shell at about 50,000 or more astronomical units. (One astronomical unit is the distance from the earth to the sun.) This distant cloud, containing perhaps some 10^{13} objects, envelops the solar system.

The Oort cloud reaches a fifth of the distance to the nearest star, Alpha Centauri. Inhabitants of this shell are thus loosely bound to the sun and readily disturbed by events beyond the solar system. If the sun passes by another star or a massive molecular cloud, some of these cometary orbits are jarred. The planetesimal might then swing into a narrow elliptical orbit that brings it toward the inner solar system. As it nears the sun, the heat vaporizes its volatile materials, which spew forth as if from a geyser. In ancient cultures this celestial spectacle was sometimes an ominous event.

Some visitors from the Oort cloud are never seen again; others have periods that get shorter with each successive pass. The best known of these comets are those that return regularly, such as Halley's, with a period of 76 years. The chance that such a comet will collide with the earth is exceedingly small, because it comes by so infrequently. But the patterns of their orbits suggest that in the next millennia, comet Halley or Swift-Tuttle (with a period of 130 years) will sometimes swing by too close for comfort.

In 1951 Gerard P. Kuiper, then at Yerkes Observatory of the University of Chicago, surmised that another belt of comets exists, just beyond Neptune's orbit, much nearer than the Oort cloud. Working at the University of Hawaii, David C. Jewitt and Jane Luu discovered the first of these objects in 1992 after a persistent search; by now some 31 bodies belonging to the Kuiper belt have been found. In fact Pluto, with its unusually elliptical orbit, is now considered to be the largest of these objects; Clyde Tombaugh, who discovered Pluto in 1930, calls it the "King of the Kuiper belt."

Comets belonging to the Kuiper belt are not directly disturbed by rival stars. Instead they can stray close to Neptune, which may either help stabilize them or, conversely, throw them out of orbit. (An as-yet-unknown 10th planet may also be stirring the comets' path, but the evidence for its existence is inconclusive.) The comets may then come very close to the sun. Although those from the Kuiper belt tend to have shorter periods than those from the Oort cloud, both types of comets can be captured in tight orbits around the sun. It is therefore impossible to tell where a particular comet—such as Tempel-Tuttle, which sweeps by at 72 kilometers per second every 33 years—originated from.

Some comets are bound into small orbits and have short periods, on the order of 10 years. These comets pose more of a concern than the ones that come by only every century or so. A collision with such a short-period comet might occur once in some three million years.

However infrequent a cometary collision might be, the consequences would be calamitous. The orbits of comets are often steeply inclined to the earth's; occasionally a comet is even going in the opposite direction. Thus, comets typically pass the earth with a high relative velocity. For example, Swift-Tuttle, which is about 25 kilometers across, flies by at 60 kilometers per second. It would impact with cataclysmic effect.

Unless it runs into something, a comet probably remains active, emitting gases and dust for some 500 passages by the sun. Eventually the volatile materials are used up, and the comet fades away as a dead object, indistinguishable from an asteroid. Up to half of the nearest asteroids might in fact be dead, short-period comets.

■ FALLING ROCKS

Indeed, most of the danger to the earth comes from asteroids. Like comets, asteroids have solar orbits that are normally circular and stable. But there are so many of them in the asteroid belt that they can collide with one another.

The debris from such collisions can end up in unstable orbits that resonate with the orbit of Jupiter. By virtue of its immense mass, Jupiter competes with the sun for control of the motions of these fragments, especially if an asteroid's orbit "beats," or resonates, with that of the giant planet. So, for instance, if the asteroid goes around the sun thrice in the same time that Jupiter orbits once, the planet's gravitational influence on the rock is greatly enhanced. Just as a child on a swing flies ever higher if someone pushes her each time the swing returns, Jupiter's rhythmic nudges ultimately cause the asteroid to veer out of its original orbit into an increasingly eccentric one.

The asteroid may either leave the solar system or move in toward the terrestrial, rocky planets. Eventually, such vagrants collide with Mars, the earth-moon system, Venus, Mercury, or even the sun. A major fragment enters the inner solar system once in roughly 10 million years and survives for about as long.

To estimate the chances of such a rock hitting the earth, the asteroids have first to be sorted according to size. The smallest ones we can observe, which are less than a few tens of meters across, rarely make it through the earth's atmosphere; friction with air generates enough heat to vaporize them. The asteroids that are roughly 100 meters and larger in diameter do pose a threat. There are 100,000 or so of these that penetrate the inner solar system deeper than the orbit of Mars. They are called near-earth asteroids.

In 1908 one such object, a loose conglomerate of silicates about 60 meters wide, entered the atmosphere and burst apart above the Tunguska Valley in Siberia. The explosion was heard as far away as London. Although the fragments did not leave a crater, the area below the explosion is still marked by burned trees laid out in a region roughly 50 kilometers across. The identity of the Tunguska object inspired a lot of nonsensical speculation for decades, and some highly imaginative suggestions were made, including that it was a mini black hole or an alien spacecraft. Scientists, however, have always understood that it was a comet or asteroid.

Events such as the Tunguska explosion may occur once a century, and it is most likely that they would occur over the oceans or remote land areas. But they would be devastating if they happened near a populated area. If one exploded over London for instance, not only the city but also its suburbs would be laid waste.

Of the smaller asteroids, the few metallic ones are tough enough to penetrate the atmosphere and carve out a crater. The 1.2-kilometer-wide Meteor Crater in northern Arizona is an example; it came from a metallic asteroid about 30 meters in diameter that fell some 50,000 years ago.

An even greater peril is posed by the 1,000 or 2,000 medium near-earth asteroids that are roughly one kilometer and larger in size. One of these asteroids is thought to collide with the earth once in about 300,000 years. Note that this estimate is only a statistical average. Such a collision can happen at any time—a year from now, in 20 years, or not in a million years.

■ FRIGHTFUL DARKNESS

The energies liberated by an impact with such an object would be tremendous. The kinetic energy can be calculated from $1/2\ mv^2$, where m is the mass of the object, and v is the incoming velocity. Assuming a density of about three grams per cubic centimeter, as known from meteorites, and an average velocity of 20 kilometers per second, a 1-kilometer-wide object would strike with a shock equivalent to tens of billions of tons of TNT—millions of times the energy released at Hiroshima in 1945.

Granted, asteroids do not emit the nuclear radiation that caused the particular horrors of Hiroshima. Still, an explosion of millions of Hiroshimas would do more than destroy a few cities or some countries. The earth's atmosphere would be globally disrupted, creating the equivalent of a nuclear winter. Large clouds of dust would explode into the atmosphere to obscure the sun, leading to prolonged darkness, subzero temperatures, and violent windstorms.

Even more dangerous are the largest near-earth asteroids, which are about 10 kilometers in diameter. Fortunately there are only a few such threatening objects, perhaps just 10. (Even more fortunately they happen to be mere fragments of the objects in the asteroid belt, which can be as large as 1,000 kilometers across.) An asteroid of this size collides with the earth only once in 100 million years or so.

One such event is evident in the fossil record. The impact of a celestial object marks the end of the Cretaceous geologic period and the beginning of the Tertiary, 65 million years ago. After years of searching, the crater from that event—a depression about 170 kilometers in diameter—has been identified in the Yucatán peninsula of Mexico. Although the crater cannot be directly seen, it has fortuitously been identified by drillings for oil and in images taken from the spade shuttle Endeavour. The depression resulted from the explosive impact of an object perhaps 10 to 20 kilometers in diameter.

Studies of the effects of that explosion paint a frightening picture. An enormous fireball ejected rocks and steam into the atmosphere, jarred the earth's crust, and triggered earthquakes and tsunamis around the globe. Vast clouds of dust, from the earth and the asteroid, erupted into the stratosphere and beyond. There ensued total darkness, which lasted for months.

Acid rain began to fall, and slowly the dust settled, creating a layer of sediment a few centimeters thick over the earth's surface. Below this thin sheet we see evidence of dinosaurs. Above it they are missing, as are three fourths of the other species. The darkness following the explosion must have initially plunged the atmosphere into a freeze.

Over many centuries the reverse effect—a slow greenhouse warming, by as much as 15 degrees Celsius—had an equally devastating outcome. The asteroid had struck the earth in a vulnerable place, slicing into a rare region with a deep layer of limestone. (Less than 2 percent of the earth's crust has so much limestone; Australia's Great Barrier Reef is an example.) The explosion ejected the carbon dioxide from the limestone into the atmosphere, where, along with other gases, it helped to trap the earth's heat. Jan Smit of the Free University, Amsterdam, has proposed that the severe warming, rather than the initial freeze, killed the dinosaurs—there is some evidence that they died off slowly.

■ SPACEWATCH

So—are we going to be hit? To begin with, the answer lies in the domain of planetary astronomy. The dangerous objects have to be located, as soon as possible, to diminish the chances of our unexpected demise. Furthermore, they have to be tracked on the succeeding nights, weeks, months, and even years so that their orbits can be accurately extrapolated into the future.

In the early 1970s a .46-meter photographic camera at the Palomar Observatory in southern California was dedicated to the research for near-earth objects. Eleanor Helin of the Jet Propulsion Laboratory in Pasadena, California, led one of the teams of astronomers, and Eugene M. and Carolyn S. Shoemaker of the U.S. Geological Survey led the other. The scientists photographed the same large areas of the sky at half-hour intervals. As asteroids orbit the sun, they move with respect to the background stars. If near to the earth, the asteroid is seen to travel relatively fast; the motion is easily recognized from the multiple exposures.

Since the pioneering efforts at Palomar, other observers have become interested in near-earth asteroids. At Siding Spring in the mountains of eastern Australia, a dedicated group of scientists uses a 1.2-meter photographic camera to hunt for these rocks. In 1994 observers in California and Australia, with their photographic methods, jointly discovered 16 near-earth asteroids. (At the end of that year, the Palomar project closed as more modern techniques were developed elsewhere.)

About 15 years ago Robert S. McMillan, also at the University of Arizona, and I began to realize that at this rate, it would take more than a century to map the 1,000 or more asteroids that are larger than one kilometer across. By taking advantage of electronic detection devices and fast computers, the rate of finding asteroids could be greatly increased. Spacewatch, a project dedicated to the study of comets and asteroids, was born in Tucson. A .9-meter telescope at the University of Arizona's Steward Observatory on Kitt Peak, 70 kilometers west of Tucson, is now dedicated to Spacewatch. Robert Jedicke, James V. Scotti, several students, and I, all from Tucson, use this facility regularly for finding comets and asteroids. McMillan, Marcus L. Perry, Toni L. Moore, and others, also from Tucson, use it for finding planets around other stars.

Instead of photographic plates, our electronic light detectors are charge-coupled devices, or CCDs. These are finely divided arrays of semiconductor picture elements, or pixels. When light hits a pixel, its energy causes positive and negative electrical charges to separate. The electrons from all the pixels provide an image of the light pattern at the focal plane of the telescope. A computer then compares images of the same patch of sky scanned at different times, marking the objects that have moved.

In this manner Spacewatch observers may find as many as 600 asteroids a night. Most of these are in the asteroid belt; only occasionally does an object move against the star field so fast that it must be close to the earth. (Similarly, an airplane high above in the sky seems to move slower than one coming in low for a landing.) In 1994 Scotti found an asteroid that passed within 105,000 kilometers of the earth. Also in that year, Spacewatch reported 77,000 precise measurements of comet and asteroid positions. One gratifying aspect of Spacewatch is that it has private and corporate supporters (currently 235) in addition to the U.S. Air Force Office of Scientific Research, the National Aeronautics and Space Administration, the Clementine space program, the National Science Foundation, and other governmental organizations.

Spacewatch has discovered an abundance of small asteroids, those in the range of tens of meters. The numbers of these objects exceed predictions by a factor of 40, but we do not as yet understand their origins. These asteroids we call the Arjunas, after the legendary Indian prince who was enjoined to persist on his charted course. Military reconnaissance satellites have since also observed the Arjunas. The data, once routinely discarded but now stored and declassified, show the continuous showering of the planet by small asteroids. Because of the atmosphere, these rocks burn up with little consequence, even though similar ones scar the airless moon.

The next step for Spacewatch is to install our new telescope, which was built with an existing 1.8-meter mirror, so that we can find fainter and more distant objects. This state-of-the-art instrument, the largest in the field of asteroid observation, should serve generations of explorers to come. Meanwhile, at Côte d'Azur Observatory in southern France, Alain Maury is about to bring a telescope into operation with an electronic detection system. Duncan Steel and his colleagues in Australia are switching to electronics as well, although this project has funding problems perhaps more severe than ours. Next to join the electronic age might be Lowell Observatory near Flagstaff, Arizona, under the supervision of Edward Bowell. The U.S. Air Force is also planning to use one of its one-meter telescopes to this end; Helin and her associates already use the one on Maui in Hawaii. And amateur astronomers are coming on-line with electronic detectors on their telescopes.

If there is an asteroid out there with our name on it, we should know by about the year 2008.

■ DEFLECTING AN ASTEROID

And what if we find a large object headed our way? If we have only five years' notice, we can say good-bye to one another and regret that we did not start surveying earlier. If we have 10 years or so, our chances are still slim. If we have 50 years' notice or more, a spacecraft could deploy a rocket that would explode near the asteroid. Perhaps the most powerful intercontinental ballistic missiles could blast a small object out of the way. (That, incidentally, would also be a good means of getting rid of these relics from the cold war.)

It seems likely, however, that we will have more than 100 years to prepare. Given that much time, a modest chemical explosion near an asteroid might be enough to deflect it. The explosion will need to change the asteroid's trajectory by only a small amount so that by the time the asteroid reaches the earth's vicinity, it will have deviated from its original course enough to bypass the planet.

Present technology for aiming and guiding rockets is close to miraculous. I once overheard two scientists arguing about why *Pioneer 11* had arrived 20 seconds late at Saturn—after a journey of six years. But the detonation will have to be carefully designed. If the asteroid is made of loosely aggregated material, it might disintegrate when shaken by an explosion. The pieces could rain down on the earth, causing even greater damage than the intact asteroid, as hunters who use buckshot know. A "standoff" explosion, at some distance from the surface, may be the most effective in that case. Earth-based radar, telescopes, and possibly space missions will be needed to determine the composition of an asteroid and how it might break up.

Farther into the future laser or microwave devices might become suitable. Gentler alternatives, such as solar sails and reflectors planted on the asteroid's surface—to harness the sun's radiation in pushing the asteroid off course—have also been suggested. A few scientists are studying the feasibility of nuclear devices to deflect very massive asteroids that show up at short notice.

Comets and asteroids remind me of Shiva, the Hindu deity who destroys and recreates. These celestial bodies allowed life to be born, but they also killed our predecessors, the dinosaurs. Now for the first time, the earth's inhabitants have acquired the ability to envision their own extinction—and the power to stop this cycle of destruction and creation.

—March 1996

THE KUIPER BELT

*Rather than ending abruptly at the orbit of Pluto, the outer
solar system contains an extended belt of small bodies*

Jane X. Luu and David C. Jewitt

After the discovery of Pluto in 1930, many astronomers became intrigued by the possibility of finding a 10th planet circling the sun. Cloaked by the vast distances of interplanetary space, the mysterious "Planet X" might have remained hidden from even the best telescopic sight, or so these scientists reasoned. Yet decades passed without detection, and most researchers began to accept that the solar system was restricted to the familiar set of nine planets.

But many scientists began seriously rethinking their notions of the solar system in 1992, when we identified a small celestial body—just a few hundred kilometers across—sited farther from the sun than any of the known planets. Since that time, we have identified nearly three dozen such objects circling through the outer solar system. A host of similar objects is likely to be traveling with them, making up the so-called Kuiper belt, a region named for Dutch-American astronomer Gerard P. Kuiper, who, in 1951, championed the idea that the solar system, contains this distant family.

What led Kuiper, nearly half a century ago, to believe the disk of the solar system was populated with numerous small bodies orbiting at great distances from the sun? His conviction grew from a fundamental knowledge of the behavior of certain comets—masses of ice and rock that on a regular schedule plunge from the outer reaches of the solar system inward toward the sun. Many of these comparatively small objects periodically provide spectacular appearances when the sun's rays warm them enough to drive dust and gas off their surfaces into luminous halos (creating large "comae") and elongate tails.

Astronomers have long realized that such active comets must be relatively new members of the inner solar system. A body such as Halley's comet, which swings into view every 76 years, loses about one ten-thousandth of its mass on each visit near the sun. That comet will survive for only about 10,000 orbits, lasting perhaps half a million years in all. Such comets were created during the formation of the solar system 4.5 billion years ago and should have completely lost their volatile constituents by now, leaving behind either inactive, rocky nuclei or diffuse streams of dust. Why then are so many comets still around to dazzle onlookers with their displays?

■ GUIDING LIGHTS

The comets that are currently active formed in the earliest days of the solar system, but they have since been stored in an inactive state—most of them preserved within a celestial deep freeze called the Oort cloud. The Dutch astronomer Jan H. Oort proposed the existence of this sphere of cometary material in 1950. He believed that this cloud had a diameter of about 100,000 astronomical units (AU—a distance defined as the average separation between the earth and the sun, about 150 million kilometers) and that it contained several hundred billion individual comets. In Oort's conception the random gravitational jostling of stars passing nearby knocks some of the outer comets in the cloud from their stable orbits and gradually deflects their paths to dip toward the sun.

For most of the past half a century, Oort's hypothesis neatly explained the size and orientation of the trajectories that the so-called long-period comets (those that take more than 200 years to circle the sun) follow. Astronomers find that those bodies fall into the planetary region from random directions—as would be expected for comets originating in a spherical repository like the Oort cloud. In contrast Oort's hypothesis could not explain short-period comets that normally occupy smaller orbits tilted only slightly from the orbital plane of the earth—a plane that astronomers call the ecliptic.

Most astronomers believed that the short-period comets originally traveled in immense, randomly oriented orbits (as the long-period comets do today) but that they were diverted by the gravity of the planets—primarily Jupiter—into their current orbital configuration. Yet not all scientists subscribed to this idea. As early as 1949, Kenneth Essex Edgeworth, an Irish gentleman-scientist (who was not affiliated with any research institution), wrote a scholarly article suggesting that there would be a flat ring of comets in the outer solar system. In his 1951 paper, Kuiper also discussed such a belt of comets, but he did not refer to Edgeworth's previous work.

Kuiper and others reasoned that the disk of the solar system should not end abruptly at Neptune or Pluto (which vie with each other for the distinction of being the planet most distant from the sun). He envisioned instead a belt beyond Neptune and Pluto consisting of residual material left over from the formation of the planets. The density of matter in this outer region would be so low that large planets could not have accreted there, but smaller objects, perhaps of asteroidal dimensions, might exist. Because these scattered remnants of primordial material were so far from the sun, they would maintain low surface temperatures. It thus seemed likely that these distant objects would be composed of water ice and various frozen gases—making them quite similar (if not identical) to the nuclei of comets.

Kuiper's hypothesis languished until the 1970s, when Paul C. Joss of the Massachusetts Institute of Technology began to question whether Jupiter's gravity could in fact efficiently transform long-period comets into short-period ones. He noted that the probability of gravitational capture was so small that the large number of short-period comets that now exists simply did not make sense. Other researchers

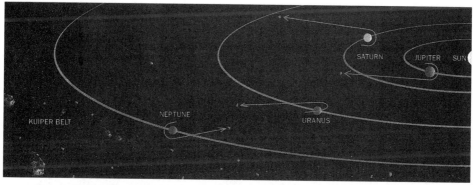

Gravity of the planets acted during the early stages of the solar system to sweep away small bodies within the orbit of Neptune. Some of these objects plummeted toward the sun; others sped outward toward the distant Oort cloud (not shown). (Peter Samek)

were, however, unable to confirm this result, and the Oort cloud remained the accepted source of the comets, long and short period alike.

But Joss had sown a seed of doubt, and eventually other astronomers started to question the accepted view. In 1980 Julio A. Fernández (then at the Max Planck Institute for Aeronomy in Katlenburg-Lindau) had, for example, done calculations that suggested that short-period comets could come from Kuiper's proposed trans-Neptunian source. In 1988 Martin J. Duncan of the University of Toronto, and Thomas Quinn and Scott D. Tremaine (both at the Canadian Institute for Theoretical Astrophysics), used computer simulations to investigate how the giant gaseous planets could capture comets. Like Joss, they found that the process worked rather poorly, raising doubts about the veracity of this well-established concept for the origin of short-period comets. Indeed, their studies sounded a new alarm because they noted that the few comets that could be drawn from the Oort cloud by the gravitational tug of the major planets should be traveling in a spherical swarm, whereas the orbits of the short-period comets tend to lie in planes close to the ecliptic.

Duncan, Quinn, and Tremaine reasoned that short-period comets must have been captured from original orbits that were canted only slightly from the ecliptic, perhaps from a flattened belt of comets in the outer solar system. But their so-called Kuiper belt hypothesis was not beyond question. In order to make their calculations tractable they had exaggerated the masses of the outer planets as much as 40 times (thereby increasing the amount of gravitational attraction and speeding up the orbital evolution they desired to examine). Other astrophysicists wondered whether this computational sleight of hand might have led to an incorrect conclusion.

■ WHY NOT JUST LOOK?

Even before Duncan, Quinn, and Tremaine published their work, we wondered whether the outer solar system was truly empty or instead full of small, unseen bodies. In 1987 we began a telescopic survey intended to address exactly that question. Our plan was to look for any objects that might be present in the outer solar system using the meager amount of sunlight that would be reflected back from such great distances. Although our initial efforts employed photographic plates, we soon decided that a more promising approach was to use an electronic detector (a charge-coupled device, or CCD) attached to one of the larger telescopes.

We conducted the bulk of our survey using the University of Hawaii's 2.2-meter telescope on Mauna Kea. Our strategy was to use a CCD array with this instrument to take four sequential, 15-minute exposures of a particular segment of the sky. We then enlisted a computer to display the images in the sequence in quick succession—a process astronomers call "blinking." An object that shifts slightly in the image against the background of stars (which appear fixed) will reveal itself as a member of the solar system.

For five years we continued the search with only negative results. But the technology available to us was improving so rapidly that it was easy to maintain enthusiasm (if not funds) in the continuing hunt for our elusive quarry. On August 30, 1992, we were taking the third of a four-exposure sequence while blinking the first two images on a computer. We noticed that the position of one faint "star" appeared to move slightly between the successive frames. We both fell silent. The motion was quite subtle, but it seemed definite. When we compared the first two images with the third, we realized that we had indeed found something out of the ordinary. Its slow motion across the sky indicated that the newly discovered object could be traveling beyond even the outer reaches of Pluto's distant orbit. Still, we were suspicious that the mysterious object might be a near-earth asteroid moving parallel with the earth (which might also cause a slow apparent motion). But further measurements ruled out that possibility.

We observed the curious body again on the next two nights and obtained accurate measurements of its position, brightness, and color. We then communicated these data to Brian G. Marsden, director of the International Astronomical Union's Central Bureau of Astronomical Telegrams at the Smithsonian Astrophysical Observatory in Cambridge, Massachusetts. His calculations indicated that the object we had discovered was indeed orbiting the sun at a vast distance (40 AU)—only slightly less remote than we had first supposed. He assigned the newly discovered body a formal, if somewhat drab, name based on the date of discovery: he christened it "1992 QB$_1$." (We preferred to call it "Smiley," after John Le Carré's fictional spy, but that name did not take hold within the conservative astronomical community.)

Our observations showed that QB$_1$ reflects light that is quite rich in red hues compared with the sunlight that illuminates it. This odd coloring matched only one other object in the solar system—a peculiar asteroid or comet called 5145 Pholus. Planetary astronomers attribute the red color of 5145 Pholus to the presence of dark, carbon-rich material on its surface. The similarity between QB$_1$ and 5145 Pholus thus heightened our excitement during the first days after the discovery. Perhaps the object we had just located was coated by some kind of red material abundant in organic compounds. How big was this ruddy new world? From our first series of measurements we estimated that QB$_1$ was between 200 and 250 kilometers across—about 15 times the size of the nucleus of Halley's comet.

Some astronomers initially doubted whether our discovery of QB$_1$ truly signified the existence of a population of objects in the outer solar system, as Kuiper and others had hypothesized. But such questioning began to fade when we found a second body in March 1993. This object is as far from the sun as QB$_1$ but is located on the opposite side of the solar system. During the past three years several other re-

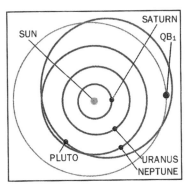

Sequential CCD exposures revealed Kuiper belt object QB₁ clearly against the background of fixed stars. The images were analyzed in order to correctly identify QB₁ and determine its orbit. (Peter Samek)

search groups have joined the effort, and a steady stream of discoveries has ensued. The current count of trans-Neptunian, Kuiper belt objects is 32.

The known members of the Kuiper belt share a number of characteristics. They are, for example, all located beyond the orbit of Neptune, suggesting that the inner edge of the belt may be defined by this planet. All these newly found celestial bodies travel in orbits that are only slightly tilted from the ecliptic—an observation consistent with the existence of a flat belt of comets. Each of the Kuiper belt objects is millions of times fainter than can be seen with the naked eye. The 32 objects range in diameter from 100 to 400 kilometers, making them considerably smaller than both Pluto (which is about 2,300 kilometers wide) and its satellite, Charon (which measures about 1,100 kilometers across).

The current sampling is still quite modest, but the number of new solar-system bodies found so far is sufficient to establish beyond doubt the existence of the Kuiper belt. It is also clear that the belt's total population must be substantial. We estimate that the Kuiper belt contains at least 35,000 objects larger than 100 kilometers in diameter. Hence, the Kuiper belt probably has a total mass that is hundreds of times larger than the well-known asteroid belt between the orbits of Mars and Jupiter.

■ COLD STORAGE FOR COMETS

The Kuiper belt may be rich in material, but can it in fact serve as the supply source for the rapidly consumed short-period comets? Matthew J. Holman and Jack L. Wisdom, both then at M.I.T., addressed this problem using computer simulations.

They showed that within a span of 100,000 years the gravitational influence of the giant gaseous planets (Jupiter, Saturn, Uranus, and Neptune) ejects comets orbiting in their vicinity, sending them out to the farthest reaches of the solar system. But a substantial percentage of trans-Neptunian comets can escape this fate and remain in the belt even after 4.5 billion years. Hence, Kuiper belt objects located more than 40 AU from the sun are likely to have held in stable orbits since the formation of the solar system.

Astronomers also believe there has been sufficient mass in the Kuiper belt to supply all the short-period comets that have ever been formed. So the Kuiper belt seems to be a good candidate for a cometary storehouse. And the mechanics of the transfer out of storage are now well understood. Computer simulations have shown that Neptune's gravity slowly erodes the inner edge of the Kuiper belt (the region within 40 AU of the sun), launching objects from that zone into the inner solar system. Ultimately many of these small bodies slowly burn up as comets. Some—such as Comet Shoemaker-Levy 9, which collided with Jupiter in July 1994—may end their lives suddenly by striking a planet (or perhaps the sun). Others will be caught in a gravitational slingshot that ejects them into the far reaches of interstellar space.

If the Kuiper belt is the source of short-period comets, another obvious question emerges: Are any comets now on their way from the Kuiper belt into the inner solar system? The answer may lie in the Centaurs, a group of objects that includes the extremely red 5145 Pholus. Centaurs travel in huge planet-crossing orbits that are fundamentally unstable. They can remain among the giant planets for only a few million years before gravitational interactions either send them out of the solar system or transfer them into tighter orbits.

With orbital lifetimes that are far shorter than the age of the solar system, the Centaurs could not have formed where they currently are found. Yet the nature of their orbits makes it practically impossible to deduce their place of origin with certainty. Nevertheless, the nearest (and most likely) reservoir is the Kuiper belt. The Centaurs may thus be "transition comets," former Kuiper belt objects heading toward short but showy lives within the inner solar system. The strongest evidence supporting this hypothesis comes from one particular Centaur—2060 Chiron. Although its discoverers first thought it was just an unusual asteroid, 2060 Chiron is now firmly established as an active comet with a weak but persistent coma.

As astronomers continue to study the Kuiper belt, some have started to wonder whether this reservoir might have yielded more than just comets. Is it coincidence that Pluto, its satellite, Charon, and the Neptunian satellite Triton lie in the vicinity of the Kuiper belt? This question stems from the realization that Pluto, Charon, and Triton share similarities in their own basic properties but differ drastically from their neighbors.

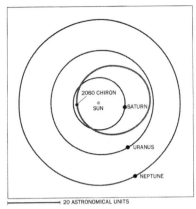

2060 CHIRON

SUN

SATURN

URANUS

NEPTUNE

20 ASTRONOMICAL UNITS

2060 Chiron may have escaped from the Kuiper belt into its current planet-crossing orbit. (Peter Samek)

■ A PECULIAR TRIO

The densities of both Pluto and Triton, for instance, are much higher than any of the giant gaseous planets of the outer solar system. The orbital motions of these bodies are also quite strange. Triton revolves around Neptune in the "retrograde" direction—opposite to the orbital direction of all planets and most satellites. Pluto's orbit slants highly from the ecliptic, and it is so far from circular that it actually crosses the orbit of Neptune. Pluto is, however, protected from possible collision with the larger planet by a special orbital relationship known as a 3:2 mean-motion resonance. Simply put, for every three orbits of Neptune around the sun, Pluto completes two.

The pieces of the celestial puzzle may fit together if one postulates that Pluto, Charon, and Triton are the last survivors of a once much larger set of similarly sized objects. S. Alan Stern of the Southwest Research Institute in Boulder first suggested this idea in 1991. These three bodies may have been swept up by Neptune, which captured Triton and locked Pluto—perhaps with Charon in tow—into its present orbital resonance.

Interestingly, orbital resonances appear to influence the position of many Kuiper belt objects as well. Up to one half of the newly discovered bodies have the same 3:2 mean-motion resonance as Pluto and, like that planet, may orbit serenely for billions of years. (The resonance prevents Neptune from approaching too closely and disturbing the orbit of the smaller body.) We have dubbed such Kuiper belt objects Plutinos— "little Plutos." Judging from the small part of the sky we have examined, we estimate that there must be several thousand Plutinos larger than 100 kilometers across.

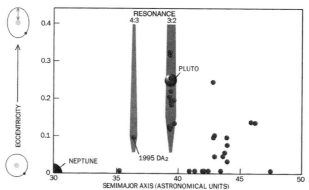

Mean-motion resonance governs the size and shape of the orbits of many Kuiper belt objects. Orbits are described by eccentricity (deviation from circularity) and semimajor axis. Like Pluto, about half the known Kuiper belt bodies circle the sun twice while Neptune completes three orbits—a 3:2 resonance. The object 1995 DA₂ orbits in other resonances. Renu Malhorra of the Lunar and Planetary Institute in Houston suggests that this pattern reflects the early evolution of the solar system, when many small bodies were ejected and the major planets migrated away from the sun. During these outward movements, Neptune could have drawn Pluto and a variety of smaller bodies into the resonant orbits that are now observed. (Peter Samek)

The recent discoveries of objects in the Kuiper belt provide a new perspective on the outer solar system. Pluto now appears special only because it is larger than any other member of the Kuiper belt. One might even question whether Pluto deserves the status of a full-fledged planet. Strangely, a line of research that began with attempts to find a 10th planet may, in a sense, have succeeded in reducing the final count to eight. This irony, along with the many intriguing observations we have made of Kuiper belt objects, reminds us that our solar system contains countless surprises.

—May 1996

THE OORT CLOUD

On the outskirts of the solar system swarms a vast cloud of comets, influenced almost as much by other stars as by our sun. The dynamics of this cloud may help explain such matters as mass extinctions on earth

Paul R. Weissman

It is common to think of the solar system as ending at the orbit of the most distant known planet, Pluto. But the sun's gravitational influence extends more than 3,000 times farther, halfway to the nearest stars. And that space is not empty—it is filled with a giant reservoir of comets, leftover material from the formation of the solar system. That reservoir is called the Oort cloud.

The Oort cloud is the Siberia of the solar system, a vast, cold frontier filled with exiles of the sun's inner empire and only barely under the sway of the central authority. Typical noontime temperatures are a frigid four degrees Celsius above absolute zero, and neighboring comets are typically tens of millions of kilometers apart. The sun, while still the brightest star in the sky, is only about as bright as Venus in the evening sky on Earth.

We have never actually "seen" the Oort cloud. But no one has ever seen an electron, either. We infer the existence and properties of the Oort cloud and the electron from the physical effects we can observe. In the case of the former, those effects are the steady trickle of long-period comets into the planetary system. The existence of the Oort cloud answers questions that people have asked since antiquity: What are comets, and where do they come from?

Aristotle speculated in the fourth century B.C. that comets were clouds of luminous gas high in the earth's atmosphere. But the Roman philosopher Seneca suggested in the first century A.D. that they were heavenly bodies, traveling along their own paths through the firmament. Fifteen centuries passed before his hypothesis was confirmed by Danish astronomer Tycho Brahe, who compared observations of the comet of 1577 made from several different locations in Europe. If the comet had been close by, then from each location it would have had a slightly different position against the stars. Brahe could not detect any difference and concluded that the comet was farther away than the moon.

Just how much farther started to become clear only when astronomers began determining the comets' orbits. In 1705 the English astronomer Edmond Halley compiled the first catalog of 24 comets. The observations were fairly crude, and Halley could fit only rough parabolas to each comet's path. Nevertheless, he argued that the orbits might be very long ellipses around the sun:

> For so their Number will be determinate and, perhaps, not so very great. Besides, the Space between the Sun and the fix'd Stars is so immense that there is Room enough for a Comet to revolve, tho' the Period of its Revolution be vastly long.

In a sense Halley's description of comets circulating in orbits stretching between the stars anticipated the discovery of the Oort cloud two and a half centuries later. Halley also noticed that the comets of 1531, 1607, and 1682 had very similar orbits and were spaced at roughly 76-year intervals. These seemingly different comets, he suggested, were actually the same comet returning at regular intervals. That body, now known as Halley's comet, last visited the region of the inner planets in 1986.

Since Halley's time astronomers have divided comets into two groups according to the time it takes them to orbit the sun (which is directly related to the comets' average distance from the sun). Long-period comets, such as the recent bright comets Hyakutake and Hale-Bopp, have orbital periods greater than 200 years; short-period comets, less than 200 years. In the past decade astronomers have further divided the short-period comets into two groups: Jupiter-family comets, such as comets Encke and Tempel 2, which have periods less than 20 years; and intermediate-period, or Halley-type, comets, with periods between 20 and 200 years.

These definitions are somewhat arbitrary but reflect real differences. The inter-mediate- and long-period comets enter the planetary region randomly from all di-rections, whereas the Jupiter-family comets have orbits whose planes are typically inclined no more than 40 degrees from the ecliptic plane, the plane of the earth's orbit. (The orbits of the other planets are also very close to the ecliptic plane.) The intermediate- and long-period comets appear to come from the Oort cloud, whereas the Jupiter-family comets are now thought to originate in the Kuiper belt, a region in the ecliptic beyond the orbit of Neptune.

■ THE NETHERWORLD BEYOND PLUTO

By the early 20th century enough long-period cometary orbits were available to study their statistical distribution *(see illustration on page 66)*. A problem emerged. About one third of all the "osculating" orbits—that is, the orbits the comets were following at the point of their closest approach to the sun—were hyperbolic. Hy-perbolic orbits would originate in and return to interstellar space, as opposed to el-liptical orbits, which are bound by gravity to the sun. The hyperbolic orbits led some astronomers to suggest that comets were captured from interstellar space by encounters with the planets.

To examine this hypothesis, celestial-mechanics researchers extrapolated, or "in-tegrated," the orbits of the long-period comets backward in time. They found that because of distant gravitational tugs from the planets, the osculating orbits did not represent the comets' original orbits *(see illustration on page 67)*. When the effects of the planets were accounted for—by integrating far enough back in time and orient-ing the orbits not in relation to the sun but in relation to the center of mass of the solar system (the sun of the sun and all the planets)—almost all the orbits became elliptical. Thus, the comets were members of the solar system, rather than inter-stellar vagabonds.

In addition, although two thirds of these orbits still appeared to be uniformly dis-tributed, fully one third had orbital energies that fell within a narrow spike. That spike represented orbits that extend to very large distances—20,000 astronomical units (20,000 times the distance of the earth from the sun) or more. Such orbits have periods exceeding one million years.

Why were so many comets coming from so far away? In the late 1940s Dutch as-tronomer Adrianus F. van Woerkom showed that the uniform distribution could be explained by planetary perturbations, which scatter comets randomly to both larger and smaller orbits. But what about the spike of comets with million-year periods?

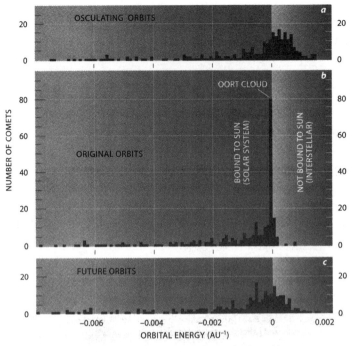

Orbital energy of known long-period comets, as shown in these histograms, reveals the Oort cloud. Astronomers first calculate the osculating orbits of the comets—the orbits they would take if their motion were entirely caused by the sun's gravity. One third of these orbits have a positive energy, making them appear interstellar (a). But when corrected for the influence of the planets and extrapolated backward in time, the energy is slightly negative—indicating that the comets came from the edge of the solar system (b). A few comets still seem to be interstellar, but this is probably the result of small observational errors. As the planets continue to exert their influence, some comets will return to the Oort cloud, some will escape from the solar system, and the rest will revisit the inner solar system (c). Technically, the orbital energy is proportional to the reciprocal of the semi-major axis, expressed in units of inverse astronomical units (AU^{-1}). (Paul R. Weissman)

In 1950 Dutch astronomer Jan H. Oort, already famous for having determined the rotation of the Milky Way galaxy in the 1920s, became interested in the problem. He recognized that the million-year spike must represent the source of the long-period comets: a vast spherical cloud surrounding the planetary system and extending halfway to the nearest stars.

Oort showed that the comets in this cloud are so weakly bound to the sun that random passing stars can readily change their orbits. About a dozen stars pass within one parsec (206,000 astronomical units) of the sun every one million years. These close encounters are enough to stir the cometary orbits, randomizing their

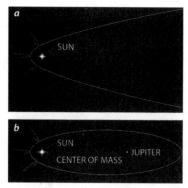

Long-period comet is so weakly bound to the sun that the planets have a decisive influence on it. Astronomers can usually see the comet only while it swings by the sun. When they apply Kepler's laws of celestial motion to plot its course—it's "osculating," or apparent orbit—the comet often seems to be on a hyperbolic trajectory, implying that it came from interstellar space and will return there (a). A more sophisticated calculation, which accounts from the planets (especially the most massive planet, Jupiter), finds that the orbit is actually elliptical (b). The orbit changes shape on each pass through the inner solar system. (Michael Goodman)

inclinations and sending a steady trickle of comets into the inner solar system on very long elliptical orbits *(see illustration on page 68)*. As they enter the planetary system for the first time the comets are scattered by the planets, gaining or losing orbital energy. Some escape the solar system altogether. The remainder return and are observed again as members of the uniform distribution. Oort described the cloud as "a garden, gently raked by stellar perturbations."

A few comets still appeared to come from interstellar space. But this was probably an incorrect impression given by small errors in the determination of their orbits. Moreover, comets can shift their orbits because jets of gas and dust from their icy surfaces act like small rocket engines as the comets approach the sun. Such nongravitational forces can make the orbits appear hyperbolic when they are actually elliptical.

■ SHAKEN, NOT STIRRED

Oort's accomplishment in correctly interpreting the orbital distribution of the long-period comets is even more impressive when one considers that he had only 19 well-measured orbits to work with. Today astronomers have more than 15 times as many. They now know that long-period comets entering the planetary region for the first time come from an average distance of 44,000 astronomical units. Such orbits have periods of 3.3 million years.

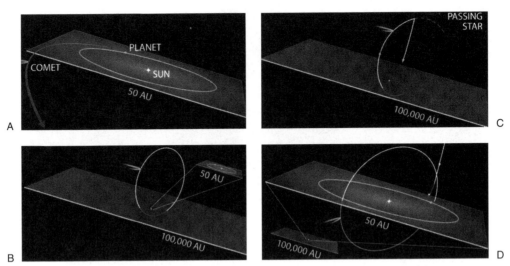

A. History of a long-period comet begins when it forms near the planets and is cata-
pulted by them into a wide orbit.
B. There the comet is susceptible to the gravitational forces of random passing stars and
giant molecular clouds, as well as the tidal forces of the galactic disk and core. These
forces randomly tilt the orbital plane of the comet and gradually pull it farther out.
C. Beyond a distance of about 20,000 astronomical units (20,000 times the earth-sun
distance), the various outside influences are capable of throwing the comet back toward the planets.
D. Once the comet reenters the inner solar system, the planets may pull it to a new orbit so that it reappears on a regular basis. (Michael Goodman)

Astronomers have also realized that stellar perturbations are not always gentle.
Occasionally a star comes so close to the sun that it passes right through the Oort
cloud, violently disrupting the cometary orbits along its path. Statistically a star is
expected to pass within 10,000 astronomical units of the sun every 36 million years
and within 3,000 astronomical units every 400 million years. Comets close to the
star's path are thrown out to interstellar space, while the orbits of comets throughout the cloud undergo substantial changes.

Although close stellar encounters have no direct effect on the planets—the closest expected approach of any star over the history of the solar system is 900 astronomical units from the sun—they might have devastating indirect consequences. In
1981 Jack G. Hills, now at Los Alamos National Laboratory, suggested that a close
stellar passage could send a "shower" of comets toward the planets, raising the rate
of cometary impacts on the planets and possibly even causing a biological mass extinction on the earth. According to computer simulations I performed in 1985 with
Piet Hut, then at the Institute for Advanced Study in Princeton, New Jersey, the
frequency of comet passages during a shower could reach 300 times the normal
rate. The shower would last two to three million years.

Recently Kenneth A. Farley and his colleagues at the California Institute of Technology found evidence for just such a comet shower. Using the rare helium 3 isotope as a marker for extraterrestrial material, they plotted the accumulation of interplanetary dust particles in ocean sediments over time. The rate of dust accumulation is thought to reflect the number of comets passing through the planetary region; each comet sheds dust along its path. Farley discovered that this rate increased sharply at the end of the Eocene epoch, about 36 million years ago, and decreased slowly over two to three million years, just as theoretical models of comet showers would predict. The late Eocene is identified with a moderate biological-extinction event, and several impact craters have been dated to this time. Geologists have also found other traces of impacts in terrestrial sediments, such as iridium layers and microtektites.

Is the earth in danger of a comet shower now? Fortunately not. Joan Garcia-Sanchez of the University of Barcelona, Robert A. Preston and Dayton L. Jones of the Jet Propulsion Laboratory in Pasadena, California, and I have been using the positions and velocities of stars, measured by the Hipparcos satellite, to reconstruct the trajectories of stars near the solar system. We have found evidence that a star has passed close to the sun in the past one million years. The next close passage of a star will occur in 1.4 million years, and that is a small red dwarf called Gliese 710, which will pass through the outer Oort cloud about 70,000 astronomical units from the sun. At that distance Gliese 710 might increase the frequency of comet passages through the inner solar system by 50 percent—a sprinkle perhaps, but certainly no shower.

In addition to random passing stars, the Oort cloud is now known to be disturbed by two other effects. First, the cloud is sufficiently large that it feels tidal forces generated by the disk of the Milky Way and, to a lesser extent, the galactic core. These tides arise because the sun and a comet in the cloud are at slightly different distances from the midplane of the disk or from the galactic center and thus feel a slightly different gravitational tug (see illustration on page 70). The tides help to feed new long-period comets into the planetary region.

Second, giant molecular clouds in the galaxy can perturb the Oort cloud, as Ludwig Biermann of the Max Planck Institute for Physics and Astrophysics in Munich suggested in 1978. These massive clouds of cold hydrogen, the birthplaces of stars and planetary systems, are 100,000 to 1 million times as massive as the sun. When the solar system comes close to one, the gravitational perturbations rip comets from their orbits and fling them into interstellar space. These encounters, though violent, are infrequent—only once every 300 million to 500 million years. In 1985 Hut and Scott D. Tremaine, now at Princeton University, showed that over the history of the solar system, molecular clouds have had about the same cumulative effect as all passing stars.

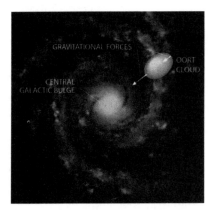

Tidal forces arise because gravity becomes weaker with distance. Therefore, the central bulge of our galaxy—a concentration of stars at the hub of the spiral pattern—pulls more on the near side of the Oort cloud (not to scale) than on the far side. The galactic plane exerts a similar force in a different direction. The galactic tides are analogous to lunar tides, which arise because the side of the earth closest to the moon feels a stronger gravitational pull than the antipode does. (Michael Goodman)

■ INNER CORE

Currently three main questions concern Oort-cloud researches. First, what is the cloud's structure? In 1987 Tremaine, Martin J. Duncan, now at Queen's University in Ontario, and Thomas R. Quinn, now at the University of Washington, studied how stellar and molecular-cloud perturbations redistribute comets within the Oort cloud. Comets at its outer edge are rapidly lost, either to interstellar space or the inner solar system, because of the perturbations. But deeper inside, there probably exists a relatively dense core that slowly replenishes the outer reaches.

Tremaine, Duncan, and Quinn also showed that as comets fall in from the Oort cloud, their orbital inclinations tend not to change. This is a major reason why astronomers now think the Kuiper belt, rather than the Oort cloud, accounts for the low-inclination, Jupiter-family comets. Still, the Oort cloud is the most likely source of the higher-inclination, intermediate-period comets, such as Halley and Swift-Tuttle. They were probably once long-period comets that the planets pulled into shorter-period orbits.

The second main question is: How many comets inhabit the Oort cloud? The number depends on how fast comets leak from the cloud into interplanetary space. To account for the observed number of long-period comets, astronomers now estimate the cloud has six trillion comets, making Oort-cloud comets the most abundant substantial bodies in the solar system. Only a sixth of them are in the outer, dynamically active cloud first described by Oort; the remainder are in the relatively dense core. If the best estimate for the average mass of a comet—about 40 billion metric tons—is applied, the total mass of comets in the Oort cloud at present is about 40 times that of the earth.

Finally, from where did the Oort-cloud comets originally come? They could not have formed at their current position, because material at those distances is too sparse to coalesce. Nor could they have originated in interstellar space; capture of comets by the sun is very inefficient. The only place left is the planetary system. Oort speculated that the comets were created in the asteroid belt and ejected by the giant planets during the formation of the solar system. But comets are icy bodies, essentially big, dirty snowballs, and the asteroid belt was too warm for ices to condense.

A year after Oort's 1950 paper, astronomer Gerard P. Kuiper of the University of Chicago proposed that comets coalesced farther from the sun, among the giant planets. (The Kuiper belt is named for him because he suggested that some comets also formed beyond the farthest planetary orbits.) Comets probably originated throughout the giant planets' region, but researchers used to argue that those near Jupiter and Saturn, the two most massive planets, would have been ejected to interstellar space rather than to the Oort cloud. Uranus and Neptune, with their lower masses, could not easily throw so many comets into escape trajectories. But more recent dynamical studies have cast some doubt on this scenario. Jupiter and particularly Saturn do place a significant fraction of their comets into the Oort cloud. Although this fraction may be smaller than that of Uranus and Neptune, it may have been offset by the greater amount of material initially in the larger planets' zones.

Therefore, the Oort-cloud comets may have come from a wide range of formation temperatures. This fact may help explain some of the compositional diversity observed in comets. Indeed, recent work I have done with Harold F. Levison of the Southwest Research Institute in Boulder, Colorado, has shown that the cloud may even contain asteroids from the inner planets' region. These objects, made of rock rather than ice, may constitute 2 to 3 percent of the total Oort-cloud population.

The key to these ideas is the presence of the giant planets, which hurl the comets outward and modify their orbits if they ever reenter the planetary region. If other stars have giant planets, as observations over the past few years suggest, they may have Oort clouds, too. If each star has its own cloud, then as stars pass by the sun, their Oort clouds will pass through our cloud. Even so, collisions between comets will be rare because the typical space between comets is an astronomical unit or more.

The Oort clouds around each star may slowly be leaking comets into interstellar space. These interstellar comets should be easily recognizable if they were to pass close to the sun, because they would approach the solar system at much higher velocities than the comets from our own Oort cloud. To date, no such interstellar comets have ever been detected. This fact is not surprising: because the solar system is a very small target in the vastness of interstellar space, there is at best a 50:50 chance that people should have seen one interstellar comet by now.

The Oort cloud continues to fascinate astronomers. Through the good fortunes of celestial mechanics, nature has preserved a sample of material from the formation of the solar system in this distant reservoir. By studying it and the cosmochemical record frozen in its icy members, researchers are learning valuable clues about the origin of the solar system.

Several space missions are now being readied to unlock those secrets. The Stardust spacecraft, due for launch next year, will fly through the coma of comet Wild 2, collect samples of cometary dust, and return them to the earth for laboratory analysis. A few years later the CONTOUR probe will fly by three comets and compare their compositions. The Deep Space 4/Champollion mission will send an orbiter and a lander to comet Tempel 1, and the Rosetta mission will do the same for comet Wirtanen. The new millennium is going to be a wonderful time for studying comets.

—September 1998

III

Planets

THE PIONEER MISSION TO VENUS

This multipart spacecraft spent 14 years scrutinizing the atmosphere, clouds, and environs of the nearest planet. The results clarify the stunningly divergent evolutionary histories of Venus and the earth

Janet G. Luhmann, James B. Pollack, and Lawrence Colin

Venus is sometimes referred to as the earth's "twin" because it resembles the earth in size and in distance from the sun. Over its 14 years of operation the National Aeronautics and Space Administration's Pioneer Venus mission revealed that the relation between the two worlds is more analogous to Dr. Jekyll and Mr. Hyde. The surface of Venus bakes under a dense carbon dioxide atmosphere, the overlying clouds consist of noxious sulfuric acid, and the planet's lack of a magnetic field exposes the upper atmosphere to the continuous hail of charged particles from the sun. Our opportunity to explore the hostile Venusian environment came to an abrupt close in October 1992, when the Pioneer Venus Orbiter burned up like a meteor in the thick Venusian atmosphere. The craft's demise marked the end of an era for the U.S. space program; in the present climate of fiscal austerity, there is no telling when humans will next get a good look at the earth's nearest planetary neighbor.

The information gleaned by Pioneer Venus complements the well-publicized radar images recently sent back by the Magellan spacecraft. Magellan concentrated on studies of Venus's surface geology and interior structure. Pioneer Venus, in comparison, gathered data on the composition and dynamics of the planet's atmosphere and interplanetary surroundings. These findings illustrate how seemingly small differences in physical conditions have sent Venus and the earth hurtling down very different evolutionary paths. Such knowledge will help scientists intelligently evaluate how human activity may be changing the environment on the earth.

Pioneer Venus consisted of two components, the Orbiter and the Multiprobe. The Multiprobe carried four craft (one large probe and three small identical ones) designed to plunge into the Venusian atmosphere, sending back data on the local conditions along the way. The Orbiter bristled with a dozen instruments with which to examine the composition and physical nature of Venus's upper atmosphere and ionosphere, the electrically charged layer between the atmosphere and outer space.

The Multiprobe was launched in August 1978 and reached Venus on December 9 of that year. Twenty-four days before its arrival the Multiprobe carrier, or "bus," released the large probe; about five days later the bus freed the three small probes to begin their own independent courses. The probes approached the planet from both high latitudes and low ones and from both the daylit and nighttime sides. In this way, information relayed by the probes during their descents enabled scientists to piece together a comprehensive picture of the atmospheric structure of Venus.

The Orbiter left the earth in May 1978, but it followed a longer trajectory than the Multiprobe, so it arrived only five days earlier, on December 4. At that time the spacecraft entered a highly eccentric orbit that looped to within 150 to 200 kilometers from the planet's surface but carried it out to a distance of 66,900 kilometers. During its closest approaches to Venus the Orbiter's instruments could directly sample the planet's ionosphere and upper atmosphere. Twelve hours later the Orbiter would have receded far enough from Venus so that the craft's remote-sensing equipment could obtain global images of the planet and could measure its near-space environment.

The gravitational pull of the sun acted to change the shape of the planet's orbit. Starting in 1986 solar gravity caused the Orbiter to pass ever closer to the planet. When the spacecraft's thrusters ran out of fuel, Pioneer Venus dove deeper into the Venusian atmosphere on each successive orbit until it met its fiery end.

Well before the arrival of Pioneer Venus, astronomers had learned that Venus does not live up to its image as the earth's near-twin. Whereas the earth maintains conditions ideal for liquid water and life, Venus is the planetary equivalent of hell. Its surface temperature of 450 degrees Celsius is hotter than the melting point of lead. Atmospheric pressure at the ground is some 93 times that at sea level on the earth.

Even aside from the heat and the pressure, the air on Venus would be utterly unbreathable to humans. The earth's atmosphere is about 78 percent nitrogen and 21 percent oxygen. Venus's much thicker atmosphere, in contrast, is composed almost entirely of carbon dioxide. Nitrogen, the next most abundant gas, makes up only about 3.5 percent of the gas molecules. Both planets possess about the same total amount of gaseous nitrogen, but Venus's atmosphere contains some 30,000 times as much carbon dioxide as does the earth's. In fact the earth does hold a quantity of carbon dioxide comparable to that in the Venusian atmosphere. On the earth, however, the carbon dioxide is locked away in carbonate rocks, not in gaseous form in the air. This crucial distinction is responsible for many of the drastic environmental differences that exist between the two planets.

The large Pioneer Venus atmospheric probe carried a mass spectrometer and gas chromatograph, devices that measured the exact composition of the atmosphere of Venus. One of the most stunning aspects of the Venusian atmosphere is that it is extremely dry. It possesses only a hundred thousandth as much water as the earth has in its oceans. If all of Venus's water could somehow be condensed onto the surface, it would make a global puddle only a couple of centimeters deep.

Unlike the earth, Venus harbors little if any molecular oxygen in its lower atmosphere. The abundant oxygen in the earth's atmosphere is a by-product of photosynthesis by plants; if not for the activity of living things, the earth's atmosphere also would be oxygen poor. The atmosphere of Venus is far richer than the earth's in sulfur-containing gases, primarily sulfur dioxide. On the earth, rain efficiently removes similar sulfur gases from the atmosphere.

Minor constituents of the Venusian atmosphere that were detected by Pioneer Venus offer clues about the internal history of the planet. The inert gas argon 40, for instance, is produced by the decay of radioactive potassium 40, which is present in nearly all rocks. As a planet's interior circulates, argon 40 that is trapped in deep rocks works its way to the surface and into the atmosphere, where it accumulates over the eons. Pioneer Venus found significantly less argon 40 in Venus's atmosphere than exists in the earth's. That disparity reflects a profound difference in how mass and heat are transported from each planet's interior to its surface. Magellan recently found evidence of earlier widespread volcanism on Venus but no signatures of the plate tectonics that keep the earth's surface geologically active and young.

Pioneer Venus revealed other ways in which Venus is more primeval than the earth. Venus's atmosphere contains higher concentrations of inert, or noble, gases—especially neon and other isotopes of argon—that have been present since the time the planets were born. This difference suggests that Venus has held on to a far greater fraction of its earliest atmosphere. Much of the earth's primitive atmosphere may have been stripped away and lost into space when our world was struck by a Mars-size body. Many planetary scientists now think the moon formed out of the cloud of debris that resulted from such a gigantic impact.

Venus's thick, carbon dioxide–dominated atmosphere is directly responsible for the inhospitable conditions on the planet's surface. On an airless body like the moon, the surface temperature depends simply on the balance between the amount of sunlight the ground absorbs and the amount of heat it emits back into space. The presence of an atmosphere complicates the situation. An atmosphere blocks some sunlight from reaching the surface and helps to carry heat upward. But more significantly, the atmospheric gases absorb infrared (thermal) radiation from the ground and reemit it back. The resultant warming of the surface is called the greenhouse effect because the atmosphere functions like a greenhouse: sunlight can get in, but infrared rays cannot get out, causing temperatures to rise.

The intensity of the greenhouse effect depends on how thoroughly the atmospheric gases capture infrared radiation. The principal greenhouse gases on the earth—carbon dioxide and water vapor—absorb complementary parts of the infrared spectrum. Adding more of these gases to the air would, in theory, increase the efficiency of the greenhouse effect, which is why people worry about the climatic impact of carbon dioxide released by human activities. The earth's atmosphere is largely transparent to infrared rays having wavelengths between 8 and 13 microns, or millionths of a meter (although ozone, methane, freon, and other gases do absorb rays in narrow portions of this band). This open "window" in the atmospheric greenhouse limits the amount of warming that the earth experiences.

Pioneer Venus showed that the greenhouse effect operates much more efficiently on Venus. Data from the four atmospheric probes enabled workers to construct a mathematical model that closely matches the observed temperatures at various altitudes. From that model it was deduced that carbon dioxide is the most significant greenhouse gas on Venus but that its action is enhanced by the presence of water vapor, clouds, sulfur dioxide, and carbon monoxide. The mixture of gases and particles in the Venusian atmosphere blocks thermal radiation at virtually all wavelengths, preventing heat from escaping into space and yielding torrid surface temperatures. These results emphasize the importance of learning more about how human-generated greenhouse gases might affect the terrestrial climate.

Astronomers have long wondered how Venus turned out so hot and dry compared with the earth, especially given that Venus and the earth probably started out with similar overall compositions. According to present theory, the two planets grew by colliding with and absorbing smaller bodies. In the process each protoplanet would have scattered some smaller bodies into orbits that would have crossed the other protoplanet's path. Hence, the earth and Venus should have accumulated comparable quantities of water-rich bodies even if, at first, water was irregularly distributed through the infant solar system. The roughly equal quantities of carbon dioxide and nitrogen on the two planets support the notion that they once had comparable amounts of water as well.

The young earth and Venus quickly developed thick atmospheres consisting of gases expelled from their interiors and of the vaporized remains of icy, impacting bodies. Water in the earth's atmosphere condensed into lakes and oceans, which proved crucial to the planet's climatic development. Much of the airborne carbon dioxide was quickly sequestered into solid carbonate, a process that occurs through the chemical weathering of rocks in the presence of liquid water.

Venus, too, may have had broad oceans during its youth. The newborn sun was about 30 percent less luminous than it is at present, so temperatures on Venus could have been well below the boiling point of water. (Venus orbits at .72 times the earth's distance from the sun.) As the sun brightened, however, surface temperatures on Venus eventually rose above boiling. From then on, any carbon dioxide exhaled by volcanoes or delivered by impacts on Venus could no longer be removed from the atmosphere by chemical weathering. As carbon dioxide accumulated in the atmosphere, the greenhouse effect grew ever more intense. The ultimate result was the sizzling, carbon dioxide–dominated world of today.

After the oceans boiled the atmosphere of Venus should have been full of water vapor—in clear contrast to the data. Where has all the water gone?

Pioneer Venus has helped answer that question. The spacecraft documented that even now Venus continues to lose water. Water molecules that wander above the cloud tops react with solar radiation and other molecules. In the process the water molecules split into their oxygen and hydrogen components. The lightweight hydrogen atoms may escape into space by interacting with energetic atoms and ions in the upper atmosphere or with the solar wind, a flow of charged particles that issues from the sun. The leftover oxygen atoms may combine with minerals on the surface, or they, too, may escape by interacting with the solar wind.

A few billion years ago Venus's upper atmosphere contained much more water than it does now; the early sun also emitted far more energetic ultraviolet rays. Both factors greatly hastened the rate at which Venus's water was destroyed and carried off into space. Calculations indicate that over the 4.5-billion-year lifetime of the solar system, Venus could have lost as much water as resides in the earth's oceans.

The earth never experienced such large losses of water because of its moderate surface temperature. Water on the earth stays mostly on the ground or in the lower atmosphere, very little reaches the upper atmosphere, where it may disappear forever. Once the oceans of Venus boiled, in contrast, the planet's atmosphere grew ever hotter, driving more and more water vapor into the upper reaches of the atmosphere.

And yet some water remains. Observations of Venus's upper atmosphere made by the Pioneer Venus Orbiter imply that the planet now loses about 5×10^{25} hydrogen atoms and ions each second. At that rate the entire amount of water in the atmosphere would be gone in about 200 million years. Venus is more than 20 times that old, so some mechanism must replenish the water that Venus is constantly losing. The water most likely derives from a mix of external sources (such as the impact of comets and icy asteroids) and internal ones (through volcanic eruptions or more widespread and steady outgasing to the surface). Because the understanding of Venus's water loss is still quite sketchy, it is possible that Pioneer Venus might actually have observed the last trickle from the planet's water-rich early atmosphere.

Despite its lack of water, Venus is cloaked in thick clouds that conceal its surface from conventional telescopes. The nature of those clouds has intrigued astronomers for centuries. By the time of the Pioneer Venus mission, planetary scientists had accumulated strong evidence that the clouds were largely composed of concentrated solutions of sulfuric acid and water. They could not determine, however, the source of the sulfur from which the cloud droplets arose.

Pioneer Venus finally settled the question. As the Orbiter circled the planet, it scrutinized the tops of the clouds using its ultraviolet spectrometer, which identifies the characteristic pattern of emission and absorption from various atoms and molecules. Also, the gas chromatograph on the large probe measured the composition of the region below the main cloud deck. The results of these studies show that the sulfuric acid in the clouds derives from sulfur dioxide in the atmosphere.

Near the top of the clouds, some 60 to 70 kilometers above the ground, ultraviolet rays from the sun split sulfur dioxide into molecular fragments known as radicals. These radicals undergo a series of chemical reactions with radicals derived from water, ultimately producing tiny droplets of sulfuric acid. Gravity and air currents cause the droplets to migrate downward. As the droplets fall, they grow by colliding with one another and by accumulating sulfuric acid vapor from the air. At and below the base of the clouds, sulfuric acid particles dissociate back into sulfur dioxide and water vapor.

Instruments on the Pioneer Venus probes detected tiny particles (less than a thousandth of a millimeter across) at altitudes between 48 and 30 kilometers, just below the base of the cloud deck. Atmospheric motions carry these particles, along with sulfuric acid vapor, to higher, colder altitudes. There the sulfuric acid rapidly condenses onto the particles, producing much larger cloud particles that are concentrated toward the clouds' base. The density of those particles varies from place to place in the lower cloud region, probably because of irregularities in the upwelling and downwelling motions.

A related *Pioneer Venus* observation has stirred great excitement and controversy. In the course of exploring Venus's sulfur chemistry the Orbiter detected an apparent steady decrease in the concentration of sulfur dioxide near the tops of the clouds. Some workers interpreted that measurement as evidence that a giant volcanic eruption spewed sulfur into the atmosphere just about the time that Pioneer Venus arrived—a tantalizing sign that Venus might have active, explosive volcanoes. Once the eruption ceased the sulfur levels would have begun to drop, as observed. Other investigators have argued that the changes in composition could have resulted from normal variations in the atmospheric circulation. The issue remains frustratingly unsettled.

Although it could not resolve the sulfur dioxide puzzle, Pioneer Venus has provided many other intriguing details about the circulation of the Venusian atmosphere. Such information is a tremendous boon for scientists attempting to understand atmospheric dynamics because it shows how weather patterns operate on a planet that differs from the earth in several crucial aspects.

Venus rotates extremely slowly; the earth completes 243 daily rotations in the time it takes Venus to turn once with respect to the stars. Also, because of the dense atmosphere, the surface temperature on Venus is nearly constant from equator to pole. One might naively assume, therefore, that winds on Venus would be very sluggish.

Pioneer Venus proved that assumption false. On the earth, winds at low latitudes move more slowly than the rotation of the planet, whereas those at higher latitudes exceed the speed of the surface, a state known as superrotation. The atmosphere of Venus superrotates at all latitudes and at all heights from close to the surface to at least 90 kilometers above the surface. The winds attain their peak velocity near the cloud tops, where they blow at an unexpectedly rapid 100 meters a second, about 60 times as fast as the rotation of the underlying surface.

Winds in the atmospheres of the earth and the other terrestrial planets are driven by local imbalances between the amount of incoming solar energy and the amount of outgoing, radiated heat. In general, low latitudes, which receive the most sun-

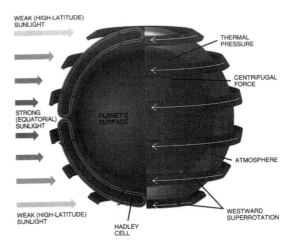

WEAK (HIGH-LATITUDE) SUNLIGHT

THERMAL PRESSURE

CENTRIFUGAL FORCE

STRONG (EQUATORIAL) SUNLIGHT

PLANET'S SURFACE

ATMOSPHERE

WEAK (HIGH-LATITUDE) SUNLIGHT

HADLEY CELL

WESTWARD SUPERROTATION

Rapid winds at Venus's cloud tops move 60 times as fast as the body of the planet. Atmospheric circulation is driven by solar radiation, which produces a north-south flow known as a Hadley cell (see above). The rotation of the atmosphere transforms Hadley cells into predominantly westward zonal winds, which may be amplified by eddies. (Schneidman/JSD)

light, experience a net heating, whereas high latitudes, which receive the least incident solar energy, undergo a net cooling. As a result the atmosphere develops a large-scale circulation pattern called a Hadley cell. In this pattern hot air rises near the equator and travels toward the poles, where it sinks and returns toward the equator.

The spinning of a planet on its axis deflects the north-south (meridional) winds sideways, however, giving rise to east-west, or zonal, winds. Surprisingly, zonal winds almost always end up being much stronger than the north-south winds from which they derived. On the earth Hadley circulation dominates atmospheric motions at low latitudes. Zonal winds close to the equator move slower than the earth's rotation (and hence are called easterlies); those closer to the poles form superrotating westerlies, culminating in the rapid flow of the jet stream. What is so odd about Venus's zonal winds is that they superrotate at almost all latitudes in the lower atmosphere.

Even now planetary scientists do not fully understand why the entire lower atmosphere of Venus superrotates. The large fraction of solar energy that is absorbed high in the atmosphere, near the tops of the clouds, probably contributes to the brisk winds. The high-altitude heating of the atmosphere may set up a circulation system that is much less influenced by frictional interaction with the surface than is the case on the earth. The atmosphere of Venus might therefore be highly susceptible to the formation of eddies that can efficiently transport angular momentum. Such eddies could counteract the ability of the Hadley circulation to prevent superrotation at low latitudes. Cloud images taken by the Pioneer Venus Orbiter provide evidence of small-scale, eddylike variations in the winds.

High above the superrotating layers of the Venusian atmosphere lies the ionosphere, an extended zone of electrically charged atoms and molecules, or ions. The ions arise when high-energy ultraviolet rays from the sun knock electrons free from atmospheric gases. Every planet that has a substantial atmosphere possesses an ionosphere, but the one on Venus has a number of unusual traits.

The Pioneer Venus Orbiter monitored the passage of radio waves through the ionosphere and, during close approaches to the planet, measured its temperature, density, and composition directly. As one might expect, Venus's ionosphere is densest in the center of the dayside hemisphere, near the equator, where the incoming sunlight is most direct. Because of the abundant chemical reactions occurring among the particles, Venus's ionosphere consists primarily of oxygen ions, even though carbon dioxide is the dominant gas at low levels.

Unlike the earth and most other planets, Venus has no significant global magnetic field, for reasons not fully understood. The absence of a magnetic field significantly affects the structure of Venus's ionosphere. The Orbiter detected a weak ionosphere that extends beyond the day-night boundary. This finding was intriguing because, in the darkness, ions and free electrons should quickly recombine into neutral atoms. An instrument on the Orbiter found that on Venus ions from the dayside are able to migrate to the nightside. On the earth the planetary magnetic field in the ionosphere inhibits such horizontal flow.

Images of the planet in ultraviolet radiation, obtained using the Orbiter's ultraviolet spectrometer, detected a previously unknown, patchy aurora on Venus's shadowed hemisphere. Scientists attribute the aurora to energetic particles, probably fast-moving electrons, that crash into the atmosphere on the nightside. When these particles strike gas molecules in the atmosphere, they excite and ionize the molecules, further contributing to Venus's nighttime ionosphere. The excited molecules soon return to their normal, low-energy state by emitting radiation, which shows up as the aurora.

As is the case for terrestrial auroras, the particles that cause the auroras on Venus derive their energy from the solar wind. The solar wind is the sun's extended, rarefied outer atmosphere. It consists of plasma, or charged particles (primarily protons and electrons), racing from the sun at supersonic speeds. At the orbit of Venus the solar wind has a density of 15 protons and electrons per cubic centimeter and a velocity of 400 kilometers per second. As the solar wind blows past the planets, it carries part of the sun's magnetic field with it.

The intrinsic magnetic fields around the earth and other planets act as obstacles to the electrically charged solar wind. The wind flows around those fields along a surface (the magnetopause) where the pressure of the wind equals the opposing magnetic pressure. The extent of the deflection depends on the strength of the planetary magnetic field. Venus, which has virtually no field at all, creates an obstacle scarcely bigger than the planet itself.

Nevertheless, the spacecraft found that solar-wind plasma was clearly being diverted around Venus. That discovery confirmed theoretical predictions that a planet's ionosphere can effectively block the solar wind even in the absence of a substantial magnetic field. Like a magnetic field, the ionosphere exerts pressure against the wind, but in this case it is the thermal pressure of the charged gas that counters the force of the solar wind. On average the balance point lies at a 300-kilometer altitude near Venus's noontime equator and at 800 to 1,000 kilometers above the day-night boundary.

The deflection of the solar-wind flow around large obstacles (such as planets) is preceded by a "bow shock," a sharp boundary closely analogous to the shock that forms in front of a supersonic aircraft. During most of its lifetime the Pioneer Venus Orbiter crossed the bow shock twice each lap, enabling it to monitor continuous changes in the magnetic environment around Venus. The craft found that the bow shock expands and contracts in step with the 11-year cycle of solar activity. The radius of the shock in the plane of the day-night boundary ranges in size from about 14,500 kilometers at solar maximum to 12,500 kilometers at solar

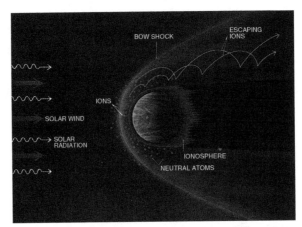

Solar wind interacts directly with the upper atmosphere of Venus because the planet has no substantial magnetic field. Where the solar wind skirts around the planet, a bow shock forms. Some neutral atoms at the top of the Venusian atmosphere become electrically charged ions that are then carried off in the solar wind. Radiation from the sun also gives rise to a permanent, charged layer, the ionosphere, which forms at lower altitudes around the planet. (Tomo Narashima)

minimum. The expansion and contraction probably result from changes in Venus's upper atmosphere associated with the varying radiation flux from the sun.

Just downstream of the bow shock the solar wind grows more dense, slows down, and changes is direction of flow. Magnetic field lines are frozen into the solar wind because it is completely ionized. After the solar wind passes through the shock, the frozen-in interplanetary magnetic field piles up.

Pioneer Venus mapped the large-scale magnetic field geometry around Venus. These data give the impression that the magnetic field lines eventually slip around the obstacle and into the wake that it creates in the solar wind. Researchers refer to this wake structure as an induced magnetotail because it derives from the interplanetary magnetic field rather than from the planet's own field, as is the case for the earth's own, much larger magnetic tail.

Because of its lack of a significant internal field, Venus interacts more directly with the solar wind than does the earth. Over the age of the solar system that interaction has affected the atmosphere of Venus. The planet's upper atmosphere, where atomic oxygen predominates, extends well above the point at which the solar wind is diverted around the planet. This gas remains largely unaffected by the solar-wind plasma as long as it remains electrically neutral. If an oxygen atom is struck by an ultraviolet ray or if it collides with a particle in the solar wind, it can become ionized. The oxygen ion couples to the flowing plasma, which may carry it away from the planet and out of the solar system.

Instruments on board the Pioneer Venus Orbiter confirm that the solar wind truly does scavenge Venus's upper atmosphere. Measurements of the density of the Venusian ionosphere indicate that the uppermost layers—those above the deduced height of the solar-wind obstacle—appear to be missing. Evidently the ions created above the obstacle have been carried off in the manner just described. The Orbiter has also detected the oxygen ions escaping tailward in the solar wind. In essence Pioneer Venus has captured a snapshot of one of the processes by which Venus evolved into a world so unlike the earth.

The extensive archive of data generated by Pioneer Venus has proved a wonderful resource for scientists studying the planet's atmosphere and near-space environs. Those data are all the more precious given that no nation has any plans for a follow-up venture. At present, major basic science projects face tight funding prospects in all developed countries. Nevertheless, the intriguing questions raised by Pioneer Venus have inspired studies for a possible return to the earth's cloud-enshrouded neighbor.

Even relatively inexpensive missions to Venus could deliver valuable results. A simple chemical composition probe, for example, could elucidate the nature of the atmospheric chemistry at various altitudes. A series of small craft deployed simultaneously all around Venus could yield a sharper picture of the planet's global weather patterns. A specialized orbiter could carry instruments to search for lightning storms and to measure in more detail the ions and atoms that escape from the planet. These scientific goals might be accomplished under the auspices of NASA'S upcoming series of fast, low-cost "Discovery-class" planetary expeditions.

A better understanding of the global environment on Venus could be considered a worthwhile goal in itself. It also provides a perspective on the nature of the environment on the earth and on the delicate balance of physical processes that keep our world habitable. Planetary missions such as Pioneer Venus clarify the earth's unique place among the worlds of the solar system. For this reason, and to satisfy the fundamental human drive to explore, we hope such missions continue to receive support in the U.S. and abroad.

—April 1994

THE CASE FOR RELIC LIFE ON MARS

A meteorite found in Antarctica offers strong evidence that Mars has had—and may still have—microbial life

Everett K. Gibson, Jr., David S. McKay,
Kathie Thomas-Keprta, and Christopher S. Romanek

Of all the scientific subjects that have seized the public psyche, few have held on as tightly as the idea of life on Mars. Starting not long after the invention of the telescope and continuing for a good part of the past three centuries, the subject has inspired innumerable studies ranging from the scientific to the speculative. But common to them all was recognition of the fact that in our solar system, if a planet other than the earth harbors life, it is almost certainly Mars.

Interest in Martian life has tended to coincide with new discoveries about the mysterious red world. Historically, these discoveries have often occurred after one of the periodic close approaches between the two planets. Every 15 years Mars comes within about 56 million kilometers of the earth (the next approach will occur in the summer of 2003). Typically, life on Mars was assumed to be as intelligent and sophisticated as that of *Homo sapiens,* if not more so. (Even less explicably, Martian beings have been popularly portrayed as green and diminutive.)

It was after one of the close approaches in the late 19th century that Italian astronomer Giovanni V. Schiaparelli announced that he had seen great lines stretching across the planet's surface, which he called *canali.* At the turn of the century U.S. astronomer Percival Lowell insisted that the features were canals constructed by an advanced civilization. In the 1960s and 1970s, however, any lingering theories about the lines and elaborate civilizations were put to rest after the U.S. and the Soviet Union sent the first space probes to the planet. The orbiters showed that there were in fact no canals, although there were long, huge canyons. Within a decade landers found no evidence of life, let alone intelligent life and civilization.

Although the debate about intelligent life was essentially over, the discussions about microbial life on the planet—particularly life that may have existed on the warmer, wetter Mars of billions of years ago—were just beginning. In August 1996 this subject was thrust into the spotlight when we and a number of our colleagues at the National Aeronautics and Space Administration Johnson Space Center and at Stanford University announced that unusual characteristics in a meteorite known to have come from Mars could most reasonably be interpreted as the vestiges of ancient Martian bacterial life. The 1.9-kilogram, potato-size meteorite, designated ALH84001, had been found in Antarctica in 1984.

Our theory was by no means universally embraced. Some researchers insisted that there were nonbiological explanations for the meteorite's peculiarities and that these rationales were more plausible than our biological explanation. We remain convinced that the facts and analyses that we will outline in this article point to the existence of a primitive form of life. Moreover, such life-forms may still exist on Mars if, as some researchers have theorized, pore spaces and cracks in rocks below the surface of the planet contain liquid water.

Why should researchers even care about the possible existence of such a simple form of life billions of years ago on the red planet? Certainly, the prevalence of life in the universe is among the most profound scientific questions. Yet almost no hard data exist that can be used to theorize on that issue. Confirmation that primitive life once flourished on Mars would be extremely useful to those studying the range of conditions under which a planet can generate the complex chemistry from which life evolves. Then, too, the information could help us understand the origin of life on the earth. Ultimately, these kinds of insights could elucidate various hypotheses—which are currently little more than guesses—about how common life is in the universe.

■ INHOSPITABLE PLANET

Conditions on Mars today are not hospitable to life as we know it. The planet's atmosphere consists of 95 percent carbon dioxide, 2.7 percent nitrogen, 1.6 percent argon, and only trace amounts of oxygen and water vapor. Surface pressure is less than 1 percent of the earth's, and daily temperatures rarely exceed zero degrees Celsius, even in the planet's warmest regions in the summer. Most important, one of life's most fundamental necessities, liquid water, seems not to exist on the planet's surface.

Given these realities, it is perhaps not surprising that the two Viking space probes that settled on the planet's surface, in July and September of 1976, failed to find any evidence of life. The results cast doubt on—but did not completely rule out—the possibility that there is life on Mars. The landers, which were equipped to detect organic compounds at a sensitivity level of one part per billion, found none, either at the surface or in the soil several centimeters down. Similarly, three other experiments found no evidence of microbial organisms. Ultimately researchers concluded that the possibility of life on Mars was quite low and that a more definite statement on the issue would have to await the analysis of more samples by future landers—and, it was hoped, the return of some samples from the red planet for detailed study on the earth.

Although the landers found no evidence of life on present-day Mars, photographs of the planet taken from orbit by the Viking craft, as well as earlier images made by the Mariner 9 probe, strongly suggest that great volumes of water had sculpted the planet's surface a few billion years ago and perhaps as recently as several hundred million years ago.

In addition, various meteorites found on the earth and known to be of Martian origin—including ALH84001 itself—offer tangible proof of Mars's watery past because they show unambiguous signs of having been altered by water. Specifically, some of these meteorites have been found to contain carbonates, sulfates, hydrates, and clays, which can be formed, so far as planetary scientists know, only when water comes into contact with other minerals in the rock.

Of course the entire argument hinges on ALH84001's having come from the red planet. Of this, at least, we can be certain. It is one of several meteorites found since the mid-1970s in meteorite-rich regions in Antarctica. In the early 1980s Donald D. Bogard and Pratt Johnson of the NASA Johnson Space Center began studying a group of meteorites found to contain minute bubbles of gas trapped within glass inside the rock. The glass is believed to have formed during impacts with meteoroids or comets while the rock was on the surface of Mars. Some of these glass-producing impacts apparently imparted enough energy to eject fragments out into space; from there, some of these rocks were captured by the earth's gravitational field. This impact scenario is the only one that planetary scientists believe can account for the existence on our world of bits of Mars.

Bogard and Johnson found that the tiny samples of gas trapped in the glass of some of the meteorites had the exact chemical and isotopic compositions as gases in the atmosphere of Mars, which had been measured by the Viking landers in 1976. The 1:1 correlation between the two gas samples—over a range of nine orders of magnitude—strongly suggests that these meteorites are from Mars. In all, five meteorites have been shown to contain samples of trapped Martian atmosphere. ALH84001 was not among the five so analyzed; however, its distribution of oxygen isotopes, mineralogy, and other characteristics place it in the same group with the other five Martian rocks.

The distribution of oxygen isotopes within a group of meteorites has been the most convincing piece of evidence establishing that the rocks—including ALH-84001—come from Mars. In the early 1970s Robert N. Clayton and his coworkers at the University of Chicago showed that the isotopes oxygen 16, oxygen 17, and oxygen 18 in the silicate materials within various types of meteorites have unique relative abundances. The finding was significant because it demonstrates that the bodies of our solar system formed from distinct regions of the solar nebula and thus have unique oxygen isotopic compositions. Using this isotopic "fingerprint," Clayton helped to show that a group of 12 meteorites, including ALH84001, are indeed closely related. The combination of trapped Martian atmospheric gases and the specific distribution of oxygen isotopes has led researchers to conclude that the meteorites must have come from Mars.

■ INVADER FROM MARS

Other analyses, mainly of radio-isotopes, have enabled researchers to outline ALH84001's history from its origins on the red planet to the present day. The three key time periods of interest are the age of the rock (the length of time since it crystallized on Mars), how long the meteorite traveled in space, and how long it has been on the earth. Analysis of three different sets of radioactive isotopes in the meteorite have established each of these time periods.

The length of time since the rock solidified from molten materials—the so-called crystallization age of the material—has been determined through the use of three different dating techniques. One uses isotopes of rubidium and strontium, another, neodymium and samarium, and the third, argon. All three methods indicated that the rock is 4.5 billion years old. By geologic standards the rock is extremely old; the 4.5-billion-year figure means that it crystallized within the first 1 percent of Mars's history. In comparison the other 11 Martian meteorites that have been analyzed are all between 1.3 billion years old and 165 million years old. It is remarkable that a rock so old, and so little altered on Mars or during its residence in the Antarctic ice, became available for scientists to study.

The duration of the meteorite's space odyssey was determined through the analysis of still other isotopes, namely helium 3, neon 21, and argon 38. While a meteorite is in space, it is bombarded by cosmic rays and other high-energy particles. The particles interact with the nuclei of certain atoms in the meteorite, producing the three isotopes listed above. By studying the abundances and production rates of these cosmogenically produced isotopes, scientists can determine how long the meteorite was exposed to the high-energy flux and, therefore, how long the specimen was in space. Using this approach, researchers concluded that after being torn free from the planet, ALH84001 spent 16 million years in space before falling in the Antarctic.

To determine how long the meteorite lay in the Antarctic ice, A. J. Timothy Jull of the University of Arizona used carbon-14 dating. When silicates are exposed to cosmic rays in space, carbon 14 is produced. In time the rates of production and decay of carbon 14 balance, and the meteorite becomes saturated with the isotope. The balance is upset when the meteorite falls from space and production of carbon 14 ceases. The decay goes on, however, reducing the amount in the rock by one half every 5,700 years. By determining the difference between the saturation level and the amount measured in the silicates, researchers can determine how long the meteorite has been on the earth. Jull's finding was that ALH84001 fell from space 13,000 years ago.

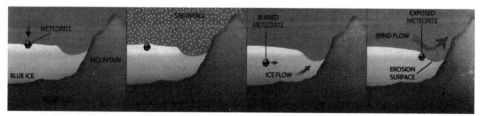

A combination of geologic and meteorological phenomena gathers meteorites at the bases of Antarctica's mountains. After landing, the meteorites become buried in compressed snow, which eventually becomes ice. Sheets of ice move toward the edges of the continent, carrying the meteorites with them. If a mountain blocks horizontal movement of the meteorites, they will in time become exposed near the mountain. The reason is that the winds slowly but continuously "ablate" the ice above the meteorites, turning it into a gas. Ablation exposes areas of ice that had been buried deep under the surface, so meteorites are found on ice that is generally more than 10,000 years old and is bluish in color. (Laurie Grace)

From the very moment it was discovered, the meteorite now known as ALH84001 proved unusual and intriguing. In 1984 U.S. geologist Roberta Score found the meteorite in the Far Western Icefield of the Allan Hills Region. Score recognized that the rock was unique because of its pale greenish gray color. The sample turned out to consist of 98 percent coarse-grained orthopyroxene [$(Mg,Fe)SiO_3$], a silicate mineral. There are also relatively minor amounts of feldspathic glass, which is also known as maskelynite ($NaAlSi_3O_8$), olivine [$(MgFe)_2SiO_4$], chromite ($FeCr_2O_4$), and pyrite (FeS_2), as well as carbonate phases and phyllosilicates.

■ CARBONATES ARE KEY

The most interesting aspect of ALH84001 is the carbonates, which exist as tiny discoids, like flattened spheres, 20 to 250 microns in diameter. They cover the walls of cracks in the meteorite and are oriented in such a way that they are flattened against the inside walls of the fractures. The globules were apparently deposited from a fluid saturated with carbon dioxide that percolated through the fracture after the silicates were formed. None of the other 11 meteorites known to have come from Mars has such globules.

It was within the carbonate globules that our research team found the assortment of unique features that led us to hypothesize that microbial organisms came into contact with the rock in the distant past. Basically, the case for ancient microbial life on Mars is built almost entirely around the globules.

Individually, none of the features we found is strongly indicative of life. Collectively, however—and especially within the confines of the tiny discoids—the globules can be plausibly explained as the ancient vestiges of microbial life. The features fall into several categories of evidence. One category centers on the presence of tiny iron oxides and iron sulfide grains, which resemble those formed by terrestrial bacteria. The second group revolves around the presence of organic carbon molecules in and on the globules. Finally, unusual structures found within the globules bear a striking resemblance to bacterial fossils found on the earth. Another relevant piece of evidence suggests the globules formed from a water-rich fluid below 100 degrees C.

NASA Johnson Space Center researchers, along with Monica Grady of the British Museum of Natural History and workers at the Open University in the U.K., performed the first isotopic analysis of carbon and oxygen in the carbonate globules. The carbon analysis indicates that the globules have more carbon 13 than any carbonates found on the earth but just the right amount to have come from Mars.

Most carbon on the earth is made up of 98.9 percent carbon 12 and 1.1 percent carbon 13. Various reactions, however, can alter this ratio. For example, in general a sample of carbon that has been a part of an organic chemical system—say, in plant matter—is somewhat more enriched in carbon 12, whereas carbon in limestone is relatively enriched in carbon 13. The carbon in the globules of ALH84001 is more enriched in carbon 13 than any natural materials on the earth. Moreover, the enrichment is different from that of the other 11 Martian meteorites. This fact suggests that the carbon in the globules—unlike the trace amounts of carbon seen in the other Martian meteorites—may have been derived from Mars's atmosphere.

Analysis of the distribution of oxygen isotopes in the carbonates can provide information about the temperature at which those minerals formed. The subject bears directly on the question of whether the carbonates were formed at temperatures that could support microbial life, because terrestrial organisms do not survive at temperatures above about 115 degrees C. The NASA-U.K. team analyzed the oxygen isotopes in the carbonate globules. Those findings strongly suggest that the globules formed at temperatures no higher than 100 degrees C. Earlier this year John W. Valley of the University of Wisconsin—Madison used an ion microprobe technique to confirm our finding.

It should be noted that another research group, led by Ralph P. Harvey of Case Western Reserve University, has analyzed the chemical composition of the minerals in the carbonates with an electron microprobe and concluded that the carbonates formed at 700 degrees C. In our view Harvey's findings are at odds with a growing body of evidence that the globules formed at relatively low temperatures.

We are extremely interested in the age of the carbonates, because it would allow us to estimate when microbial life left its mark on the rock that became ALH84001. Yet all we can say for sure is that the carbonates crystallized in the fractures in the meteorite some time after the rock itself crystallized. Various research groups have come up with ages ranging from 1.3 to 3.6 billion years; the data gathered so far, however, are insufficient to date the carbonate globules conclusively.

■ BIOMINERAL CLUES

The first category of evidence involves certain minerals found inside the carbonate globules; the type and arrangement of the minerals are similar, if not identical, to certain biominerals found on the earth. Inside, the globules are rich in magnesite ($MgCO_3$) and siderite ($FeCO_3$) and have small amounts of calcium and manganese carbonates. Fine-grained particles of magnetite (Fe_3O_4) and sulfides ranging in size from 10 to 100 nanometers on a side are present within the carbonate host. The magnetite crystals are cuboid, teardrop, or irregular in shape. Individual crystals have well-preserved structures with little evidence of defects or trace impurities.

An analysis of the samples conducted with high-resolution transmission electron microscopy coupled with energy-dispersive spectroscopy indicates that the size, purity, morphology, and crystal structures of all these magnetites are typical of magnetites produced by bacteria on the earth.

Terrestrial magnetite particles associated with fossilized bacteria are known as magnetofossils. These particles are found in a variety of sediments and soils and are classified, according to size, as superparamagnetic (less than 20 nanometers on an edge) or single-domain (20 to 100 nanometers). The magnetites within ALH84001 are typically 40 to 60 nanometers on an edge.

Single-domain magnetite has been reported in ancient terrestrial limestones and is generally regarded as having been produced by bacteria. Most intriguing, some of the magnetites in ALH84001 are arranged in chains, not unlike pearls in a necklace. Terrestrial bacteria often produce magnetite in precisely this pattern, because as they biologically process iron and oxygen from the water, they produce crystals that naturally align themselves with the earth's magnetic field.

■ ORGANIC CARBON MOLECULES

The presence of organic carbon molecules in ALH84001 constitutes the second group of clues. In recent years researchers have found organic molecules not only in Martian meteorites but also in ones known to have come from the asteroid belt in interplanetary space, which could hardly support life. Nevertheless, the type and relative abundance of the specific organic molecules identified in ALH84001 are suggestive of life processes. The presence of indigenous organic molecules within ALH84001 is the first proof that such molecules have existed on Mars.

On the earth, when living organisms die and decay they create hydrocarbons associated with coal, peat, and petroleum. Many of these hydrocarbons belong to a class of organic molecules known as polycyclic aromatic hydrocarbons (PAHs). There are thousands of different PAHs. Their presence in a sample does not in itself demonstrate that biological processes occurred. It is the location and association of the PAHs in the carbonate globules that make their discovery so interesting.

In ALH84001 the PAHs are always found in carbonate-rich regions, including the globules. In our view the relatively simple PAHs are the decay products of living organisms that were carried by a fluid and trapped when the globules were formed. In 1996 a team at the Open University showed that the carbon in the globules in ALH84001 has an isotopic composition suggestive of microbes that used methane as a food source. If confirmed, this finding will be one of the strongest pieces of evidence to date that the rock bears the imprint of biological activity.

In our 1996 announcement Richard N. Zare and Simon J. Clemett of Stanford used an extremely sensitive analytical technique to show that ALH84001 contains a relatively small number of different PAHs, all of which have been identified in the decay products of microbes. Most important, the PAHs were found to be located inside the meteorite, where contamination is very unlikely to have occurred. This critical finding supports the idea that the carbonates are Martian and contain the vestiges of ancient living organisms.

PAHs are a component of automobile exhaust, and they have also been found in meteorites, planetary dust particles, and even interstellar space. Significantly, ultrasensitive analysis of the distribution of the PAHs in ALH84001 indicated that the PAHs could not have come from the earth or from an extraterrestrial source—other than Mars.

Perhaps the most visually compelling pieces of evidence that at least vestiges of microbes came into contact with the rock are objects that appear to be the fossilized remains of microbes themselves. Detailed examination of the ALH84001 carbonates using high-resolution scanning electron microscopy (SEM) revealed unusual features that are similar to those seen in terrestrial samples associated with biogenic activity. Close-up SEM views show that the carbonate globules contain ovoid and tube-shaped bodies. The objects are around 380 nanometers long, which means they could very well be the fossilized remains of bacteria. To pack in all the components that are normally required for a typical terrestrial bacterium to function, sizes larger than 250 nanometers seem to be required. Additional tubelike curved structures found in the globules are 500 to 700 nanometers in length.

■ NANOBACTERIA OR APPENDAGES?

Other objects found within ALH84001 are close to the lower size limit for bacteria. These ovoids are only 40 to 80 nanometers long; other, tube-shaped bodies range from 30 to 170 nanometers in length and 20 to 40 nanometers in diameter. These sizes are about a factor of 10 smaller than the terrestrial microbes that are commonly recognized as bacteria. Still, typical cells often have appendages that are generally quite small—in fact about the same size as these features observed within ALH84001. It may be possible that some of the features are fragments or parts of larger units within the sample.

ALH84001's numerous ovoid and elongated features are essentially identical in size and morphology to those of so-called nanobacteria on the earth. So far little study has been devoted to nanobacteria or bacteria in the 20-to-400-nanometer range. But fossilized bacteria found within subsurface basalt samples from the Columbia River basin in Washington State have features that are essentially identical to some of those observed in the ovoids in ALH84001.

ALH84001 was present on Mars 4.5 billion years ago, when the planet was wetter, was warmer, and had a denser atmosphere. Therefore, we might expect to see evidence that the rock had been altered by contact with water. Yet the rock bears few traces of so-called aqueous alteration evidence. One such piece of evidence would be clay minerals, which are often produced by aqueous reactions. The meteorite does indeed contain phyllosilicate clay mineral, but only in trace amounts. It is not clear, moreover, whether the clay mineral formed on Mars or in the Antarctic.

Mars had liquid water on its surface early in its history and may still have an active groundwater system below the permafrost or cryosphere. If surface microorganisms evolved during a period when liquid water covered parts of Mars, the microbes might have spread to subsurface environments when conditions turned harsh on the surface. The surface of Mars contains abundant basalts that were undoubtedly fractured during the period of early bombardment in the first 600 million years of the planet's history. These fractures could serve as pathways for liquid water and could have harbored any biota that were adapting to the changing conditions on the planet. The situation has an analogue on the earth, where thin gaps between successive lava flows appear to serve as aquifers for the movement and containment of groundwater containing living bacteria.

Organisms may also have developed at hot springs or in underground hydrothermal systems on Mars where chemical disequilibriums can be maintained in environments somewhat analogous to those of the mineral-rich "hot smokers" on the seafloor of the earth.

Thus it is entirely possible that if organisms existed on Mars in the distant past, they may still be there. Availability of water within the pore spaces of a subsurface reservoir would facilitate their survival. If the carbonates within ALH84001 were formed as early as 3.6 billion years ago and have biological origins, they may be the remnants of the earliest Martian life.

The analyses so far of ALH84001 are consistent with the meteorite's carbonate globules containing the vestiges of ancient microbial life. Studies of the meteorite are far from over, however. Whether or not these investigations confirm or modify our hypothesis, they will be invaluable learning experiences for researchers, who may get the opportunity to put the experience to use in coming years. We hope that in 2005 a "sample-return" mission will be launched to collect Martian rocks and soil robotically and return them to the earth two and a half years later. To take off from the Martian surface for the return to the earth, this revolutionary mission may use oxygen produced on the Martian surface by breaking down carbon dioxide in the planet's atmosphere.

Through projects such as the sample return, we will finally begin to collect the kind of data that will enable us to determine conclusively whether life came into being on Mars. This kind of insight, in turn, may ultimately provide perspective on one of the greatest scientific mysteries: the prevalence of life in our universe.

—December 1997

GLOBAL CLIMATIC CHANGE ON MARS

Today a frozen world, Mars at one time may have had more temperate conditions, with flowing rivers, thawing seas, melting glaciers, and, perhaps, abundant life

Jeffrey S. Kargel and Robert G. Strom

To those of us who have spent a good part of our lives studying Mars, the newly discovered evidence that extraterrestrial microbes may have once lived in a rock cast off from that planet stirs feelings of awe. But the recent claim also evokes thoughts of Percival Lowell, a well-known American astronomer of the early 20th century, who turned his telescope toward Mars and saw a vast network of canals bordered by vegetation. His suggestion that Mars harbored such lushness had many people believing that the surface of the planet enjoyed conditions not so different from those on the earth. But in the 1960s three Mariner spacecraft flew by Mars and revealed the true harshness of its environment.

Observations from those unmanned probes indicated that Mars has an atmosphere that is thin, cold, and dry. This tenuous shroud, composed almost entirely of carbon dioxide, provides less than 1 percent of the surface pressure found at sea level on the earth. The images radioed back during those first fleeting encounters three decades ago were fuzzy and few in number, but they were decidedly more accurate than Lowell's telescopic views. The Mariner cameras showed no canals, no water, and no vegetation. They presented only a moonlike surface covered with craters. Sober scientists quickly dismissed any notion that the climate on Mars was sufficiently warm or wet to sustain life.

With its distant orbit—50 percent farther from the sun than the earth—and slim atmospheric blanket, Mars experiences frigid weather conditions. Surface temperatures typically average about −60 degrees Celsius (−76 degrees Fahrenheit) at the equator and can dip to −123 degrees C near the poles. Only the midday sun at tropical latitudes is warm enough to thaw ice on occasion, but any liquid water formed in this way would evaporate almost instantly because of the low atmospheric pressure.

Although the atmosphere holds a small amount of water, and water-ice clouds sometimes develop, most Martian weather involves blowing dust or carbon dioxide. Each winter, for example, a blizzard of frozen carbon dioxide rages over one pole, and a few meters of this dry-ice snow accumulate as previously frozen carbon dioxide evaporates from the opposite polar cap. Yet even on the summer pole, where the sun remains in the sky all day long, temperatures never warm enough to melt frozen water.

Despite the abundant evidence for cold, dry conditions, the impression of Mars as a perpetually freeze-dried world has been steadily giving way since the Mariner probes first reported their findings. Planetary scientists, who continue to examine the voluminous data from both the Mariner and the later Viking missions of the 1970s, now realize that Mars has had a complex climatic history—one that was perhaps punctuated with many relatively warm episodes. At certain times huge volumes of water flowed freely across the surface of the planet. Before considering what this astonishing fact means for the possibility of life evolving on Mars or the strategy for the next round of Martian exploration, it is instructive to review how this reversal in the way Mars is perceived came about.

■ MUDDY RECOLLECTIONS

Scrutinizing the Mariner and Viking images obtained from orbit, planetary scientists soon noticed that most old Martian craters (unlike lunar ones) are eroded and that features resembling mudflows occur around almost every large, young crater on Mars. Such muddy "ejecta" probably represent the frozen remnants of a cataclysmic moment in the past when an asteroid or comet collided with the Martian surface, melting a patch of icy permafrost (where water-saturated ground had been frozen) and excavating a large hole that tapped a zone containing liquid water deep underground. By the late 1970s planetologists concluded that a considerable amount of underground ice and water has been present below the Martian surface throughout much of the history of the planet.

Yet not all Martian craters have these muddy flows surrounding them. Smaller craters appear more like their counterparts on the moon, with just streaks of dry ejecta scattered around them. Near the equator of Mars only craters greater than about four kilometers in diameter display muddy ejecta, but closer to the Martian poles craters as small as one kilometer across also have relic mudflows. This dependence on latitude arises because the ice-free, surficial layer varies in thickness. This layer extends deeper near the equator (to about 800 meters) than near the poles because the relative warmth of the Martian tropics purges much of the subsurface of frozen water. Hence, near the equator only the impact of bigger objects (that is, those that leave relatively large craters) will burrow down through the upper layer to heat the underlying icy permafrost and release a torrent of mud.

Researchers have since found other indications that a thick substratum of frozen ground exists on Mars. They have also identified evidence that ice once formed on the surface, where it appears to have created characteristic glacial landscapes. These features include bouldery ridges of sediment left by melting glaciers at their margins and meandering lines of sand and gravel deposited beneath glaciers by streams running under the ice (so-called eskers).

Many telltale landforms on Mars resemble frosty sites on the earth. For example, the pitted terrain on Mars correspond to an earthly equivalent called thermokarst, which forms when the ice contained at shallow levels melts and the ground collapses. The apron-shaped lobes of rocky debris seen on the flanks of some Martian mountains might be rubble-covered glaciers. Or more likely they represent "rock glaciers," like the ones that form within the Alaska Range and in the Antarctic Dry Valleys on the earth. These distinctive sloping surfaces result after thousands of freeze-thaw cycles cause the top meter or so of water-soaked ground to creep slowly downhill.

Glacial features and muddy ejecta around craters are not the only examples of water shaping the Martian surface. In some places sinuous valleys one kilometer wide and many hundreds of kilometers long form large branching networks. Carl Sagan of Cornell University, Victor R. Baker of the University of Arizona, and their colleagues suggested in the 1970s that such troughs were created by running water. Other Martian valleys have blunt starting points and short tributaries, characteris-

ULTRAVIOLET
RAYS

LOST TO SPACE

$H_2O \longrightarrow 2H + O$

SNOW

EVAPORATION

OXIDATION
OF SURFACE
ROCKS

SEA ICE

OCEANUS BOREALIS

FORMATION
OF CARBONATES

G GROUNDWATER
M MELTWATER
R RUNOFF

HYDRATED CLAY
MINERALS AND SALTS

GROUNDWATER AQUIFER →   →   →

Water cycling during past wet episodes on Mars would have had many components. A thick atmosphere most likely carried a substantial amount of water evaporated from lakes and seas. That water vapor would, in turn, condense into clouds and eventually precipitate. Rain formed in this way would have created surface runoff, and much of this water would have percolated into the ground. Snowfalls might have accumulated to form glaciers, which in turn would have discharged their meltwaters into glacial lakes. Hydrothermal circulation, associated perhaps with sites of volcanism, could also have brought water to the surface from reservoirs deep underground. (Image by Edward Bell)

tics that are typical of erosion by groundwater "sapping." That process, common on the earth, results from the seepage of water from underground springs, which causes the overlying rock and soil to wash away.

Images of Mars also reveal enormous outflow channels etched on the surface. Some of these structures are more than 200 kilometers wide and can stretch for 2,000 kilometers or more. These channels emanate from what is called chaotic terrain, regions of fractured, jumbled rocks that apparently collapsed when groundwater suddenly surged outward. The ensuing floods carved the vast channels, leaving streamlined islands more than 100 kilometers long and gouging cavernous potholes several hundred meters deep. Baker compared the Martian outflow channels to similar, albeit smaller, flood features found on the earth in

parts of Oregon and Washington State. Those so-called channeled scablands of the Pacific Northwest formed after a glacier that had dammed a large lake broke open suddenly and caused a catastrophic flood.

The geometry of the Martian outflow channels indicates that water could have flowed along the surface as rapidly as 75 meters per second (170 miles per hour). Michael H. Carr of the U.S. Geological Survey estimates that the vast quantity of water necessary to create these many enormous channels would have been enough to fill a global Martian ocean that was 500 meters deep, although not all this liquid flowed at one time. One source for that great quantity of water may have been a deep lake in Valles Marineris, a region on Mars partly covered with sedimentary layers that appear to be ancient lake deposits. Water could also have gushed from a large reservoir under ice-impregnated permafrost that had been warmed by heat from the interior of the planet.

Why should such an underground accumulation of water suddenly inundate the surface? Scientists are unsure of the exact cause, but this groundwater might have started to flow after the icy permafrost capping it thinned and weakened, perhaps because of a sudden climate warming, volcanism, or tectonic uplift. Perhaps a large meteor impact or quake triggered the cataclysmic dousing. Once water broke through to the surface, carbon dioxide from saturated groundwater—a Martian seltzer of sorts—may have erupted in tremendous geysers, further undermining the stability of the saturated underground layers. The result was to produce chaotic terrain and to unleash floods and mudflows of a magnitude that has rarely, if ever, been matched by any earthly deluge.

■   AN OCEAN AWAY

Some highland areas on Mars contain extensive systems of valleys that drained into sediment-floored depressions. These lowlands were at one time full of water. The largest of these Martian lakes filled two gigantic impact basins called Hellas and Argyre.

But these lakes may not have been the largest bodies of water on the planet. Research groups led by David H. Scott and Kenneth L. Tanaka of the U.S. Geological Survey and by Jeffrey M. Moore of the National Aeronautics and Space Administration Ames Research Center independently concluded that repeated floods from the outflow channels emptied to the north and formed a succession of transient lakes and seas. We have interpreted many features bordering these ancient basins as marking where glaciers once emptied into these deep bodies of water. Tanaka and Moore believe that thick layers of sediment deposited in these seas now stretch across much of the extensive northern plains. According to several estimates, one of the larger of the northern seas on Mars could have displaced the combined volume of the Gulf of Mexico and the Mediterranean Sea.

Yet even that great body of water may not have been the supreme example: there may have been a Mars ocean. As early as 1973 the late Henry Faul of the University of Pennsylvania raised this intriguing possibility in a paper he romantically entitled "The Cliff of Nix Olympica." Understandably, given the paucity of observations then available, the paper was never accepted for publication. But during the past decade other researchers, working with information acquired during the Viking missions, have revived Faul's idea.

For instance, in 1989 Timothy J. Parker and his colleagues at the Jet Propulsion Laboratory in Pasadena, California, again proposed a northern ocean (arguing that many features in the northern plains looked as if they had resulted from coastal erosion). To enhance prospects for publication, however, they deliberately obscured the provocative thrust of their work with the mundane title "Transitional Morphology in the West Deuteronilus Mensae Region of Mars: Implications for Modification of the Lowland/Upland Boundary." In a subsequent paper these researchers ventured a more direct title to convey their ideas: "Coastal Geomorphology of the Martian Northern Plains." Motivated in part by such work, Baker and several colleagues (including us) named this hypothetical northern ocean Oceanus Borealis. We calculated that it was possibly four times a large as the Arctic Ocean on the earth, and we proposed a scenario for the actions of the water cycle on Mars that could have accounted for it.

Whereas most planetary scientists now agree that large bodies of water formed repeatedly in the northern plains on Mars, many do not accept that there was ever a true ocean there. Some envision that only a vast, muddy slurry, or mud ocean, existed. In any case it is clear that huge amounts of water once flowed over the surface of Mars. Yet the fate of that water remains unknown. Some of it may have percolated into the subsurface and frozen in permafrost. Some may have frozen in place and might now stretch across much of the floor of the northern plains, hidden by a mantle of dust and sand. Some water may simply have evaporated, to be later lost to space or deposited as snow at the poles.

■ TRUST THE OLD SALTS

Although images of the landforms left by ancient glaciers, river valleys, lakes, and seas are strong testament that Mars was once rich in water, evidence comes from other sources as well. Earth-based spectroscopic measurements of Mars reveal the presence of clay minerals. Even more directly, the two landers that set down on the surface during the Viking program analyzed Martian soil and found that it probably contains 10 to 20 percent salts. Martian rocks, like those on the earth, react to form salt and clay minerals when exposed to water. But such chemical weathering probably cannot occur under the cold and dry conditions that now reign on Mars.

Some scientists have also studied Martian rocks found here on the earth. These rare samples of the Martian surface were blasted into space by the impact of an asteroid or comet and later fell to the earth as meteorites. Allan H. Treiman of the Lunar and Planetary Institute in Houston and James L. Gooding of NASA Johnson Space Center have shown in the past several years that minerals in some of these so-called SNC meteorites were chemically altered by cool, salty water, whereas others were affected by warmer hydrothermal solutions. Their conclusions imply that Mars once had a relatively warm, wet climate and may have had hot springs. Just perhaps, conditions were right for life.

That possibility inspired David S. McKay of the NASA Johnson Space Center and his colleagues to examine an SNC meteorite for signs of ancient Martian life. Although their conclusion that fossil microbes are present is open to debate (and a vigorous one is indeed going on), the composition of the rock they studied—with fractures filled by minerals that probably precipitated from an aqueous solution—indicated that conditions on Mars a few billion years ago would have been compatible with the existence of life.

In agreement with this assessment, many atmospheric physicists had already concluded that Mars has lost immense quantities of water vapor to space over time. Their theoretical calculations are in good accord with measurements made by various Soviet space probes that showed oxygen and hydrogen atoms (derived from breakdown of atmospheric water exposed to sunlight) streaming away from Mars. The continuous loss of these elements implies that Mars must once have had all the water needed to fill an Oceanus Borealis.

But water was not the only substance lost. Recently David M. Kass and Yuk L. Yung of the California Institute of Technology examined the evolution of carbon dioxide—a potent greenhouse gas—in the atmosphere of Mars. They found that over time an enormous quantity of carbon dioxide has escaped to space. That amount of gaseous carbon dioxide would have constituted a thick Martian atmosphere with three times the pressure found at the surface of the earth. The greenhouse effect from that gas would have been sufficient to warm most of the surface of Mars above the freezing point of water. Thus, from this perspective, too, it seems quite plausible that the climate on Mars once was much warmer and wetter than it is today.

Yet many questions remain about how water might have arranged itself on the surface of Mars. Was there actually an ocean? Did water shift rapidly between different reservoirs? When and for how long was Mars wet? Although the absolute timing of these events remains unknown, most researchers believe that water sculpted the surface of Mars at many intervals throughout the history of the planet. The constant losses of water and carbon dioxide from the atmosphere suggest that early epochs on Mars (that is, billions of years ago) may have been especially warm and wet. But some balmy periods may also have been relatively recent: Timothy D. Swindle of the University of Arizona and his colleagues studied minerals in an SNC meteorite created by aqueous alteration and determined that they formed 300 million years ago—a long time by human standards but only a few percent of the age of the 4.6-billion-year-old solar system. Their result was, however, accompanied by a considerable degree of uncertainty.

The duration of the wet periods on Mars is also difficult to gauge exactly. If the eroded Martian landscapes formed under conditions typical of terrestrial glacial environments, more than a few thousand but less than about a million years of warm, wet climate were required. Had these conditions endured substantially longer, erosion would have presumably erased all but traces of a few impact craters, just as it does on the earth.

This limitation does not apply to the earliest history of the planet, billions of years ago, before the craters now visible had formed. A young Mars may well have had vigorous erosion smoothing its face. But eventually, as the planet slipped toward middle age, its visage became cold, dry, and pockmarked. Only scattered intervals of warmth have since rejuvenated the surface of the planet in certain regions. Yet the mechanism that causes Mars to switch between mild and frigid regimes remains largely mysterious. Scientists can now venture only crude explanations for how these climate changes might have occurred.

■ TURNING ON THE HEAT

One hypothesis involves shifts in obliquity, the tilt of the spin axis from its ideal position, perpendicular to the orbital plane. Mars, like the earth, is now canted by about 24 degrees, and that tilt changes regularly over time. Jihad Touma and Jack L. Wisdom of the Massachusetts Institute of Technology discovered in 1993 that, for Mars, the tilt can also change abruptly. Excursions of the tilt axis through a range of as much as 60 degrees may recur sporadically every 10 million years or so. In addition, the orientation of the tilt axis and the shape of the orbit that Mars follows both change cyclically with time.

These celestial machinations, particularly the tendency of the spin axis to tilt far over, can cause seasonal temperature extremes. Even with a thin atmosphere such as the one that exists today, summer temperatures at middle and high Martian latitudes during periods with large obliquity could have climbed above freezing for weeks on end, and Martian winters would have been even harsher than they are currently.

But with sufficient summer warming of one pole, the atmosphere may have changed drastically. Releases of gas from the warmed polar cap, from seltzer groundwater, or from carbon dioxide-rich permafrost may have thickened the atmosphere sufficiently to create a temporary greenhouse climate. Water could then have existed on the surface. Aqueous chemical reactions during such warm periods would in turn form salts and carbonate rocks. That process would slowly draw carbon dioxide from the atmosphere, thereby reducing the greenhouse effect. A return to moderate levels of obliquity might further cool the planet and precipitate dry-ice snow, thinning the atmosphere even more and returning Mars to its normal, frigid state.

This theory of climatic change needs to be tested, but new observations and fresh insights will undoubtedly come from a decade-long series of unpiloted spacecraft that will next visit Mars. The expeditions begin this month with the launch of

American and Russian probes. This program of exploration had been slated to conclude in 2005 with the return of Martian rocks. But the discovery of what may be fossil microbes in an SNC meteorite has sparked thoughts of obtaining Martian samples sooner so that scientists can better evaluate whether microorganisms existed on Mars several billion years ago—or even more recently.

The American spacecraft soon to be under way include Mars Pathfinder and Mars Global Surveyor. Pathfinder will land on a bouldery plain of an outflow channel that once fed an ancient sea. Although not equipped to test directly for signs of life, this lander will release a small roving vehicle to explore the local environs. Surveyor will take pictures from orbit that can resolve features that are just a few meters across. Measurements from this orbiter will also allow scientists to make detailed topographic maps and to search for icy deposits as well as new evidence of ancient glaciers, lakes, and rivers. Information gathered by these next missions should give scientists a clearer picture of what Mars looked like during its last episode of warmer climate, perhaps 300 million years in the past.

By 300 million years ago on the earth amphibians evolved from fish had crawled out of the sea and inhabited swampy coastlines. Might other complex creatures have flourished simultaneously along Martian shores? The basic conditions for life may have existed for a million years late in Martian history—perhaps much longer during an earlier period. Were these intervals conducive for organisms to evolve into forms that could survive the dramatic changes in climate? Could Martian organisms still survive today in underground hot springs? The next decade of concentrated exploration may provide the definitive answers, which, if positive, would mark an intellectual leap as great as any in human history.

—November 1996

Searching for Life in Our Solar System

If life evolved independently on our neighboring planets or moons, then where are the most likely places to look for evidence of extraterrestrial organisms?

Bruce M. Jakosky

Since antiquity human beings have imagined life spread far and wide in the universe. Only recently has science caught up, as we have come to understand the nature of life on the earth and the possibility that life exists elsewhere. Recent discoveries of planets orbiting other stars and of possible fossil evidence in Martian meteorites have gained considerable public acclaim. And the scientific case for life elsewhere has grown stronger during the past decade. There is now a sense that we are verging on the discovery of life on other planets.

To search for life in our solar system, we need to start at home. Because the earth is our only example of a planet endowed with life, we can use it to understand the conditions needed to spawn life elsewhere. As we define these conditions, though, we need to consider whether they are specific to life on the earth or general enough to apply anywhere.

Our geologic record tells us that life on the earth started shortly after life's existence became possible—only after protoplanets (small, planetlike objects) stopped bombarding our planet near the end of its formation. The last "earth-sterilizing" giant impact probably occurred between 4.4 and 4.0 billion years ago. Fossil microscopic cells and carbon isotopic evidence suggest that life had grown widespread some 3.5 billion years ago and may have existed before 3.85 billion years ago.

Once it became safe for life to exist, no more than half a billion years—and perhaps as little as 100 million to 200 million years—passed before life rooted itself firmly on the earth. This short time span indicates that life's origin followed a relatively straightforward process, the natural consequence of chemical reactions in a geologically active environment. Equally important, this observation tells us that life may originate along similar lines in any place with chemical and environmental conditions akin to those of the earth.

The standard wisdom of the past 40 years holds that prebiological organic molecules formed in a so-called reducing atmosphere, with energy sources such as lightning triggering chemical reactions to combine gaseous molecules. A more recent theory offers a tantalizing alternative. As water circulates through ocean-floor volcanic systems, it heats to temperatures above 400 degrees Celsius (720 degrees Fahrenheit). When that superhot water returns to the ocean, it can chemically reduce agents, facilitating the formation of organic molecules. This reducing environment also provides an energy source to help organic molecules combine into larger structures and to foster primitive metabolic reactions.

■ WHERE DID LIFE ORIGINATE?

The significance of hydrothermal systems in life's history appears in the "tree of life," constructed recently from genetic sequences in RNA molecules, which carry forward genetic information. This tree arises from differences in RNA sequences common to all of the earth's living organisms. Organisms evolving little since their separation from their last common ancestor have similar RNA base sequences. Those organisms closest to the "root"—or last common ancestor of all living organisms—are hyperthermophiles, which live in hot water, possibly as high as 115 degrees C. This relationship indicates either that terrestrial life "passed through" hydrothermal systems at some early time or that life's origin took place within such systems. Either way the earliest history of life reveals an intimate connection to hydrothermal systems.

As we consider possible occurrences of life elsewhere in the solar system, we can generalize environmental conditions required for life to emerge and flourish. We assume that liquid water is necessary—a medium through which primitive organisms can gain nutrients and disperse waste. Although other liquids, such as methane or ammonia, could serve the same function, water is likely to have been much more abundant, as well as chemically better for precipitating reactions necessary to spark biological activity.

To create the building blocks from which life can assemble itself, one needs access to biogenic elements. On the earth these elements include carbon, hydrogen, oxygen, nitrogen, sulfur, and phosphorus, among the two dozen or so others playing a pivotal role in life. Although life elsewhere might not use exactly the same elements, we would expect it to use many of them. Life on the earth utilizes carbon (over silicon, for example) because of its versatility in forming chemical bonds, rather than strictly its abundance. Carbon also exists readily as carbon dioxide, available as a gas or dissolved in water. Silicon dioxide, on the other hand, exists plentifully in neither form and would be much less accessible. Given the ubiquity of carbon-containing organic molecules throughout the universe, we would expect carbon to play a role in life anywhere.

Of course, an energy source must drive chemical disequilibrium, which fosters the reactions necessary to spawn living systems. On the earth today nearly all of life's energy comes from the sun, through photosynthesis. Yet chemical energy sources suffice—and would be more readily available to early life. These sources would include geochemical energy from hydrothermal systems near volcanoes or chemical energy from the weathering of minerals at or near a planet's surface.

■ POSSIBILITIES FOR LIFE ON MARS

Looking beyond the earth, two planets show strong evidence for having had environmental conditions suitable to originate life at some time in their history—Mars and Europa. (For this purpose we will consider Europa, a moon of Jupiter, to be a planetary body.)

Mars today is not very hospitable. Daily average temperatures rarely rise much above 220 kelvins, some 53 kelvins below water's freezing point. Despite this drawback, abundant evidence suggests that liquid water has existed on Mars's surface in the past and probably is present within its crust today.

Networks of dendritic valleys on the oldest Martian surfaces look like those on the earth formed by running water. The water may have come from atmospheric precipitation or "sapping," released from a crustal aquifer. Regardless of where it came from, liquid water undoubtedly played a role. The valleys' dendritic structure indicates that they formed gradually, meaning that water once may have flowed on Mars's surface, although we do not observe such signs today.

In addition ancient impact craters larger than about 15 kilometers (nine miles) in diameter have degraded heavily, showing no signs of ejecta blankets, the raised rims or central peaks typically present on fresh craters. Some partly eroded craters display gullies on their walls, which look water-carved. Craters smaller than about 15 kilometers have eroded away entirely. The simplest explanation holds that surface water eroded the craters.

Although the history of Mars's atmosphere is obscure, the atmosphere may have been denser during the earliest epochs, 3.5 to 4.0 billion years ago. Correspondingly, a denser atmosphere could have yielded a strong greenhouse effect, which would have warmed the planet enough to permit liquid water to remain stable. Subsequent to 3.5 billion years ago, evidence tells us that the planet's crust did contain much water. Evidently, catastrophic floods, bursting from below the planet's surface, carved out great flood channels. These floods occurred periodically over geologic time. Based on this evidence, liquid water should exist several kilometers underground, where geothermal heating would raise temperatures to the melting point of ice.

Mars also has had rich energy sources throughout time. Volcanism has supplied heat from the earliest epochs to the recent past, as have impact events. Additional energy to sustain life can come from the weathering of volcanic rocks. Oxidation of iron within basalt, for example, releases energy that organisms can use.

The plentiful availability of biogenic elements on Mars's surface completes life's requirements. Given the presence of water and energy, Mars may well have independently originated life. Moreover, even if life did not originate on Mars, life still could be present there. Just as high-velocity impacts have jettisoned Martian surface rocks into space—only to fall on the earth as Martian meteorites—rocks from the earth could similarly have landed on the red planet. Should they contain organisms that survive the journey and should they land in suitable Martian habitats, the bacteria could survive. Or for all we know, life could have originated on Mars and been transplanted subsequently to the earth.

An inventory of energy available on Mars suggests that enough is present to support life. Whether photosynthesis evolved, and thereby allowed life to move into other ecological niches, remains uncertain. Certainly data returned from the Viking spacecraft during the 1970s presented no evidence that life is widespread on Mars. Yet it is possible that some Martian life currently exists, cloistered in isolated, energy-rich, and water-laden niches—perhaps in volcanically heated, subsurface hydrothermal systems or merely underground, drawing energy from chemical interactions of liquid water and rock.

Recent analysis of Martian meteorites found on the earth led many scientists to conclude that life may have once thrived on Mars—based on fossil remnants seen within the rock. Yet this evidence does not definitively indicate biological activity; indeed, it may result from natural geochemical processes. Even if scientists determine that these rocks contain no evidence of Martian life, life on the red planet might still be possible—but in locations not yet searched. To draw a definitive conclusion we must study those places where life (or evidence of past life) will most likely appear.

■ EUROPA

Europa, on the other hand, presents a different possible scenario for life's origin. At first glance Europa seems an unlikely place for life. The largest of Jupiter's satellites, Europa is a little bit smaller than our moon, and its surface is covered with nearly pure ice. Yet Europa's interior may be less frigid, warmed by a combination of radioactive decay and tidal heating, which could raise the temperature above the melting point of ice at relatively shallow depths. Because the layer of surface ice stands 150 to 300 kilometers thick, a global, ice-covered ocean of liquid water may exist underneath.

Recent images of Europa's surface from the Galileo spacecraft reveal the possible presence of at least transient pockets of liquid water. Globally the surface appears covered with long grooves or cracks. On a smaller scale these quasilinear features show detailed structures indicating local ice-related tectonic activity and infilling from below. On the smallest scale blocks of ice are present. By tracing the crisscrossing grooves, the blocks clearly have moved with respect to the larger mass. They appear similar to sea ice on the earth—as if large ice blocks had broken off the main mass, floated a small distance away, and then frozen in place. Unfortunately, we cannot yet determine if the ice blocks floated through liquid water or slid on relatively warm, soft ice. The dearth of impact craters on the ice indicates that fresh ice continually resurfaces Europa. It is also likely that liquid water is present at least on an intermittent basis.

If Europa has liquid water at all, then that water probably exists at the interface between the ice and underlying rocky interior. Europa's rocky center probably has had volcanic activity—perhaps at a level similar to that of the earth's moon, which rumbled with volcanism until about 3.0 billion years ago. The volcanism within its core would create an energy source for possible life, as would the weathering of minerals reacting with water. Thus, Europa has all the ingredients from which to spark life. Of course less chemical energy is likely to exist on Europa than Mars, so we should not expect to see an abundance of life, if any. Although the Galileo space probe has detected organic molecules and frozen water on Callisto and Ganymede, two of Jupiter's four Galilean satellites, these moons lack the energy sources that life would require to take hold. Only Io, also a Galilean satellite, has volcanic heat—yet it has no liquid water, necessary to sustain life as we know it.

Mars and Europa stand today as the only places in our solar system that we can identify as having (or having had) all ingredients necessary to spawn life. Yet they are not the only planetary bodies in our solar system relevant to exobiology. In particular we can look at Venus and at Titan, Saturn's largest moon. Venus currently remains too hot to sustain life, with scorching surface temperatures around 750 kelvins, sustained by greenhouse warming from carbon dioxide and sulfur dioxide gases. Any liquid water has long since disappeared into space.

■ VENUS AND TITAN

Why are Venus and the earth so different? If the earth orbited the sun at the same distance that Venus does, then the earth, too, would blister with heat—causing more water vapor to fill the atmosphere and augmenting the greenhouse effect. Positive feedback would spur this cycle, with more water, greater greenhouse warming, and so on saturating the atmosphere and sending temperatures soaring. Because temperature plays such a strong role in determining the atmosphere's water content, both the earth and Venus have a temperature threshold, above which the positive feedback of an increasing greenhouse effect takes off. This feedback loop would load Venus's atmosphere with water, which in turn would catapult its temperatures to very high values. Below this threshold its climate would have been more like that of the earth.

Venus, though, may not always have been so inhospitable. Four billion years ago the sun emitted about 30 percent less energy than it does today. With less sunlight, the boundary between clement and runaway climates may have been inside Venus's orbit, and Venus may have had surface temperatures only 100 degrees C above the earth's current temperature. Life could survive quite readily at those temperatures—as we observe with certain bacteria and bioorganisms living near hot springs and undersea vents. As the sun became hotter Venus would have warmed gradually until it would have undergone a catastrophic transition to a thick hot atmosphere. It is possible that Venus originated life several billion years ago but that high temperatures and geologic activity have since obliterated all evidence of a biosphere. As the sun continues to heat up, the earth may undergo a similar catastrophic transition only a couple of billion years from now.

Titan intrigues us because of abundant evidence of organic chemical activity in its atmosphere, similar to what might have occurred on the early earth if its atmosphere had potent abilities to reduce chemical agents. Titan is about as big as Mercury, with an atmosphere thicker than the earth's, consisting predominantly of nitrogen, methane, and ethane. Methane must be continually resupplied from the surface or subsurface, because photochemical reactions in the atmosphere drive off hydrogen (which is lost to space) and convert the methane to longer chains of organic molecules. These longer-chain hydrocarbons are thought to provide the dense haze that obscures Titan's surface at visible wavelengths.

Surface temperatures on Titan stand around 94 kelvins, too cold to sustain either liquid water or nonphotochemical reactions that could produce biological activity— although Titan apparently had some liquid water during its early history. Impacts during its formation would have deposited enough heat (from the kinetic energy of the object) to melt frozen water locally. Deposits of liquid water might have persisted for thousands of years before freezing. Every part of Titan's surface probably has melted at least once. The degree to which biochemical reactions may have proceeded during such a short time interval is uncertain, however.

■ EXPLORATORY MISSIONS

Clearly, the key ingredients needed for life have been present in our solar system for a long time and may be present today outside of the earth. At one time or another four planetary bodies may have contained the necessary conditions to generate life.

We can determine life's actual existence elsewhere only empirically, and the search for life has taken center stage in the National Aeronautics and Space Administration's ongoing science missions. The Mars Surveyor series of missions, scheduled to take place during the coming decade, aims to determine if Mars ever had life. This series will culminate in a mission currently scheduled for launch in 2005, to collect Martian rocks from regions of possible biological relevance and return them to the earth for detailed analysis. The Cassini spacecraft currently is en route to Saturn. There the Huygens probe will enter Titan's atmosphere, its goal to decipher Titan's composition and chemistry. A radar instrument, too, will map Titan's surface, looking both for geologic clues to its history and evidence of exposed lakes or oceans of methane and ethane.

Moreover, the Galileo orbiter of Jupiter is focusing its extended mission on studying the surface and interior of Europa. Plans are under way to launch a spacecraft mission dedicated to Europa, to discern its geologic and geochemical history and to determine if a global ocean lies underneath its icy shell.

Of course it is possible that, as we plumb the depths of our own solar system, no evidence of life will turn up. If life assembles itself from basic building blocks as easily as we believe it does, then life should turn up elsewhere. Indeed, life's absence would lead us to question our understanding of life's origin here on the earth. Whether or not we find life, we will gain a tremendous insight into our own history and whether life is rare or widespread in our galaxy.

—*Scientific American Presents,* Spring 1998

Worlds Around Other Stars

Theory and observation imply that planetary systems like our own should be common. Astronomical searches are closing in on planets that may orbit some nearby stars

David C. Black

Do planets like our own orbit around other stars? The answer links intimately with the question of whether life exists beyond the earth. The most promising place for life to arise is on the surface of a planet, which functions as a "cosmic petri dish" where life can begin and be nurtured as it evolves to more complex states. A search for other planetary systems should therefore be a major component of any effort to understand the prospects for extraterrestrial life. Moreover, the results from such a search are essential for understanding the origin of the earth and the surrounding solar system.

Over many years astronomers have built up a detailed picture of the origin of the sun and its retine of planets. Unfortunately (but until now inseparably) the specific elements of that picture are based entirely on features found in a single planetary system—our own. Finding statistically significant evidence regarding the plentitude and nature of other planetary systems undoubtedly would provide many unexpected details about the process by which stars and planets are born.

Prospects for such a discovery have recently grown brighter. A number of researchers have possibly detected planetary companions circling other stars, although all the sightings remain highly tentative. A new generation of detectors and telescopes, along with some innovative detection techniques, should improve the situation.

The current view of the origin of the solar system has its roots in concepts developed by Immanuel Kant and Pierre Simon Laplace late in the 18th century. They proposed a nebular hypothesis of the solar system's origin, wherein the sun and planets condensed out of a large lumpy cloud. Over the decades the nebular hypothesis has been greatly refined and modified, but the basic concept remains. It shapes current thinking on how and where planets form and, therefore, how one might go about searching for them.

Some notable regularities and variations in the structure of the solar system hint at its nebular origin. The planets all orbit in nearly the same plane: the orbits of all planets except Mercury and Pluto lie within three degrees of the ecliptic, the plane of the earth's orbit. The mean orbital plane of the planets also sits within six degrees of the equatorial plane of the sun. These properties suggest that the planets formed from a common disklike structure, known as the solar nebula.

If viewed from above the earth's North Pole, the planets all revolve about the sun counterclockwise, the same direction in which the sun rotates on its axis. The planets also travel in nearly circular orbits (Mercury and Pluto again are mild exceptions). Such orderly movements fit with the notion that the ancestral disk was dynamically ordered, not chaotic or irregular, and that motions within the disk were dominated by rotation about the sun.

Another noteworthy pattern in the solar system is that planetary composition varies according to distance from the sun. The gaseous outer planets (Jupiter, Saturn, Uranus, and Neptune) primarily contain relatively light, volatile elements. This is particularly true for Jupiter, whose composition, being dominated by hydrogen and helium, is similar to that of the sun. The other outer planets contain less hydrogen but are rich in hydrogen compounds, such as ammonia.

In contrast the inner, or terrestrial, planets (Mercury, Venus, Earth, and Mars) consist mostly of heavier elements, such as silicon and iron, which cosmically are far less abundant than hydrogen. If enough light elements were added to the earth to make its composition match that of the sun, its mass would be comparable to Jupiter's. It seems that the outer planets are more massive because they were able to hold on to more light, volatile elements and compounds in the solar nebula when the planets formed.

The generally accepted explanation for this difference is that the inner part of the solar nebula was sufficiently hot that volatile elements existed only in gaseous form. Critical early stages of planet formation are most likely controlled by the accumulation of solid material into progressively larger objects. Water is cosmically abundant and condenses at high temperatures compared with the average nebular temperatures, so it probably played an important role in planetary formation. Regions of the nebula where the temperature was 170 kelvins or lower (the condensation point for water ice) should have contained an adequate supply of solid material for giant planets to form.

Therefore, one might reasonably assume that the largest planets always form in the cool, outer regions of a circumstellar disk. This hypothesis helps to define where, relative to the central star, astronomers should look for giant extrasolar planets.

Circumstellar disks seem to be a natural outgrowth of the manner in which stars are born. Stellar formation begins with the gravitational collapse of dense, cool cores of material within clouds of molecular gas and dust. As the core shrinks, random motions cancel out, and the core's overall rotation creates a flattened gaseous disk. Evidence suggests that such disks are highly stiff, or dissipative, causing angular momentum to transfer outward in the disk and mass to flow inward into the infant star.

Planets coalesce during the brief interval—no more than a few million years—between when the disk forms and when it disappears, either swallowed or expelled by the newborn star. Various evidence hints that the planets did not simply collapse out of the solar nebula. For instance, the rotational axes of the planets are not generally perpendicular to the ecliptic, which suggests that the planets grew through a complicated, chaotic process involving the accretion of smaller units. The implication is that a fundamental difference exists between stars and planets in the way that they form, an important distinction to keep in mind when searching for planetary systems.

If current understanding of stellar birth is broadly correct, then disks should be found in association with many young stars. Some fraction of these disks may then pass through evolutionary stages similar to those inferred to have occurred in the young solar system. The diffuse, low temperature and often heavily obscured nature of disks around forming stars make them difficult to study visually, and so astronomers have concentrated on observations made at radio or infrared frequencies. These longer wavelengths of electromagnetic radiation can penetrate thick clouds of dust far more effectively than visible light.

The first convincing indication that disk structures do indeed surround young stars was the discovery that these objects are sources of energetic jets or winds of gas and dust. The outflows presumably represent material ejected from an evolving disk, but they provide only circumstantial evidence of disks around forming stars.

More recent and direct studies have bolstered this evidence. When dust particles absorb light from a star, they grow hot and reradiate the light as less energetic infrared radiation. The wavelength of the infrared rays depends on the size of the particles. A number of young stellar objects (most notably the objects known as HL Tauri, R Monoceros, and L1551/IRS 5) emit abnormally large amounts of infrared radiation, indicating the presence of dust particles. Infrared observations of the above objects, both from earthbound telescopes and from the National Aeronautics and Space Administration's Infrared Astronomy Satellite (IRAS), imply that these dust particles are one to a few tens of microns across. Small as the particles may seem, they are significantly larger than those that exist in interstellar clouds, hinting that the building-up process that leads to planets may have begun around these stars. Some researchers argue that the dust is produced by collisions between cometlike bodies—the precursors of planets, perhaps—revolving around these stars.

A particularly powerful observation technique, known as speckle interferometry, freezes out blurring from the earth's atmosphere by using very short exposures from which the undistorted image is mathematically reconstructed. Even this technique cannot produce a clear picture of the dust clouds surrounding young stars, but it does permit astronomers to determine that the shape and size of the clouds is consistent with those expected for a disk. Images from IRAS confirmed that many stars, including older stars such as Vega and Beta Pictoris, are surrounded by disklike dusty structures. Observations of radio emission indicate the presence of gas in the disks, particularly carbon monoxide, which radiates prominently at radio wavelengths of a few millimeters.

Additional evidence for condensing disks around other stars comes from infrared observations of young stars known as T Tauri stars. As they age these stars radiate less at short infrared wavelengths, as if the hottest particles—those closest to the star—are being swept up or vaporized.

Efforts to search for other planetary systems employ either direct or indirect methods. Direct methods involve detecting reflected light or infrared radiation from the planets themselves. The primary difficulty with this approach is that emission from a planet tends to be drowned out by the vastly brighter emission of its nearby parent star.

Indirect methods involve scrutinizing a star for signs that it is responding to the gravitational tug of an orbiting planet. As the planet moves from one side of the star to the other, it pulls the star back and forth. This pull manifests itself as a slight wobble superposed on the star's overall motion across the sky. It can also be detected as a slight, periodic change in the star's velocity with respect to the earth.

Any motion toward or away from the earth causes the star's light to be slightly compressed or stretched. When light is compressed it becomes slightly bluer; when stretched it becomes slightly redder, a phenomenon known as the Doppler effect. Careful measurement of absorption lines in a star's spectrum can in principle reveal any periodic changes in its motion.

Indirect searches can be facilitated by some simplifying assumptions. Many researchers have guessed that giant planets around other stars will have orbital periods similar to Jupiter's, about one decade. Fixing the orbital period allows one to determine the size of the orbit as a function of stellar mass and thus to calculate an expected angular or velocity perturbation (given the distance to the star and assuming a certain mass for the planet).

An alternative possibility, which I think is more logical, is that giant planets form at distances from their stars where the temperature is at or below the condensation point of water. In this case the star's luminosity determines the size of the orbit for giant planets. One can then proceed as before to determine how much the planet is likely to affect the star's velocity and apparent path.

The size of the angular disturbance, or wobble, grows with increasing distance between planet and star but shrinks with increasing stellar mass. If planetary systems form in such a manner that giant planets always tend to have more or less similar orbital periods, then faint low-mass stars should be most strongly perturbed by planets, because giant planets would orbit far from these stars relative to the stars' mass. If, however, giant planets normally form at distances where temperatures are below the condensation point of water, the opposite should be true because giant planets would orbit far from bright, massive stars relative to the stars' mass. (In either case low-mass stars experience the largest velocity perturbation.)

If temperature is the determining factor, then typical orbital periods of giant planets may be much shorter than those commonly expected. The average star is cooler and much less luminous than the sun, so giant planets could orbit in relatively close, fast orbits. For a typical nearby star with a mass of about .3 solar mass, the orbital period for a giant planet would be less than one year, compared with Jupiter's orbital period of about 12 years.

Even with the use of simplifying assumptions, detecting other planetary systems remains an extremely difficult task. Current telescopes and instruments operate at the limits of their capabilities when they scan for stellar companions having masses comparable to or less than the mass of Jupiter. As a result the history of planetary searches is full of false leads and phantom discoveries.

Perhaps the best-known star to those who search for other planetary systems is the faint, cool, low-mass (M-type) star known as Barnard's star, named after the American astronomer E. E. Barnard, who noticed its unusual attributes in 1916. Barnard's star has the highest-known apparent motion across the sky; it is also, after the Centauri triple-star system, the second closest star system to ours, lying only six light-years away.

Peter van de Kamp, working at the Sproul Observatory, realized in 1937 that these two properties make Barnard's star ideal for indirect searches because any apparent wobble would be relatively large and well defined. Van de Kamp examined positional data on Barnard's star dating back to 1916 and began collecting his own data using the telescope at Sproul.

In the 1960s van de Kamp concluded that Barnard's star had two Jupiter-mass companions, one completing an orbit every 12 years, the other every 24 years. Yet studies during the past decade by George D. Gatewood of Allegheny Observatory in Pittsburgh and, independently, by Robert S. Harrington of the U.S. Naval Observatory find that the motion of Barnard's star is inconsistent with the existence of van de Kamp's planets. Indeed, the observations by these two researchers exclude the possibility that Barnard's star has any companion much in excess of Jupiter's mass, although lower-mass companions could still be present.

Another widely reported substellar object was the companion to the nearby star Van Biesbroeck 8, dubbed VB8-b, which was independently discovered in 1984 by Harrington and by Donald McCarthy and coworkers at the University of Arizona. This sighting seemed especially persuasive because two different techniques were used to find the object: Harrington used astrometric observations to infer a kink in the motion of VB8-b, and McCarthy observed what appeared to be a faint object close to it by means of infrared speckle interferometry techniques.

After considerable debate a consensus emerged that VB8-b was an intermediate-class object having about one twentieth the mass of the sun, or about 50 times that of Jupiter. Such an object, known as a brown dwarf, would be too small to trigger the nuclear reactions that power stars, and so it would be nearly impossible to observe at visible wavelengths. A few years after the discovery of VB8-b several research teams undertook more sensitive observations to confirm and characterize better its properties. Despite considerable effort, they found no trace of the much publicized object. Astronomers now generally think that VB8-b does not exist, at least not as described by the early observations.

An extremely precise technique for detecting tiny velocity perturbations has been developed recently by Bruce T. E. Campbell of the University of Victoria and his coworkers. This technique involves comparing the spectrum of stars with a very high-resolution reference spectrum from an unlikely and rather nasty compound, hydrogen fluoride. In principle Campbell's instrument shows far smaller Doppler shifts, and hence far slighter velocity disturbances, than previously could be detected.

Campbell and his colleagues at the Canada-France-Hawaii telescope located on Mauna Kea have measured the radial velocities of 15 stars six times a year with a stated accuracy of about 10 meters per second. (This accuracy is remarkable given that convective currents at the surfaces of sunlike stars typically move on the order of 1,000 meters per second.) Campbell's group has stopped short of overtly claiming to have discovered another planetary system, but they have found signs of long-term accelerations—a potential indicator of the presence of planets—in nearly half of the stars studied.

The most intriguing object studied by Campbell's group is the star Gamma Cephei, an old, orange subgiant star (class III–IV, spectral type K1) whose mass is estimated to be slightly greater than that of the sun. Spectral measurements collected since 1981 led Campbell to conclude that Gamma Cephei experiences a cyclic velocity variation having an amplitude of 25 meters per second and a period of 2.6 years. This period, combined with the star's estimated mass, implies that the companion orbits roughly 300 million kilometers from Gamma Cephei, or about twice the earth's distance from the sun. To produce the observed velocity perturbation, the companion must have at least 1.5 times the mass of Jupiter.

It would be surprising to find a giant planet orbiting so close to a relatively luminous star, but surprises are the brood stock of scientific inquiry. A more troubling aspect of Campbell's finding is that the more recent and accurate data from the study of Gamma Cephei show significant deviations from the motions that would be expected as the putative planet orbits its star. Not until Gamma Cephei has been observed through several 2.6-year "orbits" will the data appear truly convincing. Nevertheless, this work and similar, equally sensitive and accurate efforts being conducted at the University of Arizona by Robert S. McMillan and coworkers hold great potential.

David Latham and his collaborators at the Harvard-Smithsonian Center for Astrophysics, using less accurate but more conventional velocity measurement techniques, have collected some of the strongest evidence yet for the existence of a substellar companion. Latham's group observed the solar-type star HD114762 for more than 12 years and found that it undergoes periodic velocity variants. In this case measurements made at European observatories have confirmed the data.

Latham and his colleagues find a consistent velocity variation having a period of about 84 days and an amplitude of about 550 meters per second. Assuming the star has the same mass as the sun (which is consistent with its temperature and luminosity), the perceived period implies that a companion orbits at a mean distance of about 60 million kilometers, similar to the distance of Mercury from the sun. The lower limit to the mass of the companion is 11 times the mass of Jupiter. Recent, more accurate observations by W. D. Cochran and coworkers at the University of Texas at Austin confirm the period and imply that the orbit is eccentric. Their work also suggests that the companion's mass is significantly greater than previously estimated.

Many researchers understandably tend to view this companion as a true extrasolar planet, but I have doubts. The companion is far more massive than any planet in the solar system. It may represent not a planet but the very low-mass end of the population of starlike objects, and HD114762 may be an extreme type of binary star. Although the process by which binary stars form is not well understood, it seems unlikely that the process depends on whether the objects are large enough to ignite fusion reactions in their cores. There is no reason to suppose that nature would not make binary systems in which one member is a star and the other a brown dwarf.

Objects more than 10 or 20 times the mass of Jupiter probably form like stars, not like planets—that is, they condense directly from a gas cloud rather than from the disk around a star. In my view the different modes of formation are manifested in the fact that planetary systems (at least the one known) contain a multitude of bodies, whereas stars are predominantly double or triple systems. The reason for the cutoff between planets and brown dwarfs is mysterious but seems to be real. This information leads to an important principle of search efforts for planetary systems: the object of these searches is to discover planetary systems, not brown dwarfs. Systems composed of a star and a single substellar companion are not planetary systems.

Nevertheless, studies of brown dwarfs may elucidate how stars and planets form. Direct searches currently lack the sensitivity to detect Jupiter-size or smaller planets around other stars, but they can provide significant information about larger substellar companions. Most of these searches are conducted in the infrared part of the spectrum.

Studies using sophisticated new infrared array detector systems have revealed faint, cool companions around a small number of stars. Ben Zuckerman and Eric E. Becklin of the University of California at Los Angeles, among others, have detected dwarf companions by looking for excess infrared radiation that cannot be explained by a normal star's emission. All the objects found so far can be modeled as brown dwarfs weighing several tens of Jupiter masses. Sightings of brown dwarfs remain highly uncertain, however.

Various groups, such as McCarthy and his coworkers at Arizona and Geoffrey Marcy and colleagues at San Francisco State University, are compiling statistics on the abundance of brown dwarfs. The Arizona group has examined 27 nearby M-type, red dwarf stars. The researchers used speckle interferometry at infrared wavelengths of 1.6 and 2.2 microns (known as the H and K bands) to search for companions too dim to be detected at visible wavelengths.

The number of stars in the infrared survey grew greater with increasing faintness (larger magnitudes denote fainter stars) up to a K-band magnitude of about 10.0, at which point the numbers fell off abruptly. The Arizona group found no objects with K-band magnitudes between +10.0 and +11.5. A brightness of +11.5 (the limit of the survey) corresponds to the expected infrared luminosity of a brown dwarf 70 to 80 times the mass of Jupiter and several billion years old, as seen from a distance of five parsecs (16 light-years). The apparent absence of such objects suggests that no continuous population of objects extends all the way from stars to planets.

This view was confirmed by Marcy and K. Benitz, a student at San Francisco State, who recently surveyed 70 low-mass stars using radial velocity techniques precise to 230 meters per second. The measurements should have revealed any substellar companions down to a mass of about seven Jupiter masses, so long as the companions' orbital periods were less than four years. The survey uncovered six hitherto unknown stellar companions to the stars studied but gave no evidence for the presence of substellar companions. Indeed, the radial velocity survey, combined with results from long-term astrometric studies of stellar positions, indicates that less than 2 percent of all stars have substellar companions whose mass is greater than 10 times that of Jupiter.

In some instances the absence of a discovery can itself be an important discovery. I have long felt that such is the case with substellar objects. As a rule lightweight stars are more common than massive ones. Extrapolating downward, many theorists expected that brown dwarfs would be sprinkled liberally throughout the galaxy, like dust. The paucity of substellar companions is nature's way of telling us that it is time to rethink some of the physics associated with the formation of extremely low-mass stars.

It is clear that current facilities are inadequate to conduct a comprehensive search for other planetary systems. The detailed observations necessary to conduct this project in a scientifically adequate manner will require a new set of advanced instruments. In general, dedicated searches should be conducted with sufficient accuracy and sensitivity that a null result unambiguously advances the understanding of planetary-system formation. This principle should help with assessing facilities that might be used to search for other planetary systems. Many promising design concepts have already emerged, some ground based, others based in space.

Radial velocity (velocity perturbation) measurements are relatively impervious to the blurring caused by the earth's turbulent atmosphere. Larger, dedicated telescopes permit a careful survey of tens to hundreds of stars for the presence of planetary companions. Telescopes now in operation at the University of Arizona and the University of Texas could serve as prototypes for the next generation of radial velocity systems.

Another class of ground-based systems would use very large-aperture (7 to 10 meters) telescopes incorporating active optics that physically manipulate the mirror to compensate for atmospheric distortion. These instruments, which would be used to search for other planetary systems directly at infrared wavelengths, could revolutionize understanding of the structure and evolution of the disks that surround young stars.

Most of the techniques for planetary searches would function best in space. NASA's Space Infrared Telescope Facility, currently planned for launch in the late 1990s, will provide infrared observations of circumstellar disks and newborn planetary systems that will greatly improve on the images returned by IRAS in 1983.

Astrometry—the ultraprecise measurement of stellar position—will gain greatly by being conducted in space. The goal is to be able to measure stellar angular deflections with an accuracy of ten millionths of an arc second—the angular extent of a dime on the moon as viewed from the earth! Such accuracy would permit the detection of companions as small as 10 earth masses around any star within 10 parsecs (30 light-years) of the sun.

Telescopes with precisely ground mirrors and masks to blot out the bright light of a central star might be able to capture direct visible-light images of planets around other stars. An exciting possibility is that both astrometry and imaging could be done by a single telescope. A study of the feasibility of building such a combined instrument is now under way at the Jet Propulsion Laboratory in Pasadena, California.

The next generation of planetary search instruments undoubtedly will be optical interferometers, networks of telescopes placed far apart that combine their images to create, in effect, a single enormous telescope. Such devices could offer thousands of times the resolving power of exiting instruments. The moon would be an ideal location for an optical interferometer and many other astronomical instruments. It is conceivable that the lunar far side will be the site of a suite of scientific facilities that will search not only for other planetary systems but also for signals from intelligent life elsewhere in the universe.

Throughout the ages humans have wondered about the possible existence of worlds other than their own. The search for other planetary systems has been going on in earnest for more than half a century, during which time astronomers have edged tantalizingly close to their goal. An explosion of interest in this field of research, along with the development of more accurate instruments, promises that the next few decades will be especially exciting. The first positive sighting of a planetary system other than our own will be landmark, completing the revolution in thought begun some 450 years ago by Nicolaus Copernicus.

—January 1991

IV

Stars

COLLAPSE AND

FORMATION OF STARS

Hidden from observation, this process can nonetheless be modeled on high-speed computers. Pictures that emerge yield insight into the formation of our own solar system

Alan P. Boss

What are the early stages in the formation of a star? What determines whether a cloud of star-forming matter will evolve into one, two, or several stars? Because clouds of gas, dust, and debris largely obscure all but the initial and final stages of the birth of a star, these questions have so far not been answered by direct observation. Theoretical modeling offers a way to circumvent this obstacle, although not an easy one. Each model requires the execution of more basic calculations than were performed by the entire human race before 1940. Today, run on sophisticated computers, such models reveal the various stages through which a star passes as it evolves. They also give a preliminary picture of how our own solar system formed.

Stars form when nebulae (interstellar clouds of gas and dust), or parts of nebulae, collapse. Although these clouds are too dense for optical telescopes to penetrate, the more diffuse clouds are transparent to millimeter-wavelength radiation. A telescope sensitive to millimeter radiation can therefore be used to observe nebulae in which stars are on the verge of forming. The clouds are also partially transparent to infrared light, and so observations in the infrared wavelengths can be made of newly formed stars within the parent nebulae. These observations yield the basic data with which any theory of stellar formation must reckon: the initial conditions under which stars form and the characteristics of the newly formed stars. Unfortunately there is still a difference of a factor of almost 10^{20} between the density of a star-forming cloud and that of the young stars that can be observed with infrared radiation; it has been impossible to date to view the cloud as it collapses through this range of densities. Consequently stars cannot be observed as they form.

Since the late 1960s astrophysicists have developed increasingly sophisticated computer models of the events that take place between the two observable stages of stellar formation. Such models are based on systems of equations that describe the behavior of nebular gas and dust under the influence of many different forces; the solution of these equations can require roughly one million million basic operations. Even with a high-speed computer the calculation of one model can take several months.

Among the most important advances has been the use of increasingly realistic descriptions of the parent clouds. The early models pictured a spherically symmetric cloud with no rotation; at the next stage of complexity it was assumed that the cloud rotates but remains symmetric about its axis of rotation; in the most recent models the initial cloud rotates and is completely asymmetric.

These models have shown that a collapsing cloud will generally pass through two phases of rapid contraction (called dynamic collapse); phases during which outlying matter accumulates around a stable core follow each dynamic-collapse phase. In either phase of dynamic collapse the cloud might fragment into two or more protostars; whether or not the cloud fragments depends on such variables as its size and rate of rotation. It might also collapse in a way that produces a single protostar.

In fact such single stars are rare: in spite of the appearance of the night sky to the unaided eye, most stars are actually binary. (A binary system consists of two stars orbiting about each other. Often the members of binary systems are too close to be discerned without large telescopes or spectroscopic equipment.) Our sun, as a single star, is part of a minority population. The clouds that do not fragment are thus particularly interesting: they may represent models for the formation of our own solar system.

How a cloud fragments is one of the two fundamental characteristics of stellar evolution that a theory must be able to describe. Interstellar clouds can be as large as 100,000 times the mass of the sun—quite massive compared with stars, which are seldom larger than about 10 times the solar mass. In addition most stars in the disk of our galaxy seem to form in clusters containing about 100 stars. These two observations suggest that interstellar clouds fragment into many protostars.

The second fundamental characteristic that must be described concerns angular momentum. In rough terms the angular momentum of a spinning body is a measure of how much mass in the body is spinning, how fast that mass is spinning, and how large the body is. According to observational evidence, interstellar clouds have up to 10^5 times as much angular momentum per unit of mass as their progeny stars. Any theory of star formation must therefore describe how a cloud disposes of a considerable amount of angular momentum before it collapses to form a star or several stars.

One of the first sophisticated computer models of star formation was produced in 1968 by Richard B. Larson of Yale University. He developed a detailed model for the contraction of a spherically symmetric, nonrotating cloud. An important product of his work was a picture of the so-called dynamic-collapse phase of star formation. The dynamic-collapse phase is a period of rapid contraction that can be explained by the interplay of two major forces: gravity, which tends to contract the cloud, and thermal pressure, which is the tendency of hot gas within the cloud to expand. Larson showed that the dynamic-collapse phase is due in part to the way the relation between these two forces changes because of the flow of radiation within the cloud.

The outer shell of a very diffuse dust cloud is transparent to ultraviolet radiation from neighboring stars, and hence it tends to be heated substantially by such radiation. After gravity has compressed the cloud to the density of a dark cloud, the cloud becomes opaque to ultraviolet light, eliminating this source of heating. It is still transparent to infrared radiation, however, and so dust grains in its interior are able to radiate heat energy out of the cloud in the infrared portion of the spectrum.

Thus as the density of the cloud increases, its temperature drops, down to a minimum of about 10 degrees Kelvin (degrees Celsius above absolute zero). The cloud then enters an "isothermal phase" during which the temperature remains at 10 degrees as the cloud collapses through a wide range of densities, from approximately 10^5 to 10^{11} atoms per cubic centimeter. As the cloud grows smaller and denser, the gravitational force becomes stronger, eventually overwhelming the thermal pressure. The result is a dynamic collapse in which the gas and dust fall into the center at rapidly increasing velocities. As the gas and dust fall in, the density of the cloud's center increases.

When the central regions of the cloud become dense enough to be opaque to infrared radiation, the dynamic-collapse phase ends. The collapse of the cloud has generated a great deal of heat because of the compressional work gravitational forces perform on the gas. During the isothermal phase this heat was radiated out of the cloud as infrared light; once the radiation can no longer easily escape from the cloud, the temperature and pressure begin to rise. When the center of the cloud reaches a temperature of about 100 degrees K. and a density of about 10^{14} atoms per cubic centimeter, thermal pressure becomes great enough to overcome the gravitational force and to stop the cloud's dynamic collapse. The region in which the dynamic collapse halts has a radius of about five astronomical units (one astronomical unit, roughly 93 million miles, is the mean distance from the earth to the sun). This region is called the first core; matter in the outer regions, still transparent to infrared radiation, continues to fall inward, accumulating at the core.

The first core is in a quasi-equilibrium state: the matter deepest within the core flows alternately inward and outward, producing a periodic increase and decrease in density.

As matter from the envelope continues to build up at the core, the center of the core becomes progressively denser and hotter. Eventually the central temperature and density become high enough so that the diatomic hydrogen molecules dissociate into single atoms of hydrogen. At this stage the temperature of the cloud is about 2,000 degrees K. and its density is roughly 10^{16} atoms per cubic centimeter.

Because hydrogen absorbs energy as it dissociates, the temperature of the first core drops and there is less thermal pressure to support the mass of the cloud. Consequently the first core enters a second phase of dynamic collapse. The innermost regions fall in rapidly until the core reaches densities of approximately 10^{24} atoms per cubic centimeter (roughly the density of water) and temperatures of about 100,000 degrees K. At this point thermal pressure again becomes sufficient to counteract the force of gravity that has been pulling matter inward. A second core, smaller than the first, is therefore able to form. Initially this core contains only a small fraction of the total cloud and is a few times the size of the sun. The remainder of the cloud, however, continues to fall inward and to enter the second core. As this falling matter accumulates, the second core replaces the first core, which disappears.

After the second core is formed and the remainder of the cloud collapses around it, the protostar enters the main sequence of stellar evolution. The entire dynamic collapse has occupied roughly 100,000 years.

Larson's description of the collapse of spherically symmetric clouds, which I have outlined above, is basically consistent with observation. That is, it yields models of stars whose luminosities and surface temperatures fall within the observed ranges for young stars. Yet the picture of a perfectly spherical, nonrotating cloud is clearly a highly idealized one. Rotation and inhomogeneity within the cloud, neither of which is included in Larson's model, have important effects on the rate and type of collapse. It is worth noting that the nonrotational, spherically symmetric model cannot explain either fragmentation or the question of excess angular momentum.

The next step toward theoretical accuracy was taken in 1972 by Larson and in 1976 by David C. Black of the National Aeronautics and Space Administration's Ames Research Center and Peter H. Bodenheimer of the Lick Observatory of the University of California at Santa Cruz. These investigators studied the collapse of a rotating cloud; to keep the model relatively simple, they assumed that the cloud was symmetric with respect to its axis. They found that a rapidly rotating dense cloud may collapse, in several stages, to form a ring. Under certain conditions that ring may fragment into a system of several protostars.

In the first stage matter along the axis of rotation collapses toward the center in the same way as the matter of a nonrotating cloud collapses. Matter distant from the axis collapses more slowly, because much of the gravitational force that would ordinarily pull it toward the center is needed simply to hold it in orbit about the axis of rotation. In other words, because the cloud is spinning, matter in it feels an apparent "centrifugal force": the matter would normally tend to fly off along a straight trajectory; gravity overcomes that tendency and bends the trajectory into a circle. The faster the material is moving and the smaller the orbit is, the more gravitational force is needed to maintain that orbit and the less gravitational force is available to pull the matter inward.

Since matter along the axis collapses faster than matter farther from the axis, the once spherical cloud flattens out, forming a lozenge-shaped cloud, which grows progressively flatter and more disklike as it collapses. Eventually, for reasons that were first described by Joel E. Tohline (now at Louisiana State University at Baton Rouge) and me, the disk forms a ring.

Tohline and I showed that the ring develops because of an interplay between the forces of gravity and the law of the conservation of angular momentum. The angular momentum of a spinning body depends in part on the distance between the rotating matter within that body and the axis of rotation. Since the angular momentum of an isolated spinning body must remain constant, matter that falls in toward the center must orbit faster as it falls. This means that orbiting matter cannot fall all the way to the center: as it accelerates, more of the gravitational force is necessary to keep it from flying off. Eventually the falling matter will reach "centrifugal equilibrium," where the force of gravity is precisely sufficient to maintain the matter's orbit and hold it at a constant radius.

During the collapse of a rotating cloud some of the matter falling toward the center reaches and passes the radius of centrifugal equilibrium. Since the force of gravity is not strong enough to hold this matter in a small orbit, the matter stops collapsing inward and begins to flow away from the center (under the influence of an apparent "centrifugal force").

Meanwhile other material that is farther from the axis is still falling inward. In the resulting collision between the mass falling inward and that flowing outward a significant amount of mass accumulates away from the axial center of the cloud. If the accumulation is large enough, the gravitational force it exerts will attract the rest of the falling matter as well as the matter from the central regions. The result is a growing ring of gas and dust around an empty central region.

Thomas L. Cook of the Los Alamos National Laboratory and Michael L. Norman, of the Max Planck Institute for Physics and Astrophysics in Munich, have shown that such a ring might eventually fragment: if the ring is not perfectly symmetric about its axis, accumulations will form along the circumference, which eventually break the ring into a system of many protostars. Cook and Norman found that the spin angular momentum of each fragment will be reduced by a factor of 10 from the spin of the initial cloud. Their model shows that the rest of the cloud's angular momentum goes into the fragments' orbital motion about one another.

The next theoretical advance took place in 1979, when it became possible to model completely asymmetric, rapidly rotating clouds. Bodenheimer, Tohline, Black, and I found that some collapsing clouds will fragment without forming a ring; instead irregularities in the cloud can grow large enough to fragment it. The process would take roughly the same amount of time that an axially symmetric cloud requires to form a ring. We also found that those clouds that tend to fragment without forming rings usually evolve into binary systems rather than systems consisting of three or more members; apparently the first two accretions that form will pull the rest of the gas and dust toward themselves.

The initial fragments that form from rapidly rotating clouds typically have a mass of approximately one tenth of the initial cloud mass and, as in the case of ring fragments, their spin angular momentum is much less per unit of mass than that of the original cloud. Furthermore, each of the fragments is likely to undergo a second dynamic collapse. As each one collapses, it breaks up into yet another set of fragments. These subfragments are themselves likely to collapse and fragment.

This hierarchy of repeated collapse and fragmentation was hypothesized by Bodenheimer in 1978, before the numerical calculations of completely asymmetric clouds were possible. Its confirmation by modeling resolves both the fragmentation and the angular-momentum questions. If a cloud undergoes a cascade of many fragmentations, it could collapse to form a modest number of protostars whose spin angular momentums would be fairly close to those of some observed rapidly rotating stars. Furthermore, the hierarchical theory suggests, as observation confirms, that far more binary than single stars will form.

As useful as they are, computer models of asymmetric clouds have until recent years suffered one major flaw. Unlike Larson's model of the perfectly symmetric cloud, these models did not take thermodynamic factors into account. That is, they have not modeled the heating and cooling of various portions of the cloud as it is controlled by the flow of electromagnetic radiation. The flow of radiation is in turn dependent on the opacity and density of gas and dust particles within the cloud, factors that change as the cloud collapses. Because of this flaw, computer models were able to consider only the isothermal phase (that part of the first dynamic-collapse phase when the temperature of the cloud remains constant), in which the effects of radiation can be neglected.

My most recent work remedies this shortcoming: it consists of a detailed analysis of the thermodynamics of asymmetric clouds. More advanced methods make it possible to follow collapsing clouds through the isothermal phase and into the next one. In this phase the opacity rises, the first core is formed, and the first stage of dynamic collapse and fragmentation ends in the central region of the cloud.

My calculations have shown that there are certain types of clouds that probably will not fragment at all but instead will collapse to form single protostars. For example, a dust cloud with less than one tenth of the mass of the sun will not undergo the hierarchy of repeated collapse and fragmentation. Likewise, a slowly rotating cloud that is somewhat more massive may collapse to form a single protostar.

This kind of cloud will flatten into a disk, which will gradually take the shape of an elongated bar. Because of the conservation of angular momentum, the inner part of the protostar will rotate more rapidly than the outer part; consequently the bar eventually elongates into a pair of spiral arms.

The inner region of the spiral protostar will transfer some of its angular momentum to the slower-spinning outer region by way of gravitational torques; that is, the gravity of the slower-spinning outer part will pull on the faster-spinning center, slowing its rate of rotation. As the matter close to the axis of rotation slows it is able to collapse further. The protostar should then be able to contract the rest of the way to stellar densities without undergoing fragmentation due to excess angular momentum; this result has yet to be shown conclusively.

These developing models of general star formation can be applied to a specific case—the formation of our own sun and the solar system. There are three primary models of its preliminary stages. The first of these models, which suggests that the sun was originally part of a multiple stellar system, is the least likely. The second

and third models, which suggest respectively that the sun evolved from a decayed binary system and that it formed out of a slowly rotating single protostar,
coincide—that is, they predict essentially the same sequence of development once
the protostar is formed.

According to the first model, the sun was ejected from a system consisting of
three or more equally spaced protostars. Numerical modeling has shown that multiple systems will evolve into combinations of binary protostar systems and single
stars; it is therefore not inconceivable that the sun was originally part of a multiple
system that underwent decay.

It is unlikely that the sun formed in this manner, however. Modeling has shown
that a cloud must be rapidly rotating and relatively cold in order to form a triple
protostellar system. The three protostars formed would then also have relatively
large rates of rotation and low thermal energies; they would themselves collapse
and fragment. A rapidly spinning star with low thermal energy could avoid fragmentation only if it is relatively small (less than one tenth the mass of the sun).
Hence, according to this model, the sun would have to have formed from a protostar with very little mass and then acquired the bulk of its mass after becoming a
star, which is an unlikely history.

Apparently the solar system did not develop from a higher-order system. Perhaps
it was formed in the decay of a binary system. Suppose a cloud had collapsed to form
a system of two protostars spaced close together. If the binary system were somehow to transmit some of its angular momentum outward, the two stars would come
closer together. If the two protostars had already reached the quasi-equilibrium
phase and were no longer collapsing, the decreased separation could result in the
merging of the two to form a single protostar. This protostar would closely resemble
the protostar that would result from the third proposed model of solar evolution,
which is the collapse of a slowly rotating cloud.

As discussed above, a cloud that is rotating extremely slowly would not fragment
during the dynamic-collapse phase. Werner M. Tscharnuter of the Astronomical Institute of the University of Vienna has modeled the collapse of slowly rotating, axially symmetric clouds and shown that they will not form rings. He has therefore inferred that such clouds will not fragment. My own three-dimensional calculations
support Tscharnuter's results. They also show that even slowly rotating clouds will
collapse into protostars with the barlike shape characteristic of the faster-rotating
clouds; this means that the central portions of the protostar could transmit some of
their angular momentum outward by gravitational torques, enabling the protostar
to contract to stellar densities. The implication is that slowly rotating clouds will
form single stars. Since single stars are in the minority, it appears that slowly rotating clouds must also be relatively rare.

Both the model of a decayed binary and that of a slowly rotating cloud converge into one model on the formation of a single protostar surrounded by a cloud of gas and dust. At this stage the protostar must still contract through an increase in density by a factor of 10^{10} and undergo a second collapse before it reaches the main sequence. This phase of stellar evolution has not yet been calculated rigorously with a three-dimensional model; also certain physical properties that I have not discussed here, such as turbulence and magnetic fields, may have important effects.

At the same time that the sun is forming at the center of the nebula, the dust in the outer regions will form a flattened layer and begin the process of accumulation into a planetary system. This surrounding gas and dust may be essential for forming a single star, because it provides the protostar with a way to disperse some of the angular momentum that would otherwise impede its collapse. The formation of a planetary system may thus be a natural consequence of the formation of a single star. The recent exciting discovery of a flattened layer of dust surrounding the star Beta Pictoris seems to confirm this general picture of star and planet formation.

As astrophysicists have made increasingly realistic assumptions about the dust clouds, a clearer image of the process of stellar formation has developed. The next stage in my own research will be an attempt to extend the thermodynamic models of asymmetric clouds. Until now the model has been applied only to the first quasi-equilibrium phase, the forming of the first core; next I shall examine the second dynamic-collapse phase. I believe that no further fragmentation occurs after the first core has formed, and that the protostar contracts to stellar densities, but the definite answer can come only after rigorous modeling, which may take several more years.

—January 1985

Binary Neutron Stars

*These paired stellar remnants supply exquisite confirmations
of general relativity. Their inevitable collapse produces what
may be the strongest explosions in the universe*

Tsvi Piran

In 1967 Jocelyn Bell and Antony Hewish found the first pulsar. Their radio telescope brought in signals from a source that emitted very regular pulses every 1.34
seconds. After eliminating terrestrial sources and provisionally discarding the notion that these signals might come from extraterrestrial intelligent beings, they
were baffled. It was Thomas Gold of Cornell University who realized that the
pulses originated from a rotating neutron star, beaming radio waves into space like
a lighthouse. Researchers soon tuned in other pulsars.

Even as Bell and Hewish were making their discovery, military satellites orbiting
the earth were detecting the signature of even more exotic signals: powerful
gamma-ray bursts from outer space. The gamma rays triggered detectors intended
to monitor illicit nuclear tests, but it was not until 6 years later that the observations were made public; even then another 20 years passed before the bursts' origin
was understood. Many people now think gamma-ray bursts are emitted by twin
neutron stars in the throes of coalescence.

141

The discovery of binary neutron stars fell to Russell A. Hulse and Joseph H. Taylor, Jr., then at the University of Massachusetts at Amherst, who began a systematic pulsar survey in 1974. They used the Arecibo radio telescope in Puerto Rico, the largest in the world, and within a few months had found 40 previously unknown pulsars. Among their haul was a strange source named PSR 1913+16 (PSR denotes a pulsar, and the numbers stand for its position in the sky: 19 hours and 13 minutes longitude and a declination of 16 degrees). It emitted approximately 17 pulses per second, but the period of the pulses changed by as much as 80 microseconds from one day to the next. Pulsars are so regular that this small fluctuation stood out clearly.

Hulse and Taylor soon found that the timing of the signals varied in a regular pattern, repeating every seven hours and 45 minutes. This signature was not new; for many years astronomers have noted similar variations in the wavelength of light from binary stars (stars that are orbiting each other). The Doppler effect shortens the wavelength (and increases the frequency) of signals emitted when a source is moving toward the earth and increases wavelength (thus decreasing the frequency) when a source is moving away. Hulse and Taylor concluded that PSR 1913+16 was orbiting a companion star, even though available models of stellar evolution predicted only solitary pulsars.

The surprises did not end there. Analysis of the time delay indicated that the pulsar and its companion were separated by a mere 1.8 million kilometers. At that distance a normal star (with a radius of roughly 600,000 kilometers) would almost certainly have blocked the pulsar's signal at some point during its orbit. The companion could also not be a white dwarf (radius of about 3,000 kilometers), because tidal interactions would have perturbed the orbit in a way that contradicted the observations. Hulse and Taylor concluded that the companion to PSR 1913+16 must be a neutron star.

This finding earned the two a Nobel Prize in Physics in 1993. Astronomers have since mastered the challenge of understanding how binary neutron stars might exist at all, even as they have employed the signals these strange entities produce to conduct exceedingly fine tests of astrophysical models and of general relativity.

■ HOW A BINARY NEUTRON STAR FORMS

By all the astrophysical theories that existed before 1974, binary neutron stars should not have existed. Astronomers believed that the repeated stellar catastrophes needed to create them would disrupt any gravitational binding between two stars.

Neutron stars are the remnants of massive stars, which perish in a supernova explosion after exhausting all their nuclear fuel. The death throes begin when a star of six solar masses or more consumes the hydrogen in its center, expands, and becomes a red giant. At this stage its core is already extremely dense: several solar masses within a radius of several thousand kilometers. An extended envelope more than 100 million kilometers across contains the rest of the mass. In the core, heavier elements such as silicon undergo nuclear fusion to become iron.

When the core reaches a temperature of several billion kelvins, the iron nuclei begin to break apart, absorbing heat from their surroundings and reducing the pressure in the core drastically. Unable to support itself against its own gravitational attraction, the core collapses. As its radius decreases from several thousand kilometers to 15, electrons and protons fuse into neutrons, leaving a very dense star of 1.4 solar masses in a volume no larger than an asteroid.

Meanwhile the energy released in the collapse heats the envelope of the star, which for a few weeks emits more light than an entire galaxy. Observations of old supernovae, such as the Crab Nebula's, whose light reached the earth in A.D. 1054, reveal a neutron star surrounded by a luminous cloud of gas, still moving out into interstellar space.

More than half the stars in the sky belong to binary systems. As a result it is not surprising that at least a few massive pairs should remain bound together even after one of them undergoes a supernova explosion. The pair then becomes a massive x-ray binary, so named for the emission that the neutron star produces as it strips the outer atmosphere from its companion. Eventually the second star also explodes as a supernova and turns into a neutron star. The envelope ejected by the second supernova contains most of the mass of the binary (since the remaining neutron star contains a mere 1.4 solar masses). The ejection of such a large fraction of the total mass should therefore disrupt the binary and send the two neutron stars (the old one and the one that has just formed) flying into space with velocities of hundreds of kilometers per second.

Hulse and Taylor's discovery demonstrated, however, that some binaries survive the second supernova explosion. In retrospect astronomers realized that the second supernova explosion might be asymmetrical, propelling the newly formed neutron star into a stable orbit rather than out into the void. The second supernova also may be less disruptive if the second star loses its envelope gradually during the massive x-ray binary phase. Since then the discovery of three other neutron star binaries shows that other massive pairs have survived the second supernova.

Several years ago Ramesh Narayan of Harvard University, Amotz Shemi of Tel Aviv University, and I, along with E. Sterl Phinney of the California Institute of Technology, working independently, estimated that about 1 percent of massive

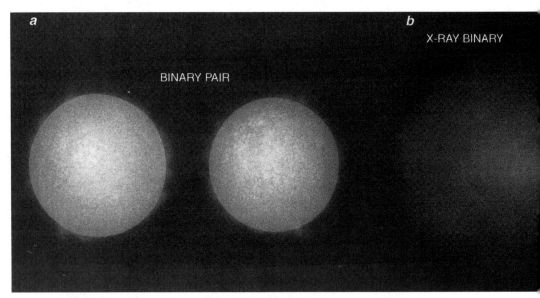

Massive binary (a) evolves through a sequence of violent events. The heavier star in the pair burns its fuel faster and undergoes a supernova explosion; if the two stars stay bound together, the result is a massive x-ray binary (b) in which the neutron star remnant of the first star strips gas from its companion and emits x-radiation. Eventually the second star also exhausts its fuel. In roughly 1 of 100 cases, the resulting explosion leaves a pair of neutron stars orbiting each other (c); in the other 99, the two drift apart (d). There are enough binary star systems that a typical galaxy contains thousands of neutron star binaries. (Al Kamajian)

x-ray binaries survive to form neutron star binaries. This figure implies that our galaxy contains a population of about 30,000 neutron star binaries. Following a similar line of argument, we also concluded that there should be a comparable number of binaries, yet unobserved, containing a neutron star and a black hole. Such a pair would form when one of the stars in a massive pair formed a supernova remnant containing more than about two solar masses and so collapsed to a singularity instead of a neutron star. Rarer, but still possible in theory, are black hole binaries, which start their lives as a pair of particularly massive stars; they should number about 300 in our galaxy.

■ TESTING GENERAL RELATIVITY

PSR 1913+16 has implications that reach far beyond the revision of theories of binary stellar evolution. Hulse and Taylor immediately realized that their dis-

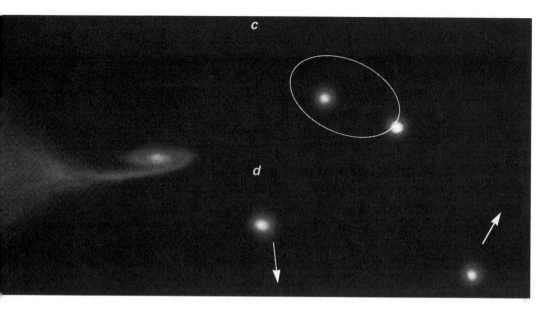

covery had provided an ideal site for testing Einstein's general theory of relativity.

Although this theory is accepted today as the only viable description of gravity, it has had only a few direct tests. Albert Einstein himself computed the precession of Mercury's orbit (the shift of the orbital axes and the point of Mercury's closest approach to the sun) and showed that observations agreed with his theory. Arthur Eddington detected the bending of light rays during a solar eclipse in 1919. In 1960 Robert V. Pound and Glen A. Rebka, Jr., then both at Harvard, first measured the gravitational redshift, the loss of energy by photons as they climb out of a powerful gravitational field. Finally, in 1964, Irwin I. Shapiro, also at Harvard, pointed out that light signals bent by a gravitational field should be delayed in comparison to those that take a straight path. He measured the delay by bouncing radar signals off other planets in the solar system. Although general relativity passed these tests with flying colors, they were all carried out in the (relativistically) weak gravitational field of the solar system. That fact left open the possibility that general relativity might break down in stronger gravitational fields.

Because a pulsar is effectively a clock orbiting in the strong gravitational field of its companion, relativity makes a range of clear predictions about how the ticks of that clock (the pulses) will appear from the earth. First, the Doppler effect causes a periodic variation in the pulses' arrival time (the pattern that first alerted Taylor and Hulse).

A "second-order" Doppler effect, resulting from time dilation caused by the pulsar's rapid motion, leads to an additional (but much smaller) variation. This second-order effect can be distinguished because it depends on the square of velocity, which varies as the pulsar moves along its elliptical orbit. The second-order Doppler shift combines with the gravitational redshift, a slowing of the pulsar's clock when it is in the stronger gravitational field closer to its companion.

Like Mercury, PSR 1913+16 precesses in its orbit about its companion. The intense gravitational fields involved, however, mean that the periastron—the nadir of the orbit—rotates by 4.2 degrees a year, compared with Mercury's perihelion shift of a mere 42 arc seconds a century. The measured effects match the predictions of relativistic theory precisely. Remarkably, the precession and other orbital information supplied by the timing of the radio pulses make it possible to calculate the masses of the pulsar and its companion: 1.442 and 1.386 solar masses, respectively, with an uncertainty of .003 solar mass. This precision is impressive for a pair of objects 15,000 light-years away.

In 1991 Alexander Wolszczan of the Arecibo observatory found another binary pulsar that is almost a twin to PSR 1913+16. Each neutron star weighs between 1.27 and 1.41 solar masses. The Shapiro time delay, which was only marginally measured in PSR 1913+16, stands out clearly in signals from the pulsar that Wolszczan discovered.

Measurements of PSR 1913+16 have also revealed a relativistic effect never seen before. In 1918, several years after the publication of his General Theory of Relativity, Einstein predicted the existence of gravitational radiation, an analogue to electromagnetic radiation. When electrically charged particles such as electrons and protons accelerate, they emit electromagnetic waves. Analogously, massive parti-

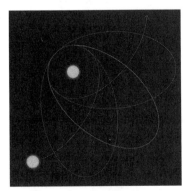

Orbital precession, the rotation of the major axis of an elliptical orbit, results from relativistic perturbations of the motion of fast-moving bodies in intense gravitational fields. It is usually almost undetectable; Mercury's orbit precesses by less than .12 of a degree every century, but that of PSR 1913+16 changes by 4.2 degrees a year. (George Retseck)

cles that move with varying acceleration emit gravitational waves, small ripples in the gravitational field that also propagate at the speed of light.

These ripples exert forces on other masses; if two objects are free to move, the distance between them will vary with the frequency of the wave. The size of the oscillation depends on the separation of the two objects and the strength of the waves. In principle all objects whose acceleration varies emit gravitational radiation. Most objects are so small and move so slowly, however, that their gravitational radiation is utterly insignificant.

Binary pulsars are one of the few exceptions. The emission of gravitational waves produces a detectable effect on the binary system. In 1941, long before the discovery of the binary pulsar, the Russian physicists Lev D. Landau and Evgenii M. Lifshitz calculated the effect of this emission on the motion of a binary. Energy conservation requires that the energy carried away by the waves come from somewhere, in this case the orbital energy of the two stars. As a result the distance between them must decrease.

PSR 1913+16 emits gravitational radiation at a rate of eight quadrillion gigawatts, about a fifth as much energy as the total radiation output of the sun. This luminosity is impressive as far as gravitational radiation sources are concerned but still too weak to be detected directly on the earth. Nevertheless, it has a noticeable effect on the pulsar's orbit. The distance between the two neutron stars decreases by a few meters a year, which suffices to produce a detectable variation in the timing of the radio pulses. By carefully monitoring the pulses from PSR 1913+16 over the years, Taylor and his collaborators have shown that the orbital separation decreases in exact agreement with the predictions of the General Theory of Relativity.

The reduction in the distance between the stars can be compared with the other general relativistic effects to arrive at a further confirmation. Just as measurements of the orbital decay produce a mathematical function relating the mass of the pulsar to the mass of its companion, so do the periastron shift and the second-order Doppler effect. All three functions intersect at precisely the same point.

■ UNDETECTABLE CATACLYSMS

At present the distance between PSR 1913+16 and its companion is decreasing only slowly. As the distance between the stars shrinks, the gravitational wave emission will increase and the orbital decay will accelerate. Eventually the neutron stars will fall toward each other at a significant fraction of the speed of light, collide, and merge. The 300 million years until PSR 1913+16 coalesces with its companion are long on a human scale but rather short on an astronomical one.

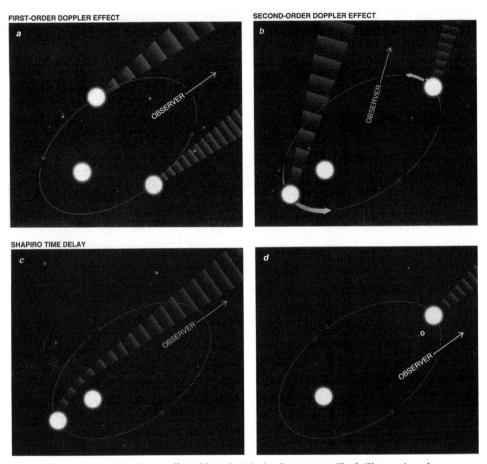

FIRST-ORDER DOPPLER EFFECT

SECOND-ORDER DOPPLER EFFECT

SHAPIRO TIME DELAY

Binary pulsar signals are affected by relativistic phenomena. (Each illustration above shows one of the effects whose combination produces the timing of the pulses that radio astronomers observe.) The Doppler effect slows the rate at which pulses reach an observer when the pulsar is moving away from the earth in its orbit and increases the rate when the pulsar is moving toward the earth (a). The second-order Doppler effect and the gravitational redshift (b) impose a similar variation because the pulsar's internal clock slows when it moves more rapidly in its orbit closest to its companion (shown by longer arrow). Most subtle is the so-called Shapiro time delay, which occurs as the gravitational field of the pulsar's companion bends signals passing near it (c). The signals travel farther than they would if they took a straight-line path (d) and so arrive later. This effect is undetectable in PSR 1913+16 but shows up clearly in a more recent discovery. (George Retseck)

Given the number of neutron star binaries in the galaxy, one pair should merge roughly every 300,000 years, a cosmological blink of the eye. Extrapolating this rate to other galaxies implies that throughout the observable universe about one neutron star merger occurs every 20 minutes—frequently enough that astronomers should consider whether they can detect such collisions.

To figure out whether such occurrences are detectable requires a solid under-standing of just what happens when two orbiting neutron stars collide. Shortly after the discovery of the first binary pulsar, Paul Clark and Douglas M. Eardley, then both at Yale University, concluded that the final outcome is a black hole. Current estimates of the maximum mass of a neutron star range between 1.4 and 2.0 solar masses. Rotation increases the maximal mass, but most models suggest that even a rapidly rotating neutron star cannot be significantly larger than 2.4 solar masses. Because the two stars together contain about 2.8 solar masses, collapse to a singu-larity is almost inevitable.

Melvyn B. Davies of Caltech, Willy Benz of the University of Arizona, Freidrich K. Thielemann of the Harvard-Smithsonian Center for Astrophysics, and I have sim-ulated the last moments of a neutron star binary in detail. The two objects are very dense and so behave effectively like point masses until they are quite close to each other. Tidal interaction between the stars becomes important only when they ap-proach to within 30 kilometers, about twice the radius of a neutron star. At that stage they begin to tear material from each other—about two tenths of a solar mass in total. Once the neutron stars touch, within a tiny fraction of a second they coa-lesce. The matter torn from the stars before the collision forms a disk around the central core and eventually spirals back into it.

What kinds of signals will this sequence of events generate? Clark and Eardley re-alized that the colliding stars will warm up and reach temperatures of several bil-lion kelvins. They figured that most of the thermal energy would be radiated as neutrinos and antineutrinos, much as it is in a supernova. Unfortunately, these weakly interacting, massless particles, which escape from the dense neutron star much more easily than do photons, are almost undetectable. When supernova 1987A exploded, the three detectors on the earth caught a total of 21 neutrinos out of the 5×10^{46} joules of radiation. Although the burst expected from a binary neutron star merger is slightly larger than that of a supernova, the typical event takes place much farther away than the mere 150,000 light-years of SN 1987A. To detect one merger a year would require picking up signals one sixteen-millionth the intensity of the 1987 event. Because current neutrino detectors must monitor interactions within thousands of tons of material, it is difficult to imagine the appa-ratus that would be required. Furthermore, supernovae are 1,000 times more fre-quent than are neutron star collisions. Even if we detected a neutrino burst from two neutron stars, it is unlikely we would be able to distinguish it among the far more numerous and far more intense supernova neutrino bursts.

Before it emits its neutrino burst, the neutron star binary sends out a similarly energetic (but not quite so undetectable) train of gravitational waves. During the 15 minutes before coalescence, the two stars cover the last 700 kilometers between them, and their orbital period shrinks from a fifth of a second to a few milliseconds.

The resulting signal is just in the optimal range for terrestrial gravity-wave detectors.

An international network of such detectors is now being built in the U.S. and in Italy. The American Caltech-M.I.T. team is building detectors for the Laser Interferometer Gravitational-wave Observatory (LIGO) near Hanford in Washington State and near Livingston in Louisiana. The French-Italian team is constructing its VIRGO facility near Pisa in Italy. The first detectors should be able to detect neutron star mergers up to 70 million light-years away; current estimates suggest that there is only one event per 100 years up to this distance. Researchers have proposed to improve their instruments dramatically over the subsequent few years; eventually they should be able to detect neutron star mergers as far away as three billion light-years—several hundred a year.

■ HIGH-ENERGY PHOTONS

For several years after the discovery of PSR 1913+16, I kept wondering whether there was a way to estimate what fraction of the coalescing stars' binding energy is emitted as electromagnetic radiation. Even if this fraction is tiny, the binding energy is so large that the resulting radiation would still be enormous. Furthermore, photons are much easier to detect than neutrinos or gravitational waves, and so mergers could be detected even from the most distant parts of the universe.

In 1987 J. Jeremy Goodman of Princeton University, Arnon Dar of the Technion, and Shmuel Nussinov of Tel Aviv University noticed that about a tenth of a percent of the neutrinos and antineutrinos emitted by a collapsing supernova core collide with one other and annihilate to produce electron-positron pairs and gamma rays. In a supernova the absorption of these gamma rays by the star's envelope plays an important role in the explosion of the outer layers.

In 1989 David Eichler of Ben Gurion University of the Negev, Mario Livio of the Technion, David N. Schramm of the University of Chicago, and I speculated that a similar fraction of the neutrinos released in a binary neutron star merger would also produce electron-positron pairs and gamma rays. The colliding neutron stars, however, have no envelope surrounding them, and so the gamma rays escape in a short, intense burst.

Gamma-ray bursts might arise from a more complex mechanism. The disk that forms during the neutron star merger falls back onto the central coalesced object within a few seconds, but during that time it, too, can trigger remissions. In 1992 Bohdan Paczyński of Princeton, Narayan of Harvard, and I suggested that the rotation of the disk could intensify the neutron star magnetic fields entangled in the disk's material, causing giant magnetic flares, a scaled-up version of the flares that rise from the surface of the sun. These short-lived magnetic disturbances could generate gamma-ray bursts in the same way that solar flares produce gamma rays and x-rays. The large variability in the observed bursts implies that both mechanisms may be at work.

■ A PUZZLE UNSCRAMBLED

Had it not been for the Limited Test Ban Treaty of 1963, we would not have known about these bursts until well into the next century. No one would have proposed a satellite to look for them, and had such a proposal been made it would surely have been turned down as too speculative. But the U.S. Department of Defense launched a series of satellites known as Vela, which carried omnidirectional x-ray and gamma-ray detectors to verify that no one was testing nuclear warheads in space.

These spacecraft never detected a nuclear explosion, but as soon as the first satellite was launched it began to detect entirely unexpected bursts of high-energy photons in the range of several hundred kiloelectron volts. Bursts lasted between a few dozen milliseconds and about 30 seconds. The lag between the arrival time of the bursts to different satellites indicated that the sources were outside the solar system. Still, the bursts were kept secret for several years, until in 1973 Ray W. Klebesadel, Ian B. Strong, and Roy A. Olson of Los Alamos National Laboratory described them in a seminal paper. Theorists proposed more than 100 models in the next 20 years; in the late 1980s a consensus formed that the bursts originated on neutron stars in our own galaxy.

A minority led by Paczyński argued that the bursts originated at cosmological distances. In the spring of 1991 the Compton Gamma Ray Observatory, which was more sensitive than any previous gamma-ray satellite, was launched by the National Aeronautics and Space Administration. It revealed two unexpected facts. First, the distribution of burst intensities is not homogeneous in the way that it would be if the bursts were nearby. Second, the bursts came from all across the sky rather than being concentrated in the plane of the Milky Way, as they would be if they originated in the galactic disk. Together these facts demonstrate that the bursts do not originate from the disk of our galaxy. A lively debate still prevails over the possibility that the bursts might originate from the distant parts of the invisible halo of our galaxy, but as the Compton Observatory collects more data, this hypothesis seems less and less likely. It seems that the minority was right.

In the fall of 1991 I analyzed the distribution of burst intensities, as did Paczyński and his colleague Shude Mao. We concluded that the most distant bursts seen by the

Compton Observatory came from several billion light-years away. Signals from such distances are redshifted (their wavelength is increased and energy decreased) by the expansion of the universe. As a result we predicted that the cosmological redshift should lead to a correlation between the intensity of the bursts, their duration, and their spectra. Fainter bursts, which tend to come from farther away, should last longer and contain a lower-energy distribution of gamma rays.

Recently a NASA team headed by Jay P. Norris of the Goddard Space Flight Center has found precisely such a correlation. The number of bursts that the Compton Observatory records also tallies quite well with our earlier estimates of the binary neutron star population. Roughly 30,000 mergers should occur every year throughout the observable universe, and the satellite's detectors can scan a sphere containing about 3 percent of that volume. Our rough estimates suggest 900 mergers a year in such a space; the Compton Observatory notes 1,000 bursts.

Although the details of how colliding neutron stars give rise to gamma rays are still being worked out, the tantalizing agreement between these data from disparate sources implies that astronomers have been detecting neutron star mergers without knowing it for the past 25 years. Researchers have proposed a few other sources that might be capable of emitting the enormous amounts of energy needed for cosmological gamma-ray bursts. The merger model, however, is the only one based on an independently observed phenomenon, the spiraling in of a neutron star binary as a result of the emission of gravity waves.

It is the only model that makes a clear prediction that can be either confirmed or refuted. If, as I expect, LIGO and VIRGO detect the unique gravitational-wave signal of spiraling neutron stars in coincidence with a gamma-ray burst, astrophysicists will have opened a new window on the final stages of stellar evolution, one that no visible-light instruments can hope to match.

—May 1995

SOHO REVEALS THE SECRETS OF THE SUN

A powerful new spacecraft—the Solar and Heliospheric Observatory, or SOHO—is now monitoring the sun around the clock, providing new clues about our nearest star

Kenneth R. Lang

From afar the sun does not look very complex. To the casual observer it is just a smooth, uniform ball of gas. Close inspection, however, shows that the star is in constant turmoil—a fact that fuels many fundamental mysteries. For instance, scientists do not understand how the sun generates its magnetic fields, which are responsible for most solar activity, including unpredictable explosions that cause magnetic storms and power blackouts here on the earth. Nor do they know why this magnetism is concentrated into so-called sunspots, dark islands on the sun's surface that are as large as the earth and thousands of times more magnetic. Furthermore, physicists cannot explain why the sun's magnetic activity varies dramatically, waning and intensifying again every 11 years or so.

To solve such puzzles—and better predict the sun's impact on our planet—the European Space Agency (ESA) and the National Aeronautics and Space Administration launched the two-ton Solar and Heliospheric Observatory (SOHO, for short) on December 2, 1995. The spacecraft reached its permanent strategic position— which is called the inner Lagrangian point and is about 1 percent of the way to the sun—on February 14, 1996. There SOHO is balanced between the pull of the earth's gravity and the sun's gravity and so orbits the sun together with the earth. Earlier spacecraft studying the sun orbited the earth, which would regularly obstruct their view. In contrast SOHO monitors the sun continuously: 12 instruments examine the sun in unprecedented detail. They downlink several thousand images a day through NASA's Deep Space Network antennae to SOHO's Experimenters' Operations Facility at the NASA Goddard Space Flight Center located in Greenbelt, Maryland.

At the Experimenters' Operations Facility solar physicists from around the world work together, watching the sun night and day from a room without windows. Many of the unique images they receive move nearly instantaneously to the SOHO home page on the World Wide Web (located at *http://sohowww.nascom.nasa.gov*). When these pictures first began to arrive, the sun was at the very bottom of its 11-year activity cycle. But SOHO carries enough fuel to continue operating for a decade or more. Thus, it will keep watch over the sun through all its tempestuous seasons—from the recent lull in magnetic activity to its next maximum, which should take place at the end of the century. Already, though, SOHO has offered some astounding findings.

■ EXPLORING UNSEEN DEPTHS

To understand the sun's cycles, we must look deep inside the star, to where its magnetism is generated. One way to explore these unseen depths is by tracing the in-and-out, heaving motions of the sun's outermost visible surface, named the photosphere from the Greek word *photos,* meaning "light." These oscillations, which can be tens of kilometers high and travel a few hundred meters per second, arise from sounds that course through the solar interior. The sounds are trapped inside the sun; they cannot propagate through the near vacuum of space. (Even if they could reach the earth, they are too low for human hearing.) Nevertheless, when these sounds strike the sun's surface and rebound back down, they disturb the gases there, causing them to rise and fall, slowly and rhythmically, with a period of about five minutes. The throbbing motions these sounds create are imperceptible to the naked eye, but SOHO instruments routinely pick them out.

The surface oscillations are the combined effect of about 10 million separate notes—each of which has a unique path of propagation and samples a well-defined section inside the sun. So to trace the star's physical landscape all the way through—from its churning convection zone, the outer 28.7 percent (by radius), into its radiative zone and core—we must determine the precise pitch of all the notes.

The dominant factor affecting each sound is its speed, which in turn depends on the temperature and composition of the solar regions through which it passes. SOHO scientists compute the expected sound speed using a numerical model. They then use relatively small discrepancies between their computer calculations and the observed sound speed to fine-tune the model and establish the sun's radial variation in temperature, density, and composition.

At present, theoretical expectations and observations made with SOHO's Michelson Doppler Imager (MDI) telescope are in close agreement, showing a maximum difference of only .2 percent. Where these discrepancies occur is, in

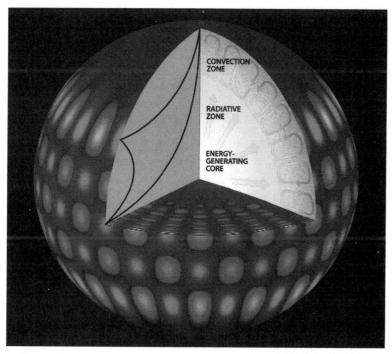

Sound waves, represented here by black lines inside the cutaway section, resonate throughout the sun. They are produced by hot gas churning in the convection zone, which lies above the radiative zone and the sun's core. As sound waves travel toward the sun's center, they gain speed and are refracted back out. At the same time, the sun's surface reflects waves traveling outward back in. Thus, the entire star throbs, with regions pulsing in and out. (Courtesy of Jack Harvey National Optical Astronomy observatories; cross sections by Michael Goodman)

fact, significant. They suggest that turbulent material is moving in and out just below the convection zone and hint that such mixing motions might occur at the boundary of the energy-generating core—concepts that could be very important for studies of stellar evolution.

For more than three centuries astronomers have known from watching sunspots that the photosphere rotates faster at the equator than at higher latitudes, and that the speed decreases evenly toward each pole. SOHO data confirm that this differential pattern persists through the convection zone. Furthermore, the rotation speed becomes uniform from pole to pole about a third of the way down. Thus, the rotation velocity changes sharply at the base of the convection zone. There the outer parts of the radiative interior, which rotates at one speed, meet the overlying convection zone, which spins faster in its equatorial middle. We now suspect that this thin base layer of rotational shear may be the source of the sun's magnetism.

The MDI telescope on board SOHO has also helped probe the sun's outer shells. Because its lenses are positioned well above the earth's obscuring atmosphere, it can continuously resolve fine detail that cannot always be seen from the ground. For this reason it has proved particularly useful in time-distance helioseismology, a new technique for revealing the motion of gases just below the photosphere. The method is quite straightforward: the telescope records small periodic changes in the wavelength of light emitted from a million points across the sun every minute. By keeping track of them, it is possible to determine how long it takes for sound waves to skim through the sun's outer layers. This travel time tells of both the temperature and gas flows along the internal path connecting two points on the visible solar surface. If the local temperature is high, sound waves move more quickly—as they do if they travel with the flow of gas.

The MDI has provided travel times for sounds crossing thousands of paths, linking myriad surface points. And SOHO scientists have used these data to chart the three-dimensional internal structure and dynamics of the sun, much in the same way that a computed tomographic (CT) scan creates an image of the inside of the brain. They fed the SOHO data to supercomputers to work out temperatures and flow directions along these intersecting paths. Using these techniques during two years of nearly continuous observations, SOHO scientists have discovered vast rivers of hot gas that circulate within the sun.

Completely unexpected currents circle the polar regions of the sun just below the photosphere. They seem to resemble the jet streams high in the atmosphere of the earth, which have a major influence on terrestrial weather. Ringing the sun at about 75 degrees latitude, the solar jet streams are totally inside the sun, 40,000 kilometers (25,000 miles) below the photosphere, and cannot be seen at the visible surface. They move about 10 percent faster than the surrounding gas—about 130 kilometers per hour faster—and they are wide enough to engulf two planet Earths.

The outer layer of the sun, to a depth of at least 25,000 kilometers, is also slowly flowing from the equator to the poles, at a speed of about 90 kilometers per hour. At this rate an object would be transported from the equator to the pole in little more than a year. Of course the sun rotates at a much faster rate of about 7,000 kilometers per hour, completing one revolution at the equator in 25.7 days. The combination of differential rotation and poleward flow has been the explanation for the stretched-out shapes of magnetic regions that have migrated toward the poles. The new SOHO MDI observations demonstrate for the first time that the poleward flow reaches deeply into the sun, penetrating at least 12 percent of the convection zone.

Researchers have also identified internal rivers of gas moving in bands near the equator at different speeds relative to each other in both the northern and southern hemispheres. The solar belts are more than 64,000 kilometers in width and move about 16 kilometers per hour faster than the gases to either side. These broad belts of higher-velocity currents remind one of the earth's equatorial tradewinds and also of Jupiter's colorful, banded atmosphere. The bands are deeply rooted, extending down approximately 19,000 kilometers into the sun. The full extent of the new-found solar meteorology could never have been seen by looking at the visible layer of the solar atmosphere.

The MDI team also investigated horizontal motions at a depth of about 1,400 kilometers and compared them with an overlying magnetic image, also taken by the MDI instrument. They found that strong magnetic concentrations tend to lie in regions where the subsurface gas flow converges. Thus, the churning gas probably forces magnetic fields together and concentrates them, thereby overcoming the outward magnetic pressure that ought to make such localized concentrations expand and disperse.

■ THE MILLION-DEGREE CORONA

SOHO is also helping scientists explain the solar atmosphere, or corona. The sun's sharp outer rim is illusory. It merely marks the level beyond which solar gas becomes transparent. The invisible corona extends beyond the planets and presents one of the most puzzling paradoxes of solar physics: it is unexpectedly hot, reaching temperatures of more than one million kelvins just above the photosphere; the sun's visible surface is only 5,780 kelvins. Heat simply should not flow outward from a cooler to a hotter region. It violates the second law of thermodynamics and all common sense as well. Thus, there must be some mechanism transporting energy from the photosphere, or below, out to the corona. Both kinetic and magnetic energy can flow from cold to hot regions. So writhing gases and shifting magnetic fields may be accountable.

For studying the corona and identifying its elusive heating mechanism, physicists look at ultraviolet (UV), extreme ultraviolet (EUV), and x-ray radiation. This is because hot material—such as that within the corona—emits most of its energy at these wavelengths. Also, the photosphere is too cold to emit intense radiation at these wavelengths, so it appears dark under the hot gas. Unfortunately, UV, EUV, and x-rays are partially or totally absorbed by the earth's atmosphere, and so they must be observed through telescopes in space. SOHO is now measuring radiation at UV and EUV wavelengths using four instruments: the Extreme-ultraviolet Imaging Telescope (EIT), the Solar Ultraviolet Measurements of Emitted Radiation (SUMER), the Coronal Diagnostic Spectrometer (DCS), and the UltraViolet Coronagraph Spectrometer (UVCS).

To map out structures across the solar disk, ranging in temperature from 6,000 to 2 million kelvins, SOHO makes use of spectral lines. These lines appear when the sun's radiation intensity is displayed as a function of wavelength. The various SOHO instruments locate regions having a specific temperature by tuning in to spectral lines emitted by the ions formed there. Atoms in a hotter gas lose more electrons through collisions, and so they become more highly ionized. Because these different ions emit spectral lines at different wavelengths, they serve as a kind of thermometer. We can also infer the speed of the material moving in these regions from the Doppler wavelength changes of the spectral lines that SOHO records.

Ultraviolet radiation has recently revealed that the sun is a vigorous, violent place even when its 11-year activity cycle is in an apparent slump—and this fact may help explain why the corona is so hot. The whole sun seems to sparkle in the UV light emitted by localized bright spots. According to SOHO measurements, these ubiquitous hot spots are formed at a temperature of a million kelvins, and they seem to originate in small, magnetic loops of hot gas found all over the sun, including both its north and south poles. Some of these spots explode and hurl material outward at speeds of hundreds of kilometers per second. SOHO scientists are now studying these bright spots to see if they play an important role in the elusive coronal heating mechanism.

SOHO has provided direct evidence for the transfer of magnetic energy from the sun's visible surface toward the corona above. Images of the photosphere's magnetism, taken with SOHO's MDI, reveal ubiquitous pairs of opposite magnetic polarity, each joined by a magnetic arch that rises above them, like bridges that connect two magnetic islands. Energy flows from these magnetic loops when they interact, producing electrical and magnetic "short circuits." The very strong electric currents in these short circuits can heat the corona to a temperature of several million degrees. Images from the EIT and CDS instruments on SOHO show the hot gases of the ever-changing corona reacting to the evolving magnetic fields rooted in the solar surface.

To explore changes at higher levels in the sun's atmosphere, SOHO relies on its UVCS and its Large Angle Spectroscopic Coronagraph (LASCO). Both instruments use occulting disks to block the photosphere's underlying glare. LASCO detects visible sunlight scattered by electrons in the corona. Initially it revealed a simple corona—one that was highly symmetrical and stable. This corona, viewed during the sun's magnetic lull, exhibited pronounced holes in the north and south. (Coronal holes are extended, low-density, low-temperature regions where EUV and x-ray emissions are abnormally low or absent.)

In contrast the equatorial regions were ringed by straight, flat streamers of outflowing matter. The sun's magnetic field shapes these streamers. At their base electrified matter is densely concentrated within magnetized loops rooted in the photosphere. Farther out in the corona, the streamers narrow into long stalks that stretch tens of millions of kilometers into space. These extensions confine material at temperatures of about two million kelvins within their elongated magnetic boundaries, creating a belt of hot gas that extends around the sun.

The streamers live up to their name: material seems to flow continuously along their open magnetic fields. Occasionally the coronagraphs record dense concentrations of material moving through an otherwise unchanging streamer—like seeing leaves floating on a moving stream. And sometimes tremendous eruptions, called coronal mass ejections, punctuate the steady outward flow. These ejections hurl billions of tons of million-degree gases into interplanetary space at speeds of hundreds of kilometers per second. This material often reaches the earth in only two or three days. To almost everyone's astonishment, LASCO found equatorial ejections emitted within hours of each other from opposite sides of the sun.

The coronagraphs have only a side view of the sun and so can barely see material moving to or from the earth. But based on what we can see, we guess that these ejections are global disturbances, extending all the way around the sun. In fact unexpectedly wide regions of the sun seem to convulse when the star releases coronal mass ejections, at least during the minimum in the 11-year activity cycle. And the coronagraphs have detected that a few days before the ejections, the streamer belt gets brighter, suggesting that more material is accruing there. The pressure and tension of this added material probably build until the streamer belt blows open in the form of an ejection. The entire process is most likely related to a large-scale global reorganization of the sun's magnetic field.

■ SOLAR WINDS AND BEYOND

The sun's hot and stormy atmosphere is forever expanding in all directions, filling the solar system with a ceaseless flow—called the solar wind—that contains electrons, ions, and magnetic fields. The million-degree corona creates an outward pressure that overcomes the sun's gravitational attraction, enabling this perpetual outward flow. The wind accelerates as it moves away from the sun, like water overflowing a dam. As the corona disperses it must be replaced by gases welling up from below to feed the wind. Earlier spacecraft measurements, as well as those from Ulysses (launched in 1990), showed that the wind has a fast and a slow component. The fast one moves at about 800 kilometers per second; the slow one travels at half that speed.

The slow component is associated with equatorial regions of the sun, now being scrutinized by LASCO and UVCS. These instruments suggest that the slow component of the solar wind flows out along the stalklike axes of equatorial coronal streamers. The high-speed component pours forth from the polar coronal holes. (Open magnetic fields there allow charged particles to escape the sun's gravitational and magnetic grasp.) SOHO is now investigating whether polar plumes— tall structures rooted in the photosphere that extend into the coronal holes—help to generate this high-speed solar wind.

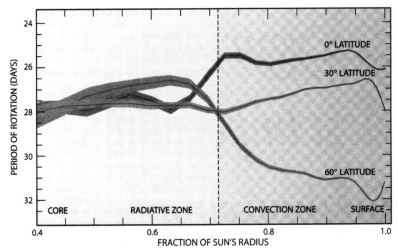

Internal rotation rate of the sun at latitudes of 0, 30 and 60 degrees has been inferred using data from the Michelson Doppler Imager. Down to the base of the convection zone, the polar regions spin more slowly than the equatorial ones do. Beyond that, uniform rotation appears to be the norm, although scientists have not yet determined rotation rates within the sun's core. (Michael Goodman. After Soho sol/MDI consortium and Alexander G. Kosovichev)

SOHO's UVCS has examined the spectral emission of hydrogen and heavily charged oxygen ions in the regions where the corona is heated and the solar wind accelerates. And these spectral-line profiles have produced surprising results, revealing a marked difference in the agitation speeds at which hydrogen and oxygen ions move. In polar coronal holes, where the fast solar wind originates, the heavier oxygen is far more agitated, with about 60 times more energy of motion; above two solar radii from the sun's center oxygen has the higher agitation speed, approaching 500 kilometers per second. Hydrogen, on the other hand, moves at only 250 kilometers per second. In contrast, within equatorial regions, where the slow-speed wind begins, the lighter hydrogen moves faster than the oxygen, as one would expect from a heat-driven wind.

Researchers are now trying to determine why the more massive oxygen ions move at greater speeds in coronal holes. One possibility is that the ions are whirling around magnetic field lines that stretch from the sun. Information about the heating and acceleration processes is probably retained within the low-density coronal holes, wherein ions rarely collide with electrons. Frequent collisions in high-density streamers might erase any signature of the relevant processes.

SOHO has obtained marvelous results to date. It has revealed features on the mysterious sun never seen before or never seen so clearly. It has provided new insights into fundamental unsolved problems, all the way from the sun's interior to Earth and out to the farthest reaches of the solar wind. Some of its instruments are now poised to resolve several other mysteries. Two of them will soon have looked at the solar oscillations long enough, and deep enough, to determine the temperature and rotation at the sun's center. Moreover, during the next few years, our home star's inner turmoil and related magnetic activity—which can directly affect our daily lives—will increase. SOHO should then offer even greater scientific returns, determining how its threatening eruptions and hot, gusty winds originate and perhaps predicting conditions in the sun's atmosphere.

—*Scientific American Presents,* Spring 1998

Accretion Disks in Interacting Binary Stars

*Disks of matter naturally organize themselves around objects
ranging from newborn stars to quasars. An unusual
class of variable stars is helping theorists
understand how these disks behave*

John K. Cannizzo and Ronald H. Kaitchuck

Disks are one of the most common structures found in the heavens. In most cases
disks surround a massive central object such as a star or a black hole. Matter in a
disk usually migrates inward and eventually accretes onto the central object; these
objects therefore are called accretion disks. Accretion disks are thought to be involved in diverse phenomena ranging from the formation of stars and planets to the
powering of quasars.

The best-studied disks reside in interacting binary star systems. We have focused our attention on a particularly intriguing, unstable class of interacting binaries, known as cataclysmic variables, that can brighten by a factor of 100 in just a few hours. These systems, fascinating in their own right, serve as a laboratory for understanding the physics of accretion disks.

One might wonder how highly organized disk shapes would develop so frequently throughout the universe. It turns out that the basic laws of physics favor such formations. Consider, for example, an irregularly shaped cloud consisting of particles moving in random orbits, the entire ensemble possessing net angular momentum. (If the cloud is gaseous, the "particle" can be thought of as a small parcel of gas.)

Each particle in the cloud responds to the combined gravitational tugs of all the other parts of the cloud. Particles passing very close to one another experience gravitational and pressure forces that deflect their paths from one of pure orbital motion. Such interactions among the particles dissipate the energy of random motions, whereas conservation of angular momentum preserves the cloud's rotational energy. The particles will eventually settle into the lowest-energy configuration: circular orbits all lying in a single plane (*see illustration below*).

Disks are naturally in irregular systems that possess net angular momentum (a). Interactions between particles in the system cancel out random motions (b), ultimately yielding a flat, circular, rotating disk (c). Viscosity effects can cause particles to spiral inward and to accrete onto more central regions of the disk. (George Retsek)

Theoreticians have developed a number of models aimed at understanding the behavior of such accretion disks. The simplest models assume that the mass flow into the disk exactly balances the accretion onto the central object. Most of these so-called static accretion disk models are based on a pioneering 1973 paper by the Russian scientists Nikolai I. Shakura of the Sternberg Astronomical Institute in Moscow and Rashid A. Sunyaev, then at the Institute of Applied Mathematics, also in Moscow.

For simplicity Shakura and Sunyaev pictured the disk as a thin, flat, gaseous object whose gravitational field is negligible compared with that of the central object. The gas therefore obeys Kepler's laws of motion, which means that the orbital velocity of each parcel of gas is inversely proportional to the square root of the distance from the central object.

Inner parts of the disk rotate faster than (and therefore slide past) outer parts. Friction between material at adjacent radii—a phenomenon known as viscosity—heats the gas and transports orbital angular momentum outward. The heated gas emits electromagnetic radiation, which escapes from the system. In the process gravitational potential energy is converted into radiant energy. As a result gas slowly drifts inward, at about one ten-thousandth of the orbital velocity.

The intensity of the coupling between adjacent radii in the disk, which determines the rate of inflow, is controlled by the strength of the viscosity of the gas. Unfortunately there is no good theory to predict this viscosity, so Shakura and Sunyaev assumed the viscosity is proportional to the pressure in the disk. In the equations describing the physical state of the disk, the strength of the viscosity appears as an adjustable parameter called alpha. All the unknown physics associated with the viscosity is hidden in alpha. Setting alpha as a constant makes it possible to construct models that predict the physical conditions in the disk. Fortunately the spectrum of radiation emitted by a disk in a steady state is independent of alpha.

The nature of the central object determines the depth of the gravitational potential well, which in turn determines the energy flux from the disk. When the central object is an ordinary star, the disk radiates primarily in the visible and in the infrared. Matter orbiting a collapsed star, such as a white dwarf or a neutron star, falls much farther before hitting the stellar surface. Disks around these objects release more total energy, and the emission peaks at ultraviolet or x-ray wavelengths.

Because of the great distances involved, it is impossible to observe directly the accretion disks of even the nearest binary star systems. Astronomers have been forced to infer the structure of stellar systems by analyzing the radiation that they produce. Light emitted by the part of a disk moving toward the viewer shifts toward the blue end of the spectrum, whereas light from the parts moving away shift in the red direction, an effect called Doppler shifting. The swirling gas in an accretion disk therefore produces a distinctive spectral signature.

In this way Arthur Wyse of Lick Observatory detected the first known accretion disk in a double star system in 1934. He was studying the spectrum of RW Tauri, a binary consisting of a hot, main-sequence star and a large, cool red companion. The orbital plane of the two stars is viewed nearly edge-on, so that once during each orbit the cool star obscures its companion. Wyse obtained a spectrum of the red star during the 90-minute interval when the companion was completely blocked from view. He observed Doppler-shifted emission lines, indicating the presence of hot, rapidly moving gas.

In the early 1940s Alfred H. Joy of Mount Wilson Observatory obtained additional spectra of RW Tauri. He found that at the beginning of an eclipse the emission lines were Doppler shifted to the red; at the end of an eclipse the lines were shifted to the blue by an equal amount. Joy inferred from his observations that the hot star is surrounded by a rapidly rotating ring of gas. As an eclipse begins, the side of the ring rotating away from the observer remains visible beyond the limb of the cool star, producing redshifted emission lines. As an eclipse draws to a close, the side rotating toward the observer is visible, yielding blueshifted lines.

Over the following years astronomers came to suspect that Joy's gaseous ring was actually a disk whose inner edge lay at the surface of the central star. In the 1940s Otto Struve of McDonald Observatory and a few others suggested that matter flows from the companion star into the disk and then accretes onto the central star, a concept now accepted by most astronomers.

The flow of matter in binary systems is sometimes a direct result of stellar evolution. When a star reaches the end of its stable life, its core becomes depleted in hydrogen fuel and so begins to contract; as the core shrinks it grows hotter and releases more energy, causing the outer parts of the star to swell. If a star in a close binary system expands beyond a certain volume, called the Roche lobe, matter moving outward from the star falls under the gravitational influence of its companion. The size and shape of the Roche lobe are largely determined by three forces felt by a particle at rest with respect to the orbiting stars: the gravitational attraction of each star and the centrifugal force produced by the orbital motion of each particle. These competing forces give the Roche lobe's surface a teardrop shape.

If the star expands beyond this surface, the gas follows the least energetic path and falls toward the companion star. The gas emerges through the point of the teardrop, called the inner Lagrangian (L1) point, where the companion's gravitational influence is greatest. Gas leaving the L1 point forms a narrow stream that moves toward the companion star. Because the gas still carries the momentum of the orbital motion of the star it just left, it does not fall directly onto the companion. Instead the stream follows a curved path toward the trailing side of the mass-gaining star.

The events that follow depend on the size of the accreting star in comparison to the orbital separation between the stars. If the accreting star is small, the stream of gas curves around it and forms a ring. The ring quickly spreads into a flat disk as a result of viscosity, which causes some gas to lose angular momentum and spiral inward while a smaller volume of gas gains the lost momentum and spirals outward.

If the accreting star is comparatively large, the stream of gas collides with the body of the mass-gaining star. Oddly enough, something resembling an accretion disk still manages to form. But the resulting disk is turbulent and unstable; it disappears quickly if the mass transfer temporarily stops, much the way the spray from a garden hose falls to the ground and disappears when the faucet is turned off. As it turns out, RW Tauri is a transient-disk binary system.

Accretion disks can act in more complicated and violent ways, as seen in a class of binary stars called cataclysmic variables. These stellar systems contain a dense, hot companion that accretes matter from its cooler neighbor. Here the mass transfer is driven by a loss of orbital angular momentum, which slowly draws the stars closer and causes the Roche lobe to shrink around the red star. The disks in cataclysmic variables seem far from stable. Because the flow of matter through these disks may not be at all steady, understanding cataclysmic variables poses a difficult challenge. But it is a challenge worth accepting: the erratic behavior of these objects holds important clues to the general nature of accretion disks.

As their names imply, some cataclysmic variables undergo severe outbursts, during which their brightness jumps as much as 100-fold in a matter of days or even hours. One subset of cataclysmic variables, called dwarf novae, flare up in a quasi-periodic fashion. Dwarf nova eruptions recur on intervals of weeks to years; each episode typically lasts from a few days to a few weeks. Dwarf novae are distinct from ordinary novae, which are thought to be powered by the nuclear fusion of hydrogen accreted onto the surface of a white dwarf star. Dwarf nova outbursts seem to draw on gravitational energy only, which explains why they are 1,000 times less energetic than ordinary novae.

Groundbreaking work by Robert P. Kraft of Mount Wilson Observatory in the 1960s revealed that cataclysmic variables are binary stars whose components orbit very close to each other. A typical cataclysmic binary has an orbital period of four hours, and a few have periods of less than 90 minutes. Such rapid orbits mean that the separation between the two stars, and hence the stars themselves, must be extremely small. In fact the typical cataclysmic system could fit inside our sun.

Studies of light spectra and radiation fluxes at a broad range of wavelengths indicate that there are three major components to a cataclysmic binary: a red dwarf star, a white dwarf star, and an accretion disk surrounding the white dwarf. The red dwarf is a cool, faint, low-mass star that is losing matter through its L1 point. The white dwarf is much hotter, more luminous, and more massive. White dwarfs are the remnant cores of evolved stars that have depleted their hydrogen fuel. Lacking an internal energy source, the core grows fantastically dense—about 10 million times as dense as water. White dwarfs are roughly as massive as the sun but are only about the size of the earth.

Because a white dwarf is so small and massive, it has a very deep gravitational potential well. Matter falling onto it from the accretion disk releases a large amount of gravitational potential energy via viscous heating in the disk. So intense is the heating that the disk becomes brighter than the stars.

Observational astronomers have deduced that enhanced mass flow through the accretion disk powers the outbursts of dwarf novae. This was demonstrated by Brian Warner of the University of Cape Town, who conducted photometric observations of the dwarf nova Z Chamaeleontis. As seen from the earth, the red star in this binary eclipses the white dwarf and its surrounding accretion disk once every orbit. During an outburst, the light loss at mideclipse is much greater than during times of quiescence, indicating that the source of the outburst cannot be the red dwarf star. Furthermore, the duration of successive eclipses lengthens as the outburst progresses, implying that the eruption is spreading across the face of the disk.

The optical spectrum of a dwarf nova changes drastically during the course of an outburst. The quiescent spectrum shows a blue continuum on which are superimposed lines of radiation emitted by hydrogen and by singly ionized helium (helium that is missing one electron). The continuum originates in the dense inner regions of the disk, whereas the emission lines are formed in the more rarefied outer regions. In systems where the orbital plane is seen nearly edge-on, the emission-line profiles appear as double-peaked curves. The peaks appear because each profile consists of two Doppler-shifted components: one from the side of the disk rotating toward the observer and one from the side rotating away.

During an outburst, the continuum radiation brightens sharply; at the same time the emission lines can become difficult or impossible to see. Broad hydrogen absorption lines often appear, the result of radiation being absorbed by cooler gas surrounding the bright part of the disk. As the outburst declines in brightness, the continuum fades, and the emission lines once again become prominent.

Dwarf nova outbursts seem to result from a sudden increase in the flow of mass through the accretion disk onto the white dwarf star. The speedier inflow could result from a surge from the mass-losing red star or from a change in the accretion disk itself. The first possibility, proposed some 20 years ago by Geoffrey T. Bath, who was then at the University of Oxford, requires that the red star periodically overflow its Roche lobe, dumping excess material into the accretion disk. The extra matter causes the rate of accretion to increase, which in turn makes the disk shine more brilliantly.

A few years later Jozef Smak of the Copernicus Astronomical Center in Poland and Yoji Osaki of the University of Tokyo independently presented a competing idea. The Smak-Osaki hypothesis holds that mass flows from the secondary star at a constant rate. Some mechanism in the accretion disk itself stores up matter and then dumps it onto the white dwarf whenever the disk's mass exceeds a critical level.

Astrophysicists largely favor the disk instability model that has grown out of the work of Smak and Osaki. Most researchers now think the instability is triggered when a critical local surface density is attained somewhere in the disk. Once the density in a local region reaches the critical value, viscosity increases tremendously, and the stored-up matter accretes onto the white dwarf star, producing an outburst. When the surface density drops below a certain level, disk viscosity plummets, and the accumulation process starts anew.

For years theorists have labored to create mathematical models that could describe accretion disk instabilities. These models predict the local surface density and temperature when values are specified for the accretion rates and for alpha at a certain radius. An increase in the accretion rate leads to higher surface density, which in turn causes increased viscous heating. Disk temperature is therefore expected to vary in proportion to surface density; a graph of local temperature versus surface density at a particular radius in an accretion disk should produce an upward-sloping curve.

The outburst models of Osaki and Smak required that at a given location in the disk there exist two possible stable values for temperature for the same value of surface density. The outburst corresponds to the high-temperature solution the quiescent state to the low-temperature solution. In this case the curve in the temperature-surface density plot must be S shaped. Osaki and Smak implicitly assumed that such an S-curve relation was physically plausible, but they did not propose a specific mechanism that would produce it.

In 1979 Reiun Hoshi of Rikkyo University in Tokyo demonstrated the existence of such a mechanism. Hoshi worked with a model that considered the vertical structure of the accretion disk (the physical conditions that prevail in a thin perpendicular slice through the disk). Although his model for the vertical structure was crude, he found that an S-curve relation naturally emerged when his model accounted for the temperatures at which hydrogen becomes partially ionized, that is, when the disk becomes so hot that some of the hydrogen atoms lose their surrounding electrons.

To see how the S-curve relation works, one must examine the physical conditions in a stable disk (said to be in a state of thermal equilibrium). In a system whose temperature remains constant, the energy escaping into space must be balanced by that released from the interior. Otherwise the system would either cool off or heat up. Furthermore, to be truly stable, the temperature of the system must return to its initial value despite small perturbations.

The sun, for example, is in a state of thermal equilibrium. If the central temperature of the sun were lowered slightly, the gas pressure would decrease, and the sun would contract. The contraction would in turn raise the central temperature and increase the nuclear-reaction rates, causing the sun to expand to its initial size. In the end, the sun would return to its original temperature and energy output.

The situation for an accretion disk is broadly similar. If the midplane temperature declines slightly for a short interval of time, the local viscosity would decrease because of reduced frictional interaction. The surface density would then increase: reduced viscosity would slow the inward flow of mass while matter from outlying radii continues to fall into the region at a constant rate. Increased surface density would result in greater viscous heating, which would return the temperature and the surface density to their initial values. Under these conditions a graph of temperature versus surface density has the expected upward slope, because regions with high temperatures require high surface densities in order to supply sufficient viscous heating.

The above mechanism can maintain thermal equilibrium in the disk provided that the opacity does not change substantially as the temperature changes. Opacity of the gas determines how freely the energy in the disk can escape into space; a substantial change in opacity can largely control the disk temperature. At a temperature of about 10,000 kelvins, hydrogen atoms begin to lose their electrons (become partially ionized), and the opacity of the hydrogen changes abruptly, increasing roughly as the tenth power of the temperature.

In regions of the disk close to the critical temperature, a light cooling leads to an enormous drop in opacity, allowing more energy to escape into space and causing the temperature to drop even farther. By the same argument, a small increase in temperature enhances opacity, impeding the flow of radiation and causing further heating. Critical regions in the disk will either heat or cool until they reach the stable branches at the top or bottom of the S curve. This thermally unstable behavior corresponds to the middle, negatively sloping portion of the S curve.

The flow of energy vertically out of the disk depends on local opacity and on the local surface density (which is proportional to the amount of gas along the line of sight). The lower branch of the S curve is thermally stable because at these low temperatures very little hydrogen is ionized, and so the opacity varies only weakly with the temperature. Likewise, the upper branch is stable because the disk is so hot that hydrogen is completely ionized; in this state opacity again changes little as a consequence of changes in temperature. On the connecting middle branch, however, the extreme sensitivity of opacity to temperature variations leads to thermally unstable conditions. Using these physical arguments, Hoshi showed that a zone of partially ionized hydrogen in the accretion disk could produce the double-valued temperature solutions that would provide a physical mechanism to account for the disk instability idea of Osaki and Smak.

Surprisingly, Hoshi's work attracted little attention. In July 1981 James E. Pringle of the University of Cambridge gave a talk at a conference at the University of California at Santa Cruz in which he discussed in general terms how an S-shaped temperature-density relation could yield the conditions that would lead to dwarf nova eruptions. In the audience were several accretion disk modelers who immediately realized the importance of calculating in detail the disk's vertical structure.

Five months later Friedrich Meyer and his wife, Emmi Meyer-Hofmeister, both of the Max Planck Institute for Astrophysics in Garching, published a short article that contained a physical basis for Pringle's suggestion. Shortly thereafter, articles on the disk instability mechanisms appeared by Smak in Poland; Pranab Ghosh, J. Craig Wheeler, and one of us (Cannizzo), then all at the University of Texas at Austin; Douglas N. C. Lin and John Faulkner of the University of California at Santa Cruz and John C. B. Papaloizou of Queen Mary College; and Shin Mineshige (now at Ibaraki University) and Osaki.

These workers all recognized that dwarf nova eruptions are a natural consequence of accretion disk instability. The conditions at each radius in the disk correspond to some point on the temperature-surface density S curve. The temperature or, equivalently, the accretion rate at a given radius is determined by the surface density and by the evolutionary history of the matter there. When local conditions move into the unstable region of the S curve, the disk switches from a quiescent state to an outburst, or vice versa (see box on next page).

The disk instability model makes a number of predictions that can be tested by observations of dwarf novae. Smak delineated many of these predictions in a landmark article that appeared in 1984. Similar studies appeared by the other researchers mentioned above at about the same time. Using computer models to simulate the time-dependent behavior of an accretion disk, Smak was able to produce artificial light cures that closely resemble the observed ones.

Several aspects of dwarf nova eruptions seem quite consistent with the disk instability model. Light curves produced by disk instabilities can show either a slow or a rapid increase in brightness. Computer models show that instabilities that begin near the outer edge of the disk produce the fast increases. In this case the radius of the disk should increase during an outburst, a phenomenon that has been observed in several dwarf novae.

In such "outside-in" eruptions, the ultraviolet emission should increase after the visible light does because the energetic ultraviolet rays come from the hot, inner regions of the disk, which should be the last ones affected by the outburst. Observations made with the International Ultraviolet Explorer (IUE) satellite have shown

HOW THE DWARF NOVA CYCLE WORKS

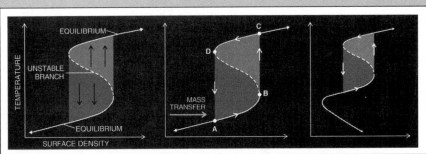

One likely model of dwarf nova eruptions holds that the accretion disk around the white dwarf experiences quasiperiodic episodes of instability. The above plots of temperature versus surface density in the disk illustrate how this model works. At any location in the disk there are two stable branches, or equilibrium states, one at high temperatures and one at low temperatures *(left)*. The midddle branch is thermally unstable.

A region of the disk where conditions correspond to the red area grows hotter until it settles at the high-temperature state. Regions corresponding to the blue area cool down until they reach the low-temperature state. The upper branch corresponds to an outburst; the lower branch, to the quiescent state.

The accretion disk surrounding the white dwarf in a dwarf nova system alternates between the two branches in the following manner. Consider a circular region, or annulus, at some fixed radius from the white dwarf star where the temperature and surface density lie at point A *(center)*. The horizontal arrow shows the rate at which matter is added from the secondary, red dwarf star. Matter is flowing into the annulus faster than it is accreting onto the white dwarf, so both surface density and temperature increase.

When the annulus reaches a state corresponding to point B, it falls out of thermal equilibrium. Viscous heating now exceeds energy losses by radiation, so the region swiftly grows hotter until it reaches the upper, thermally stable branch at point C. At these high temperatures the viscosity in the disk is quite strong. As a result the accumulated matter rapidly flows inward and accretes onto the white dwarf.

Because of the sudden draining of the disk, the local surface density and temperature decrease, so that the region moves from point C to point D. At point D thermal instability again sets in. This time the radiation losses overwhelm viscous heating, and the temperature plummets. After a period of cooling, the annulus ends up back at point A, and another cycle commences.

The above model assumes an arbitrary value for the intensity of viscosity in the disk. Cannizzo and A. G. W. Cameron of Harvard University have developed a more physically complete model in which viscous heating is assumed to arise from turbulence associated with convective motions in the accretion disk. In this case the S-shaped temperature-density relation discussed above appears to be only part of a more general, W-shaped curve *(right)*. Further work should lead to increasingly realistic accretion disk models.

Johnny Johnson

that in several dwarf novae the outburst appears at ultraviolet wavelengths about a day after it shows up in the optical. But Pringle, Frank Verbunt, and Richard A. Wade, then working at Cambridge, showed in 1986 that this behavior can be explained at least as well by Bath's secondary-star instability model.

The disk instability model makes another important prediction: when the rate of mass transfer from the red dwarf rises above a certain critical level, the flow of matter instead should occur at a steady but very high rate, corresponding to the upper stable branch of the S curve. Warner and Smak independently examined this question by inferring the rates of mass transfer for a large number of interacting binaries. They found that those systems experiencing high values of mass transfer do not exhibit eruptions, whereas those below a certain value do. Their observed dividing line fits well with the theoretically expected value, a quantity that is independent of the unknown parameter alpha. There is no reason to expect such an upper limit from Bath's model.

Observing the spectral evolution of a dwarf nova flare-up might provide important observational constraints for disk instability models. The intensity of the radiation from the regions that produce emission lines can be sensitive to small changes in the local physical conditions. Changes in the strength of the emission lines therefore might herald the coming of an eruption.

Such an effect was observed by Cathy S. Mansperger and Panayiotis Hantzios, who were then working at Ohio State University, and one of us (Kaitchuck) in the emission lines of the star RX Andromedae 18 hours before the start of an outburst. Also, the system seems to brighten a few percent during the 24 hours before the main outburst. These observations and others indicate that there is sometimes a pre-outburst stage in the dwarf nova cycle.

Mansperger and Kaitchuck have managed to observe the transition from pre-outburst to outburst in the dwarf nova TW Virginis. On one evening the star system appeared somewhat brighter than on previous nights. For the next two and a half hours the system brightened only slightly. Then, within 15 minutes, the slope of the spectrum of the continuum radiation abruptly steepened, indicating that the temperature of the disk had suddenly begun to increase. At the same time the brightening of the system accelerated. Over the remainder of the night the continuum slope steepened further, and absorption lines replaced the emission lines in the spectrum, revealing that most of the disk had become hotter than the surface of the sun. When dawn called a halt to the observations, the system had brightened by a factor of 12. By the next day it had flared to 100 times its pre-outburst brightness.

Much of the observed pre-outburst behavior was predicted in a disk model developed by Mineshige when he was at the Institute for Astronomy at Cambridge. In his model the main outburst is preceded by a warm phase that lasts for about one

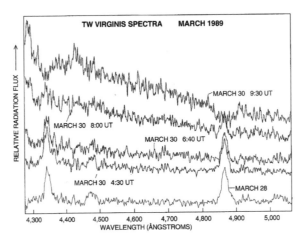

Dwarf nova spectrum changes dramatically at the start of an outburst, as seen above for the double star TW Virginis. The bottom spectrum shows the star in a quiescent state. The other spectra document the changes that occurred a few nights later. For about two hours the system was in a pre-outburst state and brightened only slightly. The star then grew rapidly brighter, especially at bluer wavelengths. By the following day TW Virginis had brightened by a factor of 100. (Johnny Johnson)

day. Then a hot front moves through the disk from the site of the thermal instability, producing the main outburst. The observed pre-outburst brightening of TW Virginis closely matches the predicted warm phase, as do the observed rates of change of the disk temperature and brightness.

If the disk instability model for dwarf nova eruptions proves correct, it should help answer many of the questions about the physics of accretion disks in general. The greatest mystery in accretion disk research is the nature of the viscosity. Clearly, a preliminary step in understanding the physical mechanism that produces viscosity involves determining empirically the magnitude of the effect and how it depends on surface density and on radius of the disk. Researchers using time-dependent computer models have tried adjusting the relation between temperature and surface density in order to make the model outbursts resemble those observed. In this way it is possible to check theories of viscous heating to make sure they are consistent with the behavior of dwarf novae.

Two main mechanisms have been proposed to account for viscous heating of the disk. One focuses on turbulence associated with convective motions that accompany the vertical transport of energy out of the disk; the other invokes energy released by the breaking and reconnecting of magnetic field lines in the disk. Cannizzo, working with A.G.W. Cameron of Harvard University, showed that convection-induced heating can reproduce some aspects of the relation between temperature and surface density needed to produce dwarf nova eruptions.

Steven Balbus and John Hawley of the Virginia Institute for Theoretical Astronomy have recently uncovered a mechanism by which motions in the ionized (and hence magnetic) gas in the disk would amplify the weak magnetic fields that should always be present. The next task for theorists is to incorporate the results from such studies into a global theory that follows the temporal evolution of the entire disk to see if the models can reproduce in detail the recorded behavior of dwarf nova eruptions.

The disk instability model may explain not only cataclysmic variables but also distant, energetic quasars. Astrophysicists have long speculated that these objects harbor accretion disks that channel matter onto central black holes having masses millions of times that of the sun. Lin and Gregory A. Shields of the University of Texas at Austin found in 1986 that the S-curve relation should operate in these giant accretion disks as well. The two researchers suggested that disk instabilities may collimate oppositely directed jets of matter when the disk is in its high state, thereby producing the sequence of knots seen in the radio-emitting jets that extend from some active galaxies.

A better understanding of accretion disks will be essential to determining the true nature of quasars and other violent phenomena that occur in the nuclei of galaxies, including the mysterious object at the center of the Milky Way. The information should also lead to new insights into the disks that seem to surround infant stars, the likely birthplace of planets such as the earth.

—January 1992

V

Galaxies

HOW THE MILKY WAY FORMED

*Its halo and disk suggest that the collapse of a gas cloud,
stellar explosions, and the capture of galactic fragments
may have all played a role*

Sidney van den Bergh and James E. Hesser

Attempts to reconstruct how the Milky Way formed and began to evolve resemble
an archaeological investigation of an ancient civilization buried below the bustling
center of an ever-changing modern city. From excavations of foundations, some
pottery shards, and a few bones, we must infer how our ancestors were born, how
they grew old and died, and how they may have helped create the living culture
above. Like archaeologists, astronomers, too, look at small, disparate clues to determine how our galaxy and others like it were born about a billion years after the
big bang and took on their current shapes. The clues consist of the ages of stars and
stellar clusters, their distribution, and their chemistry—all deduced by looking at
such features as color and luminosity. The shapes and physical properties of other
galaxies can also provide insight concerning the formation of our own.

The evidence suggests that our galaxy, the Milky Way, came into being as a consequence of the collapse of a vast gas cloud. Yet that cannot be the whole story. Recent observations have forced workers who support the hypothesis of a simple, rapid collapse to modify their idea in important ways. This new information has led other researchers to postulate that several gas cloud fragments merged to create the protogalactic Milky Way, which then collapsed. Other variations on these themes are vigorously maintained. Investigators of virtually all persuasions recognize that the births of stars and supernovae have helped shape the Milky Way. Indeed, the formation and explosion of stars are at this moment further altering the galaxy's structure and influencing its ultimate fate.

Much of the stellar archaeological information that astronomers rely on to decipher the evolution of our galaxy resides in two regions of the Milky Way: the halo and the disk. The halo is a slowly rotating, spherical region that surrounds all the other parts of the galaxy. The stars and star clusters in it are old. The rapidly rotating, equatorial region constitutes the disk, which consists of young stars and stars of intermediate age, as well as interstellar gas and dust. Embedded in the disk are the sweepingly curved arms that are characteristic of spiral galaxies such as the Milky Way. Among the middle-aged stars is our sun, which is located about 25,000 light-years from the galactic center. (When you view the night sky, the galactic center lies in the direction of Sagittarius.) The sun completes an orbit around the center in approximately 200 million years.

That the sun is part of the Milky Way was discovered less than 70 years ago. At the time, Bertil Lindblad of Sweden and the late Jan H. Oort of the Netherlands hypothesized that the Milky Way system is a flattened, differentially rotating galaxy. A few years later John S. Plaskett and Joseph A. Pearce of Dominion Astrophysical Observatory accumulated three decades' worth of data on stellar motions that confirmed the Lindblad-Oort picture.

In addition to a disk and a halo, the Milky Way contains two other subsystems: a central bulge, which consists primarily of old stars, and, within the bulge, a nucleus. Little is known about the nucleus because the dense gas clouds in the central bulge obscure it. The nuclei of some spiral galaxies, including the Milky Way, may contain a large black hole. A black hole in the nucleus of our galaxy, however, would not be as massive as those that seem to act as the powerful cores of quasars.

All four components of the Milky Way appear to be embedded in a large, dark corona of invisible material. In most spiral galaxies the mass of this invisible corona exceeds by an order of magnitude that of all the galaxy's visible gas and stars. Investigators are intensely debating what the constituents of this dark matter might be.

The clues to how the Milky Way developed lie in its components. Perhaps the only widely accepted idea is that the central bulge formed first, through the col-

| 600,000 MILES |
Ground Based
Wide Angle View

| 100,000 MILES |
HST View
Region Containing the Nuclei

| 40,000 MILES |
HST View
Closeup Near Brightest Nucleus

(a) An image of comet Shoe-maker-Levy 9 taken on March 30, 1993, with the Spacewatch Camera of the University of Arizona. Although not clearly visible in this image, the bright "streak" near the center contains many individual nuclei and their associated comae and tails. In addition, light scattered from fine dust particles can be seen extending for large distances beyond the region of the streak. (Dr. J. V. Scotti, University of Arizona. Courtesy NASA)

(b) An image taken by the Planetary Camera of the Hubble Space Telescope (HST) on July 1, 1993 shows the streak region in much greater detail. The appearance of this broken-up comet is reminiscent of a "string of pearls." (Courtesy NASA)

(c) An enlargement of the HST image in the region of the brightest nucleus. Looking carefully, one can tell that the "brightest nucleus" near the center of the frame is actually a group of at least four separate pieces that are blurred together when observed at poorer resolution. (Courtesy NASA)

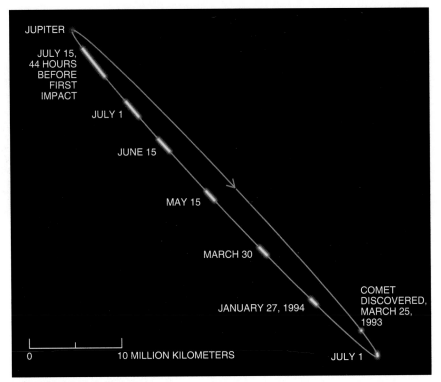

JUPITER

JULY 15,
44 HOURS
BEFORE
FIRST
IMPACT

JULY 1

JUNE 15

MAY 15

MARCH 30

JANUARY 27, 1994

COMET
DISCOVERED,
MARCH 25,
1993

0   10 MILLION KILOMETERS   JULY 1

The train of fragments of comet Shoemaker-Levy 9 grew continuously in length from the time of discovery in 1993 to the final collisions in 1994. (Jared Schneidman/JSD; Source: Paul W. Chodas)

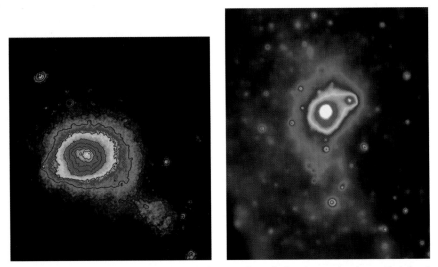

X-ray images of Coma (left) and Virgo (right) clusters show the hot intergalactic gas that dominates the luminous part of these structures. The gas in Coma has a more regular shape than that in Virgo, suggesting that the cluster has reached a more advanced stage of formation. Both clusters are surrounded by infalling material. (J. Patrick Henry, Ulrich G. Briel, and Hans Bohringer)

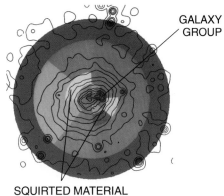

GALAXY
GROUP

SQUIRTED MATERIAL

(a) *Three galaxy clusters are at different stages in their evolution, as shown in these x-ray images (left column) and temperature maps (right column). The first cluster, Abell 2256, is busily swallowing a small group of galaxies, which is identified by its relatively low temperature. On the map red is comparatively cool, orange intermediate, and yellow hot.*

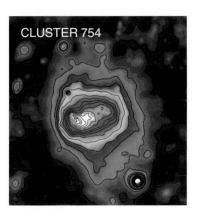
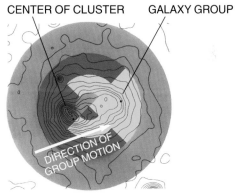

CENTER OF CLUSTER   GALAXY GROUP

DIRECTION OF
GROUP MOTION

(b) *The second cluster, Abell 754, is several hundred million years further along in its digestion of a galaxy group. The hapless group probably entered from the southeast, because the cluster is elongated in that direction. The galaxies of the group have separated from the gas and passed through the cluster.*

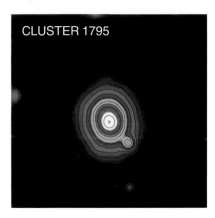
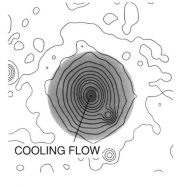

COOLING FLOW

(c) *The third cluster, Abell 1795, has gone several billion years since its last meal. Both its x-ray brightness and gas temperature are symmetrical. At the core of the cluster is a cool spot, a region of dense gas that has radiated away much of its heat. (J. Patrick Henry, Ulrich G. Briel, and Hans Bohringer)*

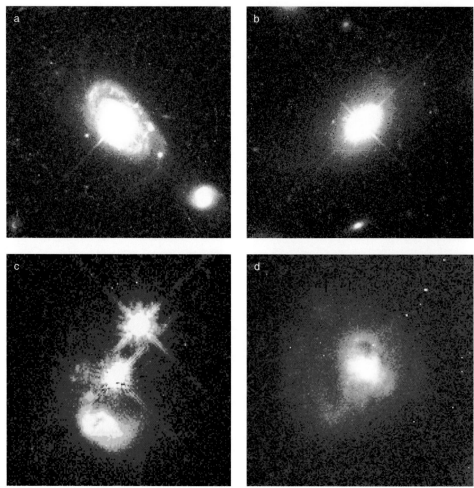

Host galaxies surround most of the quasars observed by the Hubble Space Telescope. The spiral galaxy around PG 0052+251 (a) and the elliptical galaxy around PHL 909 (b) appear to be undisturbed by collisions. But a galactic crash seems to be fueling IRAS 04505–2958 (c). A spiral ring torn from one of the galaxies is below the quasar; the object above it is a foreground star. HST has also captured evidence of a dance between two merging galaxies. The galaxies may have orbited each other several times before merging, leaving distinct loops of glowing gas around quasar IRAS13218+00552 (d). (Images a and b courtesy John Bahcall, Institute for Advanced Study. Images c and d courtesy NASA).

Gravitational lensing has enabled cosmologists to see greater detail within faint galaxies. The many blue, looped objects in the picture above are multiple images of the same distant galaxy. The yellow elliptical and spiral galaxies clustered at the center of the photograph, named 0024 + 1654 for their position in the sky, possess a tremendous gravitational field. This field bends light and so distorts, magnifies, or duplicates the images of objects behind it. (Courtesy NASA)

Spiral galaxy NGC2997, located in the southern Antlia cloud, probably resembles our own galaxy. Like all spiral galaxies, NGC2997 is embedded in an extended dark diffuse halo, whose composition remains unknown. (Photograph by David Malin; copyright Anglo-Australian Observatory)

Andromeda galaxy is a prime example of why calculating the expansion rate of the universe is difficult. Andromeda is 2.5 million light-years away from earth, but it still feels the gravitational pull of our own galaxy. Consequently, its relative motion has little to do with the expansion of the universe. By observing more distant galaxies, astronomers can detect the expansion, but they do not know its precise rate because it is difficult to measure distances to remote galaxies. (Courtesy California Institute of Technology)

Orion Nebula, a 10-light-year-wide cloud of hydrogen gas, illustrates the capabilities of the Hubble Space Telescope. The nebula has been photographed extensively using terrestrial telescopes. Hubble's view shows a snorkel-shaped feature (at bottom) and many previously unseen wisps and filaments. The jetlike formation (at left) appears to be matter flowing from a newborn star. The image is a composite, consisting of blue light emitted by ionized oxygen, red light from ionized sulfur, and green light from neutral hydrogen. (Copyright Anglo-Australian Observatory)

lapse of a gas cloud. The central bulge, after all, contains mostly massive, old stars. But determining when and how the disk and halo formed is more problematic.

In 1958 Oort proposed a model according to which the population of stars forming in the halo flattened into a thick disk, which then evolved into a thin one. Meanwhile further condensation of stars from the hydrogen left over in the halo replenished that structure. Other astronomers prefer a picture in which these populations are discrete and do not fade into one another. In particular, V. G. Berman and A. A. Suchkov of the Rostov State University in Russia have indicated how the disk and halo could have developed as separate entities.

These workers suggest a hiatus between star formation in the halo and that in the disk. According to their model, a strong wind propelled by supernova explosions interrupted star formation in the disk for a few billion years. In doing so, the wind would have ejected a significant fraction of the mass of the protogalaxy into intergalactic space. Such a process seems to have prevailed in the Large Magellanic Cloud, one of the Milky Way's small satellite galaxies. There an almost 10-billion-year interlude appears to separate the initial burst of creation of conglomerations of old stars called globular clusters and the more recent epoch of star formation in the disk. Other findings lend additional weight to the notion of distinct galactic components. The nearby spiral M33 contains a halo but no nuclear bulge. This characteristic indicates that a halo is not just an extension of the interior feature, as many thought until recently.

In 1962 a model emerged that served as a paradigm for most instigators. According to its developers—Olin J. Eggen, now at the National Optical Astronomical Observatories, Donald Lynden-Bell of the University of Cambridge, and Allan R. Sandage of the Carnegie Institution—the Milky Way formed when a large, rotating gas cloud collapsed rapidly, in about a few hundred million years. As the cloud fell inward on itself, the protogalaxy began to rotate more quickly; the rotation created the spiral arms we see today. At first the cloud consisted entirely of hydrogen and helium atoms, which were forged during the hot, dense initial stages of the big bang. Over time the protogalaxy started to form massive, short-lived stars. These stars modified the composition of galactic matter, so that the subsequent generations of stars, including our sun, contain significant amounts of elements heavier than helium.

Although the model gained wide acceptance, observations made during the past three decades have uncovered a number of problems with it. In the first place, investigators found that many of the oldest stars and star clusters in the galactic halo move in retrograde orbits—that is, they revolve around the galactic center in a direction opposite to that of most other stars. Such orbits suggest that the protogalaxy was quite clumpy and turbulent or that it captured sizable gaseous fragments whose matter was moving in different directions. Second, more refined dynamic models show that the protogalaxy would not have collapsed as smoothly as predicted by the simple model; instead the densest parts would have fallen inward much faster than more rarefied regions.

Third, the time scale of galaxy formation may have been longer than that deduced by Eggen and his colleagues. Exploding supernovae, plasma winds pouring from massive, short-lived stars, and energy from an active galactic nucleus are all possible factors. The galaxy may also have subsequently rejuvenated itself by absorbing large inflows of pristine intergalactic gas and by capturing small, gas-rich satellite galaxies.

Several investigators have attempted to develop scenarios consistent with the findings. In 1977 Alar Toomre of the Massachusetts Institute of Technology postulated that most galaxies form from the merger of several large pieces rather than from the collapse of a single gas cloud. Once merged in this way, according to Toomre, the gas cloud collapsed and evolved into the Milky Way now seen. Leonard Searle of the Carnegie Institution and Robert J. Zinn of Yale University have suggested a somewhat different picture, in which many small bits and pieces coalesced. In the scenarios proposed by Toomre and by Searle and Zinn, the ancestral fragments may have evolved in chemically unique ways. If stars began to shine and supernovae started to explode in different fragments at different times, then each ancestral fragment would have its own chemical signature. Recent work by one of us (van den Bergh) indicates that such differences do indeed appear among the halo populations.

Discussion of the history of galactic evolution did not advance significantly beyond this point until the 1980s. At that time workers became able to record more precisely than ever before extremely faint images. This ability is critically important because the physical theories of stellar energy production—and hence the lifetimes and ages of stars—are most secure for so-called main-sequence stars. Such stars burn hydrogen in their cores; in general, the more massive the star, the more quickly it completes its main-sequence life. Unfortunately, this fact means that within the halo the only remaining main-sequence stars are the extremely faint ones. The largest, most luminous ones, which have burned past their main-sequence stage, became invisible long ago. Clusters are generally used to determine age. They are crucial because their distances from the earth can be determined much more accurately than can those of individual stars.

The technology responsible for opening the study of extremely faint halo stars is the charge-coupled device (CCD). This highly sensitive detector produces images electronically by converting light intensity into current. CCDs are far superior in most respects to photographic emulsions, although extremely sophisticated software, such as that developed by Peter B. Stetson of Dominion Astrophysical Observatory, is required to take full advantage of them. So used, the charge-coupled device has yielded a tenfold increase in the precision of measurement of color and luminosity of the faint stars in globular clusters.

Among the most important results of the CCD work done so far are more precise age estimates. Relative age data based on these new techniques have revealed that clusters whose chemistries suggest they were the first to be created after the big bang have the same age to within 500 million years of one another. The ages of other clusters, however, exhibit a greater spread.

The ages measured have helped researchers determine how long it took for the galactic halo to form. For instance, Michael J. Bolte, now at Lick Observatory, carefully measured the colors and luminosities of individual stars in the globular clusters NGC 288 and NGC 362 (see illustration below). Comparison between these data and stellar evolutionary calculations shows that NGC 288 is approximately 15 billion years old and that NGC 362 is only about 12 billion years in age. This difference is greater than the uncertainties in the measurements. The observed age range indicates that the collapse of the outer halo is likely to have taken an order of magnitude longer than the amount of time first envisaged in the simple, rapid collapse model of Eggen, Lynden-Bell, and Sandage.

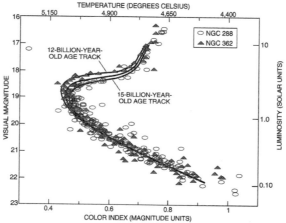

Color-luminosity diagrams can be used to determine stellar ages. The one above compares the plots of stars in globular clusters NGC 288 and NGC 362 with age tracks (black lines) generated by stellar evolution models. The color index, expressed in magnitude units, is a measure of the intensity of blue wavelengths minus visual ones. In general, the brighter the star, the lower the color index; the trend reverses for stars brighter than about visual magnitude 19. The plots suggest the clusters differ in age by about three billion years. The temperature (inversely related to the color index) and luminosity have been set to equal those of NGC 288. (Tomo Narashima)

Of course it is possible that more than one model for the formation of the galaxy is correct. The Eggen-Lynden-Bell-Sandage scenario may apply to the dense bulge and inner halo. The more rarefied outer parts of the galaxy may have developed by the merger of fragments, along the lines theorized by Toomre or by Searle and Zinn. If so, then the clusters in the inner halo would have formed before those in the more tenuous outer regions. The process would account for some of the age differences found for the globular clusters. More precise modeling may have to await the improved image quality that modifications to the Hubble Space Telescope cameras will afford.

Knowing the age of the halo is, however, insufficient to ascertain a detailed formation scenario. Investigators need to know the age of the disk as well and then to compare that age with the halo's age. Whereas globular clusters are useful in determining the age of the halo, another type of celestial body—very faint white dwarf stars—can be used to determine the age of the disk. The absence of white dwarfs in the galactic disk near the sun sets a lower limit on the disk's age. White dwarfs, which are no longer producing radiant energy, take a long time to cool, so their absence means that the population in the disk is fairly young—less than about 10 billion years. This value is significantly less than the ages of clusters in the halo and is thus consistent with the notion that the bulk of the galactic disk developed after the halo.

It is, however, not yet clear if there is a real gap between the time when formation of the galactic halo ended and when creation of the old thick disk began. To estimate the duration of such a transitional period between halo and disk, investigators have compared the ages of the oldest stars in the disk with those of the youngest ones in the halo. The oldest known star clusters in the galactic disk, NGC 188 and NGC 6791, have ages of nearly eight billion years, according to Pierre Demarque and David B. Guenther of Yale and Elizabeth M. Green of the University of Arizona. Stetson and his colleagues and Roberto Buonanno of the Astronomical Observatory in Rome and his coworkers examined globular clusters in the halo population. They found the youngest globulars—Palomar 12 and Ruprecht 106—to be about 11 billion years old. If the few billion years' difference between the disk objects and the young globular is real, then young globulars may be the missing links between the disk and halo populations of the galaxy.

At present, unfortunately, the relative ages of only a few globular clusters have been precisely estimated. As long as this is the case, one can argue that the Milky Way could have tidally captured Palomar 12 and Ruprecht 106 from the Magellanic Clouds. This scenario, proposed by Douglas N. C. Lin of the University of California at Santa Cruz and Harvey B. Richer of the University of British Columbia, would obviate the need for a long collapse time. Furthermore, the apparent age gap between disk and halo might be illusory. Undetected systematic errors may lurk in the age-dating processes. Moreover, gravitational interactions with massive interstellar clouds may have disrupted the oldest disk clusters, leaving behind only younger ones. Determining the relative ages of the halo and disk reveals much about the sequence of the formation of the galaxy. On the other hand, it leaves open the question of how old the entire galaxy actually is. The answer would provide some absolute framework by which the sequence of formation events can be discerned. Most astronomers who study star clusters favor an age of some 15 to 17 billion years for the oldest clusters (and hence the galaxy).

Confidence that those absolute age values are realistic comes from the measured abundance of radioactive isotopes in meteorites. The ratios of thorium 232 to uranium 235, of uranium 235 to uranium 238, or of uranium 238 to plutonium 244 act as chronometers. According to these isotopes, the galaxy is between 10 and 20 billion years old. Although ages determined by such isotope ratios are believed to be less accurate than those achieved by comparing stellar observations and models, the consistency of the numbers is encouraging.

Looking at the shapes of other galaxies alleviates to some extent the uncertainty of interpreting the galaxy's evolution. Specifically the study of other galaxies presents a perspective that is unavailable to us as residents of the Milky Way—an external view. We can also compare information from other galaxies to see if the processes that created the Milky Way are unique. The most immediate observation one can make is that galaxies come in several shapes. In 1925 Edwin P. Hubble found that luminous galaxies could be arranged in a linear sequence according to whether they are elliptical, spiral, or irregular *(see illustration on next page)*. From an evolutionary point of view, elliptical galaxies are the most advanced. They have used up all (or almost all) of their gas to generate stars, which probably range in age from 10 to 15 billion years. Unlike spiral galaxies, ellipticals lack disk structures. The main differences between spiral and irregular galaxies is that irregulars have neither spiral arms nor compact nuclei.

The morphological types of galaxies can be understood in terms of the speed with which gas was used to create stars. Determining the rate of gas depletion would corroborate estimates of the Milky Way's age and history. Star formation in elliptical galaxies appears to have started off rapidly and efficiently some 15 billion years ago and then declined sharply. In most irregular galaxies the birth of stars has taken place much more slowly and at a more nearly constant rate. Thus, a significant fraction of their primordial gas still remains.

The rate of star formation in spirals seems to represent a compromise between that in ellipticals and that in irregulars. Star formation in spirals began less rapidly than it did in ellipticals but continues to the present day.

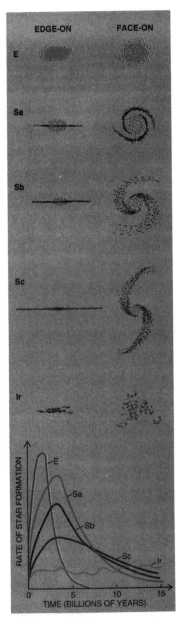

Morphological classification of galaxies (top) ranges from ellipticals (E) to spirals (subdivided into categories Sa, Sb, and Sc) and irregulars (Ir)). The history of star formation varies according to morphology (bottom). In elliptical galaxies stars developed in an initial burst. Star formation in spirals was less vigorous but continues today. In most irregular galaxies the birthrate of stars has probably remained constant. (Jared Schneidman)

Spirals are further subdivided into categories Sa, Sb, and Sc. The subdivisions refer to the relative size of the nuclear bulges and the degree to which the spiral arms coil. Objects of type Sa have the largest nuclear bulges and the most tightly coiled arms. Such spirals also contain some neutral hydrogen gas and a sprinkling of young blue stars. Sb spirals have relatively large populations of young blue stars in their spiral arms. The central bulge, containing old red stars, is less prominent than is the central bulge in spirals of type Sa. Finally, in Sc spirals the light comes mainly from the young blue stars in the spiral arms; the bulge population is inconspicuous or absent. The Milky Way is probably intermediate between types Sb and Sc.

Information from other spirals seems consistent with the data obtained for the Milky Way. Like those in our galaxy, the stars in the central bulges of other spirals arose early. The dense inner regions of gas must have collapsed first. As a result most of the primordial gas initially present near the centers has turned into stars or has been ejected by supernova-driven winds.

There is an additional kind of evidence on which to build our understanding of how the Milky Way came into existence: the chemical composition of stars. This information helps to pinpoint the relative ages of stellar populations. According to stellar models, the chemistry of a star depends on when it formed. The chemical differences exist because first-generation stars began to "pollute" the protogalaxy with elements heavier than helium. Such so-called heavy elements, or "metals," as astronomers refer to them, were created in the interiors of stars or during supernova explosions. Examining the makeup of stars can provide stellar evolutionary histories that corroborate or challenge age estimates.

Different types of stars and supernovae produce different relative abundances of these metals. Researchers believe that most "iron-peak" elements (those closest to iron in the periodic table) in the galaxy were made in supernovae of type Ia. The progenitors of such supernovae are thought to be pairs of stars, each of which has a mass a few times that of the sun. Other heavy elements—the bulk of oxygen, neon, magnesium, silicon, and calcium, among others—originated in supernovae that evolved from single or binaries of massive, short-lived stars. Such stars have initial masses of 10 to 100 solar masses and violently end their lives as supernovae of type Ib, Ic, or II.

Stars that subsequently formed incorporated some of these heavy elements. For instance, approximately 1 to 2 percent of the mass of the sun consists of elements other than hydrogen or helium. Stars in nuclear bulges generally harbor proportionally more heavy elements than do stars in the outer disks and halos. The abundance of heavy elements decreases gradually by a factor of .8 for every kiloparsec (3,300 light-years) from the center to the edge of the Milky Way disk. Some 70 percent of the 150 or so known globular clusters in the Milky Way exhibit an average metal content of about one twentieth that of the sun. The remainder shows a mean of about one third that of the sun.

Detailed studies of stellar abundances reveal that the ratio of oxygen to iron-peak elements is larger in halo stars than it is in metal-rich disk stars. This difference suggests the production of heavy elements during the halo phase of galactic evolution was dominated by supernovae of types Ib, Ic, and II. It is puzzling that iron-producing type Ia supernovae, some of which are believed to have resulted from progenitor stars with lifetimes as short as a few hundred million years, did not contribute more to the chemical mixture from which halo stars and some globular clusters formed. This failure would seem to imply that the halo collapsed very rapidly—before supernovae of type Ia could contribute their iron to the halo gas.

That idea, however, conflicts with the four-billion-year age spread observed among galactic globular clusters, which implies that the halo collapsed slowly. Perhaps supernova-driven galactic winds swept the iron-rich ejecta from type Ia supernovae into intergalactic space. Such preferential removal of the ejecta of type Ia supernovae might have occurred if supernovae of types Ib, Ic, and II exploded primarily in dense gas clouds. Most of type Ia supernovae then must have detonated in less dense regions, which are more easily swept out by the galactic wind.

Despite the quantity of data, information about metal content has proved insufficient to settle the controversy concerning the time scale of disk and halo formation. Sandage and his colleague Gary A. Fouts of Santa Monica College find evidence for a rather monolithic collapse. On the other hand, John E. Norris and his collaborators at the Australian National Observatory, among others, argue for a significant decoupling between the formation of halo and disk. They also posit a more chaotic creation of the galaxy, similar to that envisaged by Searle and Zinn.

Such differences in interpretation often reflect nearly unavoidable effects arising from the way in which particular samples of stars are selected for study. For example, some stars exhibit chemical compositions similar to those of "genuine" halo stars, yet they have kinematics that would associate them with one of the subcomponents of the disk. As vital as it is, chemical information alone does not resolve ambiguities about the formation of the galactic halo and disk. "Cats and dogs may have the same age and metallicity, but they are still cats and dogs" is the way Bernard Pagel of the Nordic Institute for Theoretical Physics in Copenhagen puts it.

As well as telling us about the past history of our galaxy, the disk and halo also provide insight into the Milky Way's probable future evolution. One can easily calculate that almost all of the existing gas will be consumed in a few billion years. This estimate is based on the rate of star formation in the disks of other spirals and on the assumption that the birth of stars will continue at its present speed. Once the gas has been depleted, no more stars will form, and the disks of spirals will then fade. Eventually the galaxy will consist of nothing more than white dwarfs and black holes encapsulated by the hypothesized dark matter corona.

Several sources of evidence exist for such an evolutionary scenario. In 1978 Harvey R. Butcher of the Kapteyn Laboratory in the Netherlands and Augustus Oemler, Jr., of Yale found that dense clusters of galaxies located about six billion light-years away still contained numerous spiral galaxies. Such spirals are, however, rare or absent in nearby clusters of galaxies. This observation shows that the disks of most spirals in dense clusters must have faded to invisibility during the past six billion years. Even more direct evidence of the swift evolution of galaxies comes from the observation of so-called blue galaxies. These galaxies are rapidly generating large stars. Such blue galaxies seem to be less common now than they were only a few billion years ago.

Of course the life of spiral galaxies can be extended. Copious infall of hydrogen from intergalactic space might replenish the gas supply. Such infall can occur if a large gas cloud or another galaxy with a substantial gas reservoir is nearby. Indeed, the Magellanic Clouds will eventually plummet into the Milky Way, briefly rejuvenating our galaxy. Yet the Milky Way will not escape its ultimate fate. Like people and civilizations, stars and galaxies leave behind only artifacts in an evolving, ever-dynamic universe.

—January 1993

THE EVOLUTION OF GALAXY CLUSTERS

The most massive objects in the universe are huge clusters of galaxies and gas that have slowly congregated over billions of years. The process of agglomeration may now be ending

J. Patrick Henry, Ulrich G. Briel, and Hans Böhringer

The royal Ferret of Comets was busy tracking his prey. On the night of April 15, 1779, Charles Messier watched from his Paris observatory as the Comet of 1779 slowly passed between the Virgo and Coma Berenices constellations on its long journey through the solar system. Messier's renown in comet spotting had inspired the furry moniker from King Louis XV, but on this night he took his place in astronomy history books for a different reason. He noticed three fuzzy patches that looked like comets yet did not move from night to night; he added them to his list of such impostors so as not to be misled by them during his real work, the search for comets. Later he commented that a small region on the Virgo-Coma border contained 13 of the 109 stationary splotches that he, with the aid of Pierre Mechain,

191

eventually identified—the Messier objects well known to amateur and professional astronomers today.

As so often happens in astronomy, Messier found something completely different from what he was seeking. He had discovered the first example of the most massive things in the universe held together by their own gravity: clusters of galaxies. Clusters are assemblages of galaxies in roughly the same way that galaxies are assemblages of stars. On the cosmic organizational chart they are the vice presidents—only one level below the universe itself. In fact they are more massive relative to a human being than a human being is relative to a subatomic particle.

In many ways clusters are the closest that astronomers can get to studying the universe from the outside. Because a cluster contains stars and galaxies of every age and type, it represents an average sample of cosmic material—including the dark matter that choreographs the movements of celestial objects yet cannot be seen by human eyes. And because a cluster is the result of gravity acting on immense scales, its structure and evolution are tied to the structure and evolution of the universe itself. Thus, the study of clusters offers clues to three of the most fundamental issues in cosmology: the composition, organization, and ultimate fate of the universe.

A few years after Messier's observations in Paris, William Herschel and his sister, Caroline, began to examine the Messier objects from their garden in England. Intrigued, they decided to search for others. Using substantially better telescopes than their French predecessor had, they found more than 2,000 fuzzy spots— including 300 in the Virgo cluster alone. Both William and his son, John, noticed the lumpy arrangement of these objects on the sky: What organized these objects (which we now know to be galaxies) into the patterns they saw?

A second question emerged in the mid-1930s, when astronomers Fritz Zwicky and Sinclair Smith measured the speeds of galaxies in the Virgo cluster and in a slightly more distant cluster in Coma. Just as the planets orbit about the center of mass of the solar system, galaxies orbit about the center of mass of their cluster. But the galaxies were orbiting so fast that their collective mass could not provide enough gravity to hold them all together. The clusters had to be nearly 100 times as heavy as the visible galaxies, or else the galaxies would have torn out of the clusters long ago. The inescapable conclusion was that the clusters were mostly made of unseen, or "dark," matter. But what was this matter?

These two mysteries—the uneven distribution of galaxies in space and the unknown nature of dark matter—continue to confound astronomers. The former became especially puzzling after the discovery in the mid-1960s of the cosmic microwave background radiation. The radiation, a snapshot of the universe after the big bang and before the formation of stars and galaxies, is almost perfectly smooth. Its tiny imperfections somehow grew to the structures that exist today; but the process is still not clear. As for dark matter, astronomers have learned a bit more about it since the days of Zwicky. But they are still in the uncomfortable position of not knowing what most of the universe is made of.

■ LIGHT FROM DARK MATTER

Impelled by these mysteries, the pace of discovery in the study of clusters has accelerated over the past 40 years. Astronomers now know of some 10,000 of them. American astronomer George Abell compiled the first large list in the early 1950s, based on photographs of the entire northern sky taken at Palomar Observatory in California. By the 1970s astronomers felt they at least understood the basic properties of clusters: they consisted of speeding galaxies bound together by huge amounts of dark matter. They were stable and immutable objects.

Then came 1970. In that year a new satellite, named Uhuru ("freedom" in Swahili) in honor of its launch from Kenya, began observing a form of radiation hitherto nearly inaccessible to astronomers: x-rays. Edwin M. Kellogg, Herbert Gursky, and their colleagues at American Science and Engineering, a small company in Massachusetts, pointed Uhuru at the Virgo and Coma clusters. They found that the clusters consist not only of galaxies but also of huge amounts of gas threading the space between the galaxies. The gas is too tenuous to be seen in visible light, but it is so hot—more than 25 million degrees Celsius—that it pours out x-rays.

In short, astronomers had found some of the dark matter—20 percent of it by mass. Although the gas is not enough to solve the dark matter mystery completely, it does account for more mass than all the galaxies put together. In a way, the term "clusters of galaxies" is inaccurate. These objects are balls of gas in which galaxies are embedded like seeds in a watermelon.

Since the early 1970s the x-ray emission has been scrutinized by other satellites, such as the Einstein X-Ray Observatory, the Roentgen Satellite (ROSAT), and the Advanced Satellite for Cosmology and Astrophysics (ASCA). Our own research mainly uses ROSAT. The first x-ray telescope to record images of the entire sky, ROSAT is well suited for observations of large diffuse objects such as clusters and is now engaged in making detailed images of these regions. With this new technology, astronomers have extended the discoveries of Messier, Zwicky and the other pioneers.

When viewed in x-rays, the Coma cluster has a mostly regular shape with a few lumps. These lumps appear to be groups of galaxies—that is, miniature clusters. One lump to the southwest is moving into the main body of the cluster, where other lumps already reside. Virgo, in comparison, has an amorphous shape. Although it has regions of extra x-ray emission these bright spots are coming from some of the Messier galaxies rather than from clumps of gas. Only the core region in the northern part of Virgo has a nearly symmetrical structure *[see color plate 2]*.

Such x-ray images have led astronomers to conclude that clusters form from the merger of groups. The lumps in the main body of the Coma cluster presumably represent groups that have already been drawn in but have not yet been fully assimilated. Virgo seems to be in an even earlier stage of formation. It is still pulling in surrounding material and, at the current rate of progress, will look like Coma after a few billion years. This dynamic view of clusters gobbling up and digesting nearby matter is in stark contrast to the static view that astronomers held just a few years ago.

■ TAKING THEIR TEMPERATURE

Ever since astronomers obtained the first good x-ray images in the early 1980s they have wanted to measure the variation of gas temperature across clusters. But making these measurements is substantially more difficult than making images, because it requires an analysis of the x-ray spectrum for each point in the cluster. Only in 1994 did the first temperature maps appear.

The maps have proved that the formation of clusters is a violent process. Images of the cluster Abell 2256, for example, show that x-ray emission has not one but rather two peaks. The western peak is slightly flattened, suggesting that a group slamming into the main cluster has swept up material just as a snowplow does. A temperature map supports this interpretation *[see color plate 3]*. The western peak, it turns out, is comparatively cool; its temperature is characteristic of the gas in a group of galaxies. Because groups are smaller than clusters, the gravitational forces within them are weaker; therefore, the speed of the gas molecules within them— that is, their temperature—is lower. A typical group is 50 trillion times as massive as the sun and has a temperature of 10 million degrees C. In comparison a typical cluster weighs 1,000 trillion suns and registers a temperature of 75 million degrees C; the heaviest known cluster is five times as massive and nearly three times as hot.

Two hot regions in Abell 2256 appear along a line perpendicular to the presumed motion of the group. The heat seems to be generated as snowplowed material squirts out the sides and smashes into the gas of the main cluster. In fact these observations match computer simulations of merging groups. The group should penetrate to the center of the cluster in several hundred million years. Thus, Abell 2256 is still in the early stages of the merger.

The late stages of a merger are apparent in another cluster, Abell 754. This cluster has two distinguishing features. First, optical photographs show that its galaxies reside in two clumps. Second, x-ray observations reveal a bar-shaped feature from which the hot cluster gas fans out. One of the galaxy clumps is in the bar region and the other is at the edge of the high-temperature region to the west.

Theorists can explain this structure with an analogy. Imagine throwing a water balloon, which also contains some pebbles, into a swimming pool. The balloon represents the merging group: the water is gas, and the pebbles are galaxies. The swimming pool is the main cluster. When the balloon hits the water in the pool, it ruptures. Its own water stays at the surface and mixes very slowly, but the pebbles can travel to the other side of the pool. A similar process apparently took place in Abell 754. The gas from the merging group was suddenly stopped by the gas of the cluster, while the group galaxies passed right through the cluster to its far edge.

A third cluster, Abell 1795, shows what a cluster looks like billions of years after a merger. The outline of this cluster is perfectly smooth and its temperature is nearly uniform, indicating that the cluster has assimilated all its groups and settled into equilibrium. The exception is the cool region at the very center. The lower temperatures occur because gas at the center is dense, and dense gas emits x-rays more efficiently than tenuous gas. If left undisturbed for two or three billion years, dense gas can radiate away much of its original energy, thereby cooling down.

As the gas cools, substantial amounts of lukewarm material build up—enough for a whole new galaxy. So where has all this material gone? Despite exhaustive searches, astronomers have yet to locate conclusively any pockets of tepid gas. That the cluster gas is now losing heat is obvious from the temperature maps. Perhaps

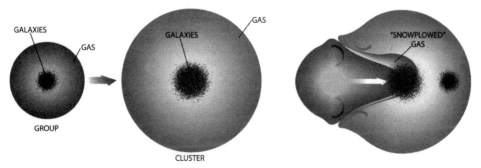

Absorption of galaxy group allows a cluster to grow to colossal size. Pulled in by gravity, the group slams into the cluster, pushing gas out the sides. The galaxies themselves pass through the cluster, their progress unimpeded by the tenuous gas. Eventually the galaxies and gas mix together, forming a unified cluster that continues to draw in other groups until no more are to be found. (Slim Films)

the heat loss started only fairly recently, or perhaps the collision of galaxy groups prevents cool gas from collecting in one spot. These so-called cooling flows remain yet another unsolved mystery.

■   BOTTOMS UP

The sequence represented by these three Abell clusters is probably undergone by every cluster as it grows. Galaxy groups occasionally join the cluster; with each, the cluster gains hot gas, bright galaxies, and dark matter. The extra mass creates stronger gravitational forces, which heat the gas and accelerate the galaxies. Most astronomers believe that almost all cosmic structures agglomerated in this bottom-up way. Star clusters merged to form galaxies, which in turn merged to form groups of galaxies, which are now merging to form clusters of galaxies. In the future it will be the clusters' turn to merge to form still larger structures. There is, however, a limit set by the expansion of the universe. Eventually, clusters will be too far apart to merge. Indeed, the cosmos may be approaching this point already.

By cosmological standards, all the above-mentioned clusters (Coma, Virgo, and Abell 2256, 754, and 1795) are nearby objects. Astronomers' efforts to understand their growth are analogous to understanding human growth from a single photograph of a crowd of people. With a little care you could sort the people in the picture into the proper age sequence. You could then deduce that as people age, they generally get taller, among other visible changes.

You could also study human growth by examining a set of photographs, each containing only people of a certain age—for example, class pictures from grade school, high school, and college. Similarly, astronomers can observe clusters at ever-increasing distances, which correspond to ever-earlier times. On average the clusters in a more distant sample are younger than those in a nearby one. Therefore, researchers can piece together "class photos" of clusters of different ages. The advantage of this approach is that it lets astronomers work with a whole sample of clusters, rather than just a few individual clusters. The disadvantage is that the younger objects are too far away to study in detail; only their average properties can be discerned.

One of us (Henry) applied this method to observations from the ASCA x-ray satellite. He found that distant, younger clusters are cooler than nearby, older ones. Such a temperature change shows that clusters become hotter and hence more massive over time—further proof of the bottom-up model. From these observations researchers have estimated the average rate of cluster evolution in the uni-

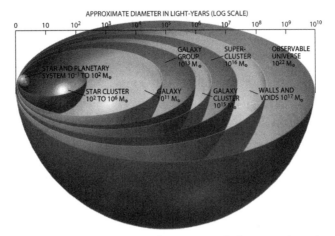

APPROXIMATE DIAMETER IN LIGHT-YEARS (LOG SCALE)

Hierarchy of cosmic structures ranges from stars and planets to the universe itself. The largest objects held together by gravity are galaxy clusters with masses up to 10^{15} times that of the sun (denoted as $M_\odot$). Although there is a higher level of organization consisting of superclusters and great walls, these patterns are not bound gravitationally. On even larger scales the universe is featureless. Astronomers think most of these structures form from the progressive agglomeration of smaller units. (Slim Films)

verse. The rate, which is related to the overall evolution of the universe and to the nature of the dark matter, implies that the universe will expand forever.

New x-ray observations may shed light on the remaining dark matter in clusters. By the end of 2000 there will be three advanced x-ray observatories in orbit: the Advanced X-Ray Astrophysics Facility from the U.S., the X-ray Multi-mirror Mission from Europe, and ASTRO-E from Japan.

In the meantime observations of another form of radiation, known as extreme ultraviolet light, are yielding mysteries of their own. The extreme ultraviolet has an energy that is only slightly lower than that of x-rays. It is heavily absorbed by material in our galaxy, so astronomers assumed that most clusters are not visible in this wavelength band. But recently Richard Lieu of the University of Alabama at Huntsville, C. Stuart Bowyer of the University of California at Berkeley, and their colleagues studied five clusters using the sensitive Extreme Ultraviolet Explorer satellite.

These clusters, they discovered, shine brightly in the extreme ultraviolet. In some ways this discovery was as unexpected as the first detection of x-rays from clusters in the early 1970s. Although some of the radiation comes from the same gas that generates the x-rays, there appears to be an additional source in at least some of the clusters. This finding is very new and has not yet been explained. Perhaps astronomers are seeing another component of the clusters' dark matter for the first time. The upcoming x-ray facilities may identify this new component.

Those of us involved in this work feel a special bond with Charles Messier as he strained to glimpse those faint patches of light in Virgo, not knowing their true significance. As advanced as our technology has become, we still strain to understand these clusters. We feel a bond with future observers as well, for science advances in a continuous process of small increments. We have been helped by those who preceded us; we share our new understanding with those who follow.

—December 1998

Colossal Galactic Explosions

Enormous outpourings of gas from the centers of nearby galaxies may ultimately help explain both star formation and the intergalactic medium

Sylvain Veilleux, Gerald Cecil,
and Jonathan Bland-Hawthorn

Millions of galaxies shine in the night sky, most made visible by the combined light of their billions of stars. In a few, however, a pointlike region in the central core dwarfs the brightness of the rest of the galaxy. The details of such galactic dynamos are too small to be resolved even with the Hubble Space Telescope. Fortunately, debris from these colossal explosions—in the form of hot gas glowing at temperatures well in excess of a million degrees—sometimes appears outside the compact core, on scales that can be seen directly from the earth.

The patterns that this superheated material traces through the interstellar gas and dust surrounding the site of the explosion provide important clues to the

nature and history of the powerful forces at work inside the galactic nucleus. Astronomers can now determine what kind of engines drive these dynamos and the
effects of their tremendous outpourings on the intergalactic medium.

Furthermore, because such cataclysms appear to have been taking place since
early in the history of the universe, they have almost certainly affected the environment in which our own Milky Way galaxy evolved. Understanding how such events
take place today may illuminate the distribution of chemical elements that has
proved crucial to formation of stars like the sun.

Astronomers have proposed two distinctly different mechanisms for galactic dynamos. The first was the brainchild of Martin J. Rees of the University of Cambridge and Roger D. Blandford, now at the California Institute of Technology.
During the early 1970s the two sought to explain the prodigious luminosity—
thousands of times that of the Milky Way—and the spectacular "radio jets" (highly
focused streams of energetic material) that stretch over millions of light-years from
the centers of some hyperactive young galaxies known as quasars. They suggested
that an ultramassive black hole—not much larger than the sun but with perhaps a
million times its mass—could power a quasar.

A black hole itself produces essentially no light, but the disk of accreted matter spiraling in toward the hole heats up and radiates as its density increases. The inner, hotter part of the disk produces ultraviolet and x-ray photons over a broad range of
energies, a small fraction of which are absorbed by the surrounding gas and reemitted as discrete spectral lines of ultraviolet and visible light. In the years since Rees and
Blandford proposed their model, astronomers have come to understand that similar
black holes may be responsible for the energy output of nearer active galaxies.

As the disk heats up, gas in its vicinity reaches temperatures of millions of degrees and expands outward from the galactic nucleus at high speed. This flow, an
enormous cousin to the solar wind that streams away from the sun or other stars,
can sweep up other interstellar gases and expel them from the nucleus. The resulting luminous shock waves can span thousands of light-years—comparable to the
visible sizes of the galaxies themselves—and can be studied from space or ground-
based observatories. Some of these galaxies also produce radio jets: thin streams of
rapidly moving gas that emit radio waves as they traverse a magnetic field that may
be anchored within the accretion disk.

Black holes are not the only engines that drive violent galactic events. Some
galaxies apparently undergo short episodes of rapid star formation in their cores:
so-called nuclear starbursts. The myriad new stars produce strong stellar winds
and, as the stars age, a rash of supernovae. The fast-moving gas ejected from the supernovae strikes the background interstellar dust and gas and heats it to millions of
degrees.

The pressure of this hot gas forms a cavity, like a steam bubble in boiling water. As the bubble expands, cooler gas and dust accumulate in a dense shell at the edge of the bubble, slowing its expansion. The transition from free flow inside the bubble to near-stasis at its boundary gives rise to a zone of turbulence that is readily visible from the earth. If the energy injected into the cavity is large enough, the bubble bursts out of the galaxy's gas disk and spews the shell's fragments and hot gas into the galaxy halo or beyond, thousands of light-years from their origins.

Roberto Terlevich of the Royal Greenwich Observatory and his collaborators have led the most recent research aimed at determining whether starbursts alone can drive the outpourings of hot gas characteristic of active galaxies. In 1985 Terlevich and Jorge Melnick, now at the European Southern Observatory, argued that many such galaxies contain unusual stars they dubbed "warmers"—extremely hot stars with temperatures higher than 100,000 degrees and very powerful stellar winds. Such stars, the two scientists proposed, arise naturally when a starburst occurs in a region enriched in heavy chemical elements from previous supernovae. Terlevich and his colleagues contend that their model explains the spectra and many other properties of certain active galaxies.

■   IDENTIFYING THE ENGINE

Both the starburst and the black-hole explanations appear plausible, but there are important differences between the two that may reveal which one is at work in a given galaxy. A black hole can convert as much as 10 percent of the infalling matter to energy. Starbursts, in contrast, rely on nuclear fusion, which can liberate only .1 percent of the reacting mass. As a result they require at least 100 times as much matter, most of which accumulates as unburned fuel. Over the lifetime of a starburst-powered quasar, the total mass accumulated in the nucleus of the galaxy could reach 100 billion times the mass of the sun, equivalent to the mass of all the stars in the Milky Way galaxy.

The more mass near the nucleus, the more rapidly the orbiting stars must move. Ground-based optical observations, which are limited by atmospheric blurring, have not placed tight constraints on the concentration of mass in galactic centers. Recent radio-telescope findings, however, have revealed an accretion disk with an inner radius of half a light-year spinning rapidly around a mass 20 million times that of the sun at the center of a nearby spiral galaxy called NGC 4258.

Several research groups are now measuring the distributions of stellar motions across galactic nuclei using the spectrograph on board the Hubble telescope. The discovery that gas in the inner core of the active galaxy M87 is moving in a manner consistent with a black-hole accretion disk has demonstrated the promise of such techniques, and the instrument will become far more efficient after astronauts upgrade it in 1997.

Starbursts and black holes also differ in the spectra of the most energetic photons they produce. Near a black hole the combination of a strong magnetic field and a dense accretion disk creates a soup of very fast particles that collide with one another and with photons to generate x-rays and gamma rays. A starburst, in contrast, produces most of its high-energy radiation from collisions between supernova ejecta and the surrounding galactic gas and dust. This impact heats gas to no more than about a billion degrees and so cannot produce any radiation more energetic than x-rays. The large numbers of gamma rays detected recently from some quasars by the Compton Gamma Ray Observatory imply that black holes are at their centers.

A final difference between black holes and starbursts lies in the forces that focus the flow of outrushing gas. The magnetic field lines attached to the accretion disk around a black hole direct outflowing matter along the rotation axis of the disk in a thin jet. The material expelled by a starburst bubble, in contrast, simply follows the path of least resistance in the surrounding environment. A powerful starburst in a spiral galaxy will spew gas perpendicular to the plane of the galaxy's disk of stars and gas, but the flow will be distributed inside an hourglass-shaped region with a wide opening. The narrow radio jets that extend millions of light-years from the core of some active galaxies clearly suggest the presence of black holes.

All that we know about galaxies—active or otherwise—comes from the radiation they emit. Our observations supply the data that astrophysicists can use to choose among competing theories. The three of us have concentrated on visible light, from which we can determine the temperatures, pressures, and concentrations of different atoms in the gas agitated by galactic explosions. We compare the wavelength and relative intensities of emission lines from excited or ionized atoms with those measured in terrestrial laboratories or derived from theoretical calculations.

■ SPECTRAL SIGNATURES

Thanks to the Doppler shift, which changes the frequency and wavelength of light emitted by moving sources, this analysis also reveals how fast the gas is moving. Approaching gas emits light shifted toward the blue end of the spectrum, and receding gas emits light shifted toward the red end.

Until recently astronomers unraveled gas behavior by means of two complementary methods: emission line imaging and long-slit spectroscopy. The first produces images through a filter that selects light of a particular wavelength emitted by an element such as hydrogen. Such images often dramatically reveal the filamentary patterns of explosions, but they cannot tell observers anything about the speed or direction of the gases' motions, because the filter does not discriminate finely enough to measure redshifts or blueshifts. Long-slit spectrometers, which disperse light into its constituent colors, provide detailed information about gas motions but only over a tiny region.

For almost a decade our group has used an instrument that combines the advantages of these two methods without the main drawbacks. The Hawaii Imaging Fabry-Perot Interferometer (HIFI) yields detailed spectral information over a large field of view. Named after the turn-of-the-century French inventors Charles Fabry and Alfred Pérot, such interferometers have found wide-ranging applications in astronomy. At their heart are two glass plates that are kept perfectly parallel while separated by less than a twentieth of a millimeter. The inner surfaces of the plates are highly reflecting, so light passing through the plates is trapped into repeated reflections. Light of all but a specific wavelength—determined by the precise separation—is attenuated by destructive interference as the light waves bounce back and forth between the plates. By adjusting the separation between the plates, we can produce a series of images that are essentially a grid of spectra obtained by the interferometer at every position over the field of view.

The HIFI takes its pictures atop the 14,000-foot dormant volcano Mauna Kea, using the 2.2-meter telescope owned by the University of Hawaii and the 3.6-meter Canada-France-Hawaii instrument. The smooth airflow at the mountaintop produces sharp images. Charge-coupled devices, which are very stable and sensitive to faint light, collect the photons. In a single night this powerful combination can generate records of up to a million spectra across the full extent of a galaxy.

■ NEARBY ACTIVE GALAXIES

We have used the HIFI to explore NGC 1068, an active spiral galaxy 46 million light-years away. As the nearest and brightest galaxy of this type visible from the Northern Hemisphere, it has been studied extensively. At radio wavelengths NGC 1068 looks like a miniature quasar: two jets extend about 900 light-years from the core, with more diffuse emission from regions farther out. Most likely, emission from gaseous plasma moving at relativistic speeds creates the radio jets, and the "radio lobes" arise where the plasma encounters matter from the galactic disk. As might a supersonic aircraft, the leading edge of the northeast jet produces a V-shaped shock front.

The same regions also emit large amounts of visible and ultraviolet light. We have found, however, that only 10 percent of the light comes from the nucleus. Another

5 percent comes from galaxy-disk gas that has piled up on the expanding edge of the northeast radio lobe. All the rest come from two fans of high-velocity gas moving outward from the center at speeds of up to 1,500 kilometers per second.

The gas flows outward in two conical regions; it is probably composed of dense filaments of matter that have been swept up by the hot wind from the accretion disk. The axis of the cones of outflowing wind is tilted above the plane of the galaxy but does not point toward the poles.

The effects of the activity within the nucleus reach out several thousand light-years, well beyond the radio lobes. The diffuse interstellar gas exhibits unusually high temperatures and a large fraction of the atoms have lost one or more electrons and become ionized. At the same time phenomena in the disk appear to influence the nucleus. Infrared images reveal an elongated bar of stars that extends more than 3,000 light-years from the nucleus. The HIFI velocity measurements suggest that the bar distorts the circular orbit of the gas in the disk, funneling material toward the center of the galaxy. This inflow of material may in fact fuel the black hole.

Another tremendous explosion is occurring in the core of one of our nearest neighbor galaxies, M82, just a few million light-years away. In contrast to NGC 1068, this cataclysm appears to be an archetypal starburst-driven event. Images exposed through a filter that passes the red light of forming hydrogen atoms reveal a web of filaments spraying outward along the galactic poles. Our spectral grids of emission from filaments perpendicular to the galactic disk reveal two main masses of gas, one receding and the other approaching. The difference in velocity between the two increases as the gas moves outward from the core, reaching about 350 kilometers per second at a distance of 3,000 light-years. At a distance of 4,500 light-years from the core the velocity separation diminishes.

The core of M82 is undergoing an intense burst of star formation, possibly triggered by a recent orbital encounter with its neighbors M81 and NGC 3077. Its infrared luminosity is 30 billion times the total luminosity of the sun, and radio astronomers have identified the remnants of large numbers of supernovae. The filamentary web visible from the earth results from two elongated bubbles oriented roughly perpendicular to the disk of M82 and straddling the nucleus. X-ray observatories in space have detected the hot wind that inflates these bubbles; their foamy appearance probably arises from instabilities in the hot gas as it cools.

■ AMBIGUOUS ACTIVITY

Unfortunately the identity of the principal source of energy in active galaxies is not always so obvious. Sometimes a starburst appears to coexist with a black-hole engine. Like M82, many of these galaxies are abnormally bright at infrared wavelengths and rich in molecular gas, the raw material of stars. Radio emission and visual spectra resembling those of a quasar, however, suggest that a black hole may also be present.

Such ambiguity plagues interpretations of the behavior of the nearby galaxy NGC 3079. This spiral galaxy appears almost edge-on from the earth—an excellent vantage point from which to study the gas expelled from the nucleus. Like M82, NGC 3079 is anomalously bright in the infrared, and it also contains a massive disk of molecular gas spanning 8,000 light-years around its core. At the same time, the core is unusually bright at radio wavelengths, and the linear shape of radio-emitting regions near the core suggests a collimated jet outflow. On a larger scale the radio-emission pattern is complex and extends more than 6,500 light-years from either side of the galactic disk.

Images made in red hydrogen light show a nearly circular ring 3,600 light-years across just east of the nucleus; velocity measurements from the HIFI confirm that the ring marks the edge of a bubble as seen from the side. The bubble resembles an egg with its pointed extremity balanced on the nucleus and its long axis aligned with the galactic pole. There is another bubble on the west side of the nucleus, but most of it is hidden behind the dusty galaxy disk.

Our spectral observations imply that the total energy of this violent outflow is probably 10 times that of the explosions in NGC 1068 or M82. The alignment of the bubble along the polar axis of the host galaxy implies that galactic dust and gas, rather than a central black hole, are collimating the outflow. Nevertheless, the evidence is fairly clear that NGC 3079 contains a massive black hole at its core.

Is the nuclear starburst solely responsible for such a gigantic explosion? We have tried to answer this question by analyzing the infrared radiation coming from the starburst area. Most of the radiation from young stars embedded in molecular clouds is absorbed and reemitted in the infrared, so the infrared luminosity of NGC 3079's nucleus may be a good indicator of the rate at which supernovae and stellar winds are injecting energy at the center of the galaxy. When we compare the predictions of the starburst model with our observations, we find that the stellar ejecta appears to have enough energy to inflate the bubble. Although the black hole presumed to exist in the core of NGC 3079 may contribute to the outflow, there is no need to invoke it as an energy source.

■ HOW ACTIVE GALAXIES FORM

Although astronomers now understand the basic principles of operation of the engines that drive active galaxies, many details remain unclear. There is a vigorous debate about the nature of the processes that ignite a starburst or form a central black hole. What is the conveyor belt that transports fuel down to the pointlike nucleus? Most likely, gravitational interactions with gas-rich galaxies redistribute gas in the host galaxy, perhaps by forming a stellar bar such as the one in NGC 1068. Computer simulations appear to indicate that the bar, once formed, may be quite stable. (Indeed, the bar must be stable, because NGC 1068 currently has no close companion.)

Researchers are also divided on which comes first, nuclear starburst or black hole. Perhaps the starburst is an early phase in the evolution of active galaxies, eventually fading to leave a dense cluster of stellar remnants that rapidly coalesce into a massive black hole.

The anomalous gas flows in the galaxies that we and others have observed are almost certainly only particularly prominent examples of widespread, but more subtle, processes that affect many more galaxies. Luminous infrared galaxies are common, and growing evidence is leading astronomers to believe that many of their cores are also the seats of explosions. These events may profoundly affect the formation of stars throughout the galactic neighborhood. The bubble in NGC 3079, for instance, is partially ruptured at the top and so probably leaks material into the outer galactic halo or even into the vast space between galaxies. Nuclear reactions in the torrent of supernovae unleashed by the starburst enrich this hot wind in heavy chemical elements. As a result the wind will not only heat its surroundings but also alter the environment's chemical composition.

The full impact of this "cosmic bubble bath" over the history of the universe is difficult to assess accurately because we currently know very little of the state of more distant galaxies. Images of distant galaxies taken by the Hubble will help clarify some of these questions. Indeed, as the light that left those galaxies billions of years ago reaches our instruments, we may be watching an equivalent of our own galactic prehistory unfolding elsewhere in the universe.

—February 1996

Dark Matter in Spiral Galaxies

It appears that much of the matter in spiral galaxies emits no light. Moreover, it is not concentrated near the center of the galaxies

Vera Rubin

After evidence was obtained (in the 1920s) that the universe is expanding it became reasonable to ask: Will the universe continue to expand indefinitely or is there enough mass in it for the mutual attraction of its constituents to retard the expansion and finally bring it to a halt? Most cosmologists agree that the universe started in a big bang 10 to 20 billion years ago from an infinitely small and dense state and that it has been expanding ever since. It can be calculated that the critical density of matter needed to brake the expansion and "close" the universe is on the order of 5×10^{-30} gram per cubic centimeter, which is equal to about three hydrogen atoms per cubic meter. The amount of luminous matter in the form of galaxies, however, comes to only about 7.5×10^{-32} gram per cubic centimeter.

Therefore if the expansion of the universe is to stop, the density of the invisible matter must exceed the density of the luminous matter by a factor of roughly 70.

With this factor in mind astronomers over the past half century have sought to determine the mass of the galaxies that populate the universe out to the limits of observation. From the luminosity of typical galaxies one can estimate that they have a mass ranging from a few billion to a few trillion times the mass of the sun. The actual stellar population of a galaxy is of course highly diverse. Some stars are 10,000 times more luminous than the sun per unit of mass; others are only a small fraction as luminous. Given this diversity one would like to know: Is the distribution of luminosity in galaxies a reliable indicator of the distribution of mass? And by extrapolation, is the distribution of luminosity in galaxies a reliable indicator of the distribution of mass in the universe?

My colleagues and I in the Department of Terrestrial Magnetism of the Carnegie Institution of Washington have sought to answer these questions by measuring the rotational velocity of selected galaxies at various distances from their center of rotation. It has been known for a long time that outside the bright nucleus of a typical spiral galaxy the luminosity of the galaxy falls off rapidly with distance from the center. If luminosity were a true indicator of mass, most of the mass would be concentrated toward the center. Outside the nucleus the rotational velocity would fall off inversely as the square root of the distance, in conformity with Kepler's law for the orbital velocity of bodies in the solar system. Instead it has been found that the rotational velocity of spiral galaxies in a diverse sample either remains constant with increasing distance from the center or rises slightly out as far as it is possible to make measurements. This unexpected result indicates that the falloff in luminous mass with distance from the center is balanced by an increase in nonluminous mass.

Our results, taken together with those of many other workers who have attacked the mass question in other ways, now make it possible to say with some confidence that the distribution of light is not a valid indicator of the distribution of mass either in galaxies or in the universe at large. As much as 90 percent of the mass of the universe is evidently not radiating at any wavelength with enough intensity to be detected on the earth. Originally astronomers described the nonluminous component as "missing matter." Today they recognize that it is not missing; it is just not visible. Such dark matter could be in the form of extremely dim stars of low masses, of large planets like Jupiter, or of black holes, either small or massive. Other candidates include neutrinos (if indeed they have mass, as recent work suggests) or such hypothetical particles as magnetic monopoles or gravitinos.

Early in the century it was reasonable for astronomers to assume that the distribution of luminous matter, wherever it was found, coincided with the distribution of mass. Nearly 50 years ago, however, Sinclair Smith and Fritz Zwicky of the California Institute of Technology discovered that in some large clusters of galaxies the individual members are moving so rapidly that their mutual gravitational attraction is insufficient to keep the clusters from flying apart. Either such clusters should be dissolving or there must be enough dark matter present to hold them together. Almost all the evidence suggests that clusters of galaxies are stable configurations. Hence the early observations of Smith and Zwicky marshaled the first evidence that such clusters harbor matter both luminous and nonluminous.

Recent work by many other astronomers has strengthened this conclusion. Studies of the dynamics of individual galaxies, including our own galaxy, of pairs of galaxies, of groups of galaxies, and of clusters of galaxies all point to a component of unobservable but ubiquitous mass. Such studies detect the presence of nonluminous mass solely by its gravitational effects.

For the past several years W. Kent Ford, Jr., Norbert Thonnard, David Burstein, and I have sought to learn about the distribution of mass in the universe by investigating the distribution of matter within galaxies with a structure similar to that of our own galaxy, namely the general class of spiral galaxies. We have adopted this approach because spiral galaxies have a geometry favorable for the identification of mass, whether it is luminous or nonluminous, and modern large telescopes equipped with image-tube spectrographs make it possible to complete an observation of a single galaxy with an exposure of about three hours. Before I describe our observations it will be helpful if I review how celestial objects respond to the gravitational force acting on them and how that response can reveal the large-scale distribution of matter.

Toward the end of the 17th century Robert Hooke suspected that the planets were subject to a gravitational force from the sun whose intensity decreased inversely as the square of the distance. Isaac Newton then recognized that all pairs of objects in the universe have a gravitational attraction for each other that is proportional to the product of their masses and inversely proportional to the square of the distance between them. In other words, if the distance between the objects is increased by, say, a factor of two, their mutual attraction decreases by a factor of four.

For planets in orbit around the sun, which embodies essentially all the mass in the solar system, the decrease in gravitational attraction with distance is exactly paralleled by a decrease in the velocity needed to hold the planet in its orbit. Therefore Mercury, lying at .39 astronomical unit from the sun (that is, .39 of the mean distance between the sun and the earth), has an orbital velocity of about 47.9 kilometers per second. Pluto, 100 times farther away at a mean distance of 39.5 astronomical units, has an orbital velocity only a tenth that of Mercury, or 4.7 kilometers per second. Spiral galaxies rotate because they retain the angular momentum and the orbital momentum of the initial clumps of gas from which they formed.

In a spiral galaxy the gas, dust, and stars in the disk of the galaxy (together with any associated planets and their satellites) are all in orbit around a common center.

Like the planets in the solar system, the gas and stars move in response to the combined gravitational attraction of all the other mass. If the galaxy is visualized as a spheroid, the gravitational attraction due to the mass M_r lying between the center and an object of mass m in an equatorial orbit at a distance r from the center is given by Newton's law GmM_r/r^2, where G is the constant of gravitation. If the galaxy is neither contracting nor expanding, the gravitational force is exactly equal to the centrifugal force on the mass at distance r: $GmM_r/r^2 = mV_r^2/r$, where V_r is the orbital velocity.

When this equation is solved for V_r, the value of m drops out and the velocity of a body at distance r from the center is determined only by the mass M_r inward from its position. If, as in the solar system, virtually all the mass is near the center, then the velocities outward from the center decrease as $1/r^2$. Such a decrease in orbital velocity is called Keplerian after Johannes Kepler, who first stated the laws of planetary motion.

In a galaxy the brightness is strongly peaked near the center and falls off rapidly with distance. Astronomers had long assumed that the mass too decreased rapidly with distance, in accordance with the distribution of luminosity. Hence it was expected that stars at increasing distances from the center would have decreasing Keplarian orbital velocities. Until recently few velocity observations had been made in the faint outer regions of galaxies, either to confirm this expectation or refute it.

Although the forms of spiral galaxies are exceedingly diverse, astronomers are able to group them into three useful classes following a scheme proposed some 60 years ago by Edwin P. Hubble. Galaxies designated Sa have a large central bulge surrounded by tightly wound smooth arms in which "knots," or bright regions, are barely resolved. Sb galaxies have a less pronounced central bulge and more open arms with more pronounced knots. Sc galaxies have a small central bulge and well-separated arms speckled with distinct luminous segments. The progression from type Sa to type Sc is one of decreasing prominence of the central bulge and increasing prominence of the disk rotating about it. That the disk is indeed rotating is assumed on simple dynamical grounds.

Within each type there are systematic variations in size and luminosity. For example, Sc galaxies range from small, low-luminosity, low-mass objects to galaxies of enormous luminosity and mass. For completeness, therefore, the study of the dynamics of galaxies should include not only objects with a range of morphological types but also objects with a range of luminosities.

Only for the closest stars in our own galaxy is it possible to detect motion by observing the changing position of the star against the background of more distant stars and galaxies on the celestial sphere. Even for the Andromeda galaxy, the large spiral galaxy closest to our own, it would take some 20,000 years for an orbital velocity of 200 kilometers per second (a velocity comparable to the sun's) to carry a star one second of arc across the sky. This is the minimum angular separation that can be detected optically from the earth. To study the motions in galaxies a different method is needed, one based on the phenomenon of the Doppler shift.

Doppler shifts are shifts in the frequency of waves from a source caused by the motion of the source toward or away from the observer. When the spectrum of the bright nucleus of a spiral galaxy is recorded, the absorption lines arising from the constituent stars are shifted toward the long-wavelength (red) end of the spectrum compared with the same lines in spectra made in laboratories on the earth. Such redshifts in the spectra of all but a few of the nearest galaxies, first observed in about 1915 by V. M. Slipher of the Lowell Observatory, provide the evidence that the universe is expanding, carrying almost all the other isolated galaxies away from ours and away from one another. As a result of Smith and Zwicky's work it is known that in pairs, groups, and clusters of galaxies the local gravitational field overcomes the general expansion, so that these denser agglomerations of matter remain bound. Although the distances between clusters of galaxies are increasing, the distances between galaxies within clusters remain about the same. Slipher also noted that the spectra of individual galaxies can yield additional information about the motions of stars and gas within the galaxy.

If the disk of a spiral galaxy is oriented so that its plane is sharply tilted with respect to the line of sight from the earth, the rotation of the galaxy will carry the stars and gas on one side of the galactic nucleus toward our galaxy and those on the other side away from it. The spectral lines of the approaching material will therefore be blueshifted, or raised in frequency, and the lines of the receding material will be redshifted, or lowered in frequency. A measurement at any point on a spectral line will therefore supply both the angular distance of that point from the galactic nucleus and the velocity along the line of sight at that distance.

It is difficult to make a spectroscopic measurements of the velocities of individual stars, which are faint even in galaxies fairly close to our own. In our work, therefore, we observe not stars but the light from the clouds of gas, rich in hydrogen and helium, that surround certain hot stars. The spectra of such clouds consist of bright emission lines that arise as an electron in an excited atom drops from a higher energy state to a lower one. In addition to emission lines of hydrogen and helium there usually are bright lines from atoms of nitrogen and sulfur that are singly ionized, or stripped of one electron. These lines are called forbidden because they arise only for atoms in the near-vacuum of space; in terrestrial laboratories such singly ionized atoms are rapidly deexcited by collisions with other atoms before the forbidden transition can occur.

Until recently it was not possible to get high-resolution optical spectra of the faint outer regions of galaxies. It is the present availability of large optical telescopes, of high-resolution, long-slit spectrography, and of efficient electronic imaging devices that have made our observing program feasible. Six years ago my colleagues and I set out to measure the rotational velocities completely across the luminous disk of suitably tilted spiral galaxies. Our aim was to study the internal dynamics and distribution of mass in individual galaxies as a function of the galaxies' morphology. We have now observed 60 spiral galaxies: 20 each of the three major types Sa, Sb, and Sc. We have selected galaxies that have a well-defined type, that are well inclined to the plane of the sky (yielding a large component of orbital velocity along the line of sight), that have an angular diameter no larger than the slit of the spectrograph, and that span a large range of luminosities within each type.

Most of the spectra have been obtained with two four-meter telescopes, the one at the Kitt Peak National Observatory in Arizona and the one at the Cerro Tololo Inter-American Observatory in Chile. A few of the spectra were recorded with the 2.5-meter telescope at the Las Campanas Observatory in Chile.

After the photons from the galactic source pass through the slit of the spectrograph and are dispersed by a diffraction grating, they are focused on a "Carnegie" image tube (RCA C33063), where they are multiplied by a factor of 10 or more before they are recorded by the photographic emulsion. Exposures of two to three hours are recorded on Kodak IIIa-J plates whose sensitivity, matched to that of the image tube, has been much increased by having previously been baked at 65 degrees Celsius for two hours in a special "forming" gas (nitrogen with an admixture of 2 percent hydrogen) and preexposed to flashes of light. Without the image tube and the plate-sensitizing methods exposure times would have been prohibitively long: from 20 to 60 hours.

Generally two exposures are made of each galaxy. In one exposure the spectrograph slit is made to coincide with the major (long) axis of the galaxy; each point on the spectrum arises from a single region of the galactic disk. The Doppler, or velocity, displacements of the emission lines are readily discerned in the developed image. A second exposure is made with the spectrograph slit aligned with the minor axis of the galactic disk. Since the orbital velocities are now perpendicular to the line of sight, no Doppler shifts are evident. The absence of line displacements with the slit along the minor axis is confirming evidence that the motions we study are indeed orbital ones.

In order to have a reference scale against which to measure the displacement of emission lines in galactic spectra, astronomers formerly recorded neon lines from a lamp along the edges of the spectrum. We have now dispensed with this procedure. Instead we measure displacements directly from the unshifted lines on each plate that are emitted by hydroxyl (OH) molecules in the earth's atmosphere. Many astronomers have adopted sophisticated plate-scanning devices to measure line positrons, particularly for faint signals. We, however, still measure the location of the emission lines with the aid of a microscope whose stage can be moved in two directions. We are able to measure positrons in each coordinate to the high accuracy of one micrometer.

In our work we define the nominal radius of a galaxy as that distance at which the surface brightness of the galaxy has fallen to the threshold of detectability on plates made with the 48-inch Schmidt telescope on Mount Palomar, a value equal to 25th magnitude per square second of arc. For establishing the distance to the objects examined, and hence their actual size, we adopt a value for the Hubble constant (which specifies the expansion rate of the universe) of 50 kilometers per second per megaparsec. (A megaparsec is 3.26 million light-years.)

From the measured velocities of the strongest emission lines we compute a smooth rotation curve by averaging together the approaching and receding velocities from the two sides of the galactic disk. Although each galaxy exhibits distinctive features in its rotational pattern, the systematic trends that emerge are impressive. With increasing luminosity galaxies are bigger, orbital velocities are higher, and the velocity gradient across the nuclear bulge is steeper. Moreover, each type of galaxy displays characteristic rotational properties. For example, the most luminous Sa galaxies rotate more than 50 percent faster at the midpoint of their radius than equally luminous Sc galaxies. Among Sc galaxies the most luminous rotate more than twice as fast at comparable radial distances as Sc galaxies that are only a hundredth as luminous.

One overwhelming conclusion emerges from our observations. Virtually all the rotation curves are either flat or rising out to the visible limits of the galaxy. There are no extensive regions where the velocities fall off with distance from the center, as would be predicted if mass were centrally concentrated. The conclusion is inescapable: mass, unlike luminosity, is not concentrated near the center of spiral galaxies. Thus the light distribution in a galaxy is not at all a guide to mass distribution.

On the basis of their rotational velocities the masses of the galaxies in our study range from 6×10^9 to 2×10^{12} times the mass of the sun inside their optical radius. We cannot yet specify the total mass of any one galaxy because we do not see any "edge" to the mass. Instead the mass inside any given radial distance is increasing linearly with distance and, contrary to what one might expect, is not converging to a limiting mass at the edge of the visible disk. The linear increase of mass with radius indicates that each successive shell of matter in the galaxy must contain just as much mass as every other shell of the same thickness. Since the volume of each successive shell increases as the square of the radius, the density of matter in successive shells must decrease as 1 over the radius squared in order for the product of the density times the volume to remain constant.

The theoretical model that least disturbs generally accepted ideas about galaxies accounts for the observed rotation curves by embedding each spiral galaxy in a spherical "halo" of matter that extends well beyond the visible limits of the galactic disk. The gravitational attraction of this unseen mass keeps the orbital velocities of the galaxies from decreasing with distance from the galactic center. It is perhaps disappointing that the observations yield almost no information on the detailed distribution of the invisible dark matter. One can nonetheless say that the dark matter is not part of the overall background density of matter in the universe but rather is strongly clumped around galaxies. This is evident because the density of nonluminous matter decreases, albeit slowly, with distance from the galactic center, and the density even at large radial distances is between 100 and 1,000 times higher than the mean density of the universe.

Although there are other models that try to account for the high orbital velocities, all are less satisfactory than a single halo of dark matter. If all the required unseen matter is put in a disk, the disk will quickly become unstable and form itself into a bar. The important finding that halos are necessary for stabilizing a disk was first elucidated by Jeremiah P. Ostriker and P. J. E. Peebles of Princeton University.

The observed dynamic effects are reproduced by models of spiral galaxies that put the mass in a nucleus, a surrounding bulge, a disk, and a halo. Particularly interesting models have been developed by John N. Bahcall and Raymond M. Soneira of the Institute for Advanced Study, Maarten Schmidt of Caltech, and S. Casertano of the Scuola Normale Superiore in Pisa. Perhaps the most radical idea for explaining the observed high rotational velocities is one advanced independently by Joel E. Tohline of Louisiana State University and M. Milgrom and J. Bekenstein of the Weizmann Institute of Science. They have proposed that at great distances the Newtonian theory of gravitation must be modified, thereby allowing rotational velocities in galaxies to remain high at such distances from the galactic center even in the absence of unseen mass.

Additional evidence on the high rotational velocities of matter in spiral galaxies is provided by the 21-centimeter radio waves emitted by the neutral (unionized) hydrogen in the galactic disk. Early studies of the 21-centimeter radiation of a few spiral galaxies by Morton S. Roberts of the National Radio Astronomy Observatory showed that the rotational velocities of the hydrogen are high. With multiple radio telescopes, notably the array at Westerbork in the Netherlands and the Very Large Array at Socorro, New Mexico, it is possible to match and even exceed the resolving power of optical telescopes and thereby to study the distribution of hydrogen in galaxies similar to those we have observed. Albert Bosma of the State University of Leiden has shown for a wide variety of galaxy types that the orbital velocities of neutral hydrogen remain high at large distances from the galactic center.

In general the apparent diameters of galaxies are similar whether they are measured by optical observations or by radio ones. For a small set of galaxies, however, hydrogen extends several times as far out from the center as the luminous stars do. For such objects it is possible to determine the gravitational potential beyond the limits of the optically visible galaxy. In several instances the hydrogen does not remain in a plane but is warped sharply near the edge of the visible disk. It is therefore not certain whether the gas velocities that have been measured at the largest distances from the center are true circular orbital velocities or whether they represent more complex motions.

Renzo Sancisi of the University of Groningen, who has studied such warped galaxies, has suggested that the orbital velocities may in fact be decreasing beyond the limits of the visible galaxy. The velocities, however, seem to decrease only slightly, perhaps by 20 kilometers per second or about 10 percent, and then hold constant at that value at larger distances. The radio observations are continuing and should offer important information on the far outer regions of galaxies.

Students of galaxies are fortunate in being able to examine the properties of galaxies a long way off and then return to the galaxy where they live and ask if it exhibits the same features as other galaxies. It was not so long ago that astronomers believed the sun, about eight kiloparsecs from the center of our galaxy, was near the edge of it and that the galaxy was only of moderate size. Now all the evidence indicates that our galaxy too extends well beyond the position of the sun and that its mass continues to increase.

The velocity of the sun in its orbit around the center of the galaxy is placed at 220 kilometers per second by James E. Gunn and Gillian R. Knapp of Princeton and Scott D. Tremaine of the Massachusetts Institute of Technology. Other estimates run as high as 260 kilometers per second. At the lower value the amount of mass between the sun and the center of the galaxy is about 10^{11} solar masses. On the evidence that substantial mass lies beyond the sun's distance from the galactic center, the galactic mass out to 100 kiloparsecs may reach 10^{12} solar masses, which would place our galaxy in a class with the largest galaxies of its type.

Some 30 years ago Jan H. Oort of the Leiden Observatory demonstrated that the observable mass of stars and gas in the galactic disk in the vicinity of the sun is too low by almost a factor of two to account for the disk's gravitational attraction on the stars far out of its central plane. This study offered the first evidence that our galaxy too harbors mass that is not luminous.

More recent evidence comes from the orbital velocities of objects in the plane of the galaxy considerably farther out than the sun. Measurements are difficult, but the velocities have been deduced for a few special cases. For example, Leo Blitz of the University of Maryland at College Park has determined the velocities of clouds of carbon monoxide at a distance of nearly 16 kiloparsecs from the galactic center. These velocities, together with the velocities of hydrogen clouds determined by Blitz and Shrinivas Kulkarni and Carl E. Heiles of the University of California at Berkeley, yield a rotation curve that continues to rise with increasing distance from the galactic center.

In order to deduce the mass at still-larger distances the velocities of globular star cluster in the halo of our galaxy, with one sample of clusters at 30 kiloparsecs from the center and another at 60 kiloparsecs, have been measured by F. D. A. Hartwick of the University of Victoria, Wallace L. W. Sargent of Caltech, Carlos Frenk of the University of Cambridge, and Simon White of the University of California at Berkeley. Their work shows that the mass continues to increase with approximate linearity to the mean distance of the clusters.

With effort and imagination it is possible to sample the gravitational potential at even more remote distances. Our galaxy is not alone in intergalactic space; it has a retinue of smaller satellite galaxies. The orbits of the two closest satellites, the Large and Small Clouds of Magellan, a little less than 60 kiloparsecs from the center of our galaxy, are highly uncertain. Model orbits have been proposed, however, by Tadayuki Murai and Mitsuaki Fujimoto of Nagoya University, D. N. C. Lin of the Lick Observatory, and Donald Lynden-Bell of the University of Cambridge. From the model orbits they deduce values of mass that are consistent with those yielded by the globular clusters.

For still-greater distances Jaan Einasto and his colleagues at the Estonian S.S.R. Academy of Sciences have relied on a combination of enormously distant globular clusters and satellite galaxies to deduce the mass to distances beyond 80 kiloparsecs. When the results from such analyses are combined, they indicate a galaxy in which orbital velocities remain in the range of 220 to 250 kilometers per second out to almost 10 times the distance of the sun from the galactic center. Such a mass distribution is mandatory if our galaxy is to resemble all the other spiral galaxies my colleagues and I have studied. It moves the sun from a relatively rural position to a much more urban one.

The broad conclusion that can be drawn from all these results is that as the disk of a spiral galaxy is scanned from the center outward, the total mass of luminous and dark material falls off slowly and the luminosity (measured in the blue region of the spectrum) falls off rapidly. As a result the ratio of the local mass density to the local (blue) luminosity density, which can be expressed for convenience as the value of the ratio M/L, increases steadily with distance from the galactic center. In the nuclear region a lot of luminosity is produced by relatively little mass, whereas at large distances little luminosity is produced by a lot of mass. If there were no invisible material clumped around galaxies, the mass distribution would simply follow the luminosity distribution and the M/L ratio would be approximately constant across the disk from its center to its edge.

If mass and luminosity are measured in units of solar mass and solar luminosity, the M/L ratio of the sun is of course $1/1$. In such units (omitting the denominator, which is simply 1) the average M/L ratio near the nucleus of a spiral galaxy has sunlike values of 1 or perhaps 2 or 3. Toward the edge of the visible disk, as luminosity decreases, the M/L value climbs to 10 or 20. Beyond the visible disk, where the luminosity falls essentially to zero and the mass remains high, the average M/L value soars into the hundreds.

In an effort to identify the constituents of the invisible halo we must ask what celestial objects have high values of M/L. Stars like the sun are clearly ruled out. The luminous hot young stars that delineate a galaxy's spiral arms are even poorer candidates; their M/L values are about 10^{-4}. At the other extreme the old red dwarf stars that populate the nuclear bulge and the outlying regions of the galaxy have both a low mass and a low blue luminosity. Their M/L values, about 20, are still far short of the values needed for the halo. Moreover, a halo consisting of very low-mass red dwarfs would reveal its presence by radiating strongly in the infrared region of the spectrum. All attempts to detect a halo by its visual, infrared, radio, or x-ray radiation have failed.

What candidates are left? Normal stars radiate energy generated by thermonuclear processes, which convert hydrogen and helium into heavier elements. Such nuclear processes are kindled only in bodies whose mass is large enough for the gravitational energy to raise the temperature at the core of the star to several million degrees Kelvin (degrees C above absolute zero). The minimum mass required is about .085 times the mass of the sun. Jupiter, the largest planet in the solar system, falls short of this value by a factor of nearly 100. A halo of planetlike bodies, perhaps protostars that failed to become stars, is at least conceivable, although rather unlikely. In sum, the only requirement for the halo is the presence of matter in any cold, dark form that meets the M/L constraint, from neutrinos to black holes.

So far I have described the rotational properties of relatively isolated normal spiral galaxies. There is additional observational evidence for high M/L ratios at large distances from the nuclei of other galaxies. Occasionally nature offers an unexpected opportunity to probe its secrets. Recently François Schweizer of the Carnegie Institution, Bradley C. Whitmore of Arizona State University, and I have been fascinated by the faint "anonymous" galaxy AO 136–0801, one of a class of spindle galaxies with polar rings. It is called anonymous because it is not listed in any of the standard galactic catalogs. Its numerical designation corresponds to its location in the sky.

Our observations of the distribution of light across the spindle show that it is a low-luminosity disk of stars viewed nearly edge-on, with little or no gas and dust and no spiral structure. Such galaxies are classed as SO galaxies and represent a significant fraction of all disk galaxies. By our usual methods we have determined the rotational properties of the disk by measuring the Doppler shift of absorption lines from its component stars. A short distance from the center of the object along the major axis of the spindle, rotational velocities reach 145 kilometers per second, a value that corresponds closely to velocities measured in type-Sa galaxies of low luminosity. Along the minor axis the orbital velocities show no line-of-sight component, confirming evidence that we are observing a rotating disk of stars.

The unusual feature of AO 136–0801 is a large ring, also seen nearly edge-on, that encircles the narrow axis of the spindle by passing almost over the disk's center of rotation. The ring is composed of gas, dust, and luminous young stars. The gas reveals itself by its emission-line spectrum, the dust by its absorbing effects where it crosses in front of the spindle, and the stellar component by its knotty, bluish appearance in photographs. The maximum diameter of the ring is several times greater than the long axis of the spindle. As a consequence the motions of the objects in the ring offer a unique opportunity to probe the gravitational field perpendicular to the galactic disk out to distances exceeding the visible radius of the disk.

Our spectrographic observations confirm that the ring is indeed rotating at right angles to the plane of rotation of the disk. It seems improbable that this dynamical configuration could have arisen in the normal evolution of an isolated disk galaxy; the configuration must be the result of some kind of event, such as an encounter with another galaxy or with a disk of gas. By measuring the displacement of emission lines we find that the ring's velocity of rotation is about 170 kilometers per second and that the velocity curve is flat or slightly rising out to a distance of almost three times the radius of the inner disk. If the velocity curves of the disk and the ring are plotted on the same velocity-distance scale, the two are seen to have nearly identical values at the same distance from the center of the galaxy. The high rotational velocity of the ring offers strong evidence for the existence of a massive halo extending at least three times farther than the visible radius of the disk. Moreover, the shape of the halo must be more nearly spherical than disklike. Calculations show that if the halo were as flat as the disk, the velocities above the plane of the disk would be smaller than those in the disk by 20 to 40 percent.

I have been describing determinations of mass made by measuring the velocity of orbiting test objects, objects in the central disk of a galaxy, and objects orbiting the pole of an unusual galaxy. Other special instances can help to shed light on the quantity of dark matter in the universe. Galaxies often exist in pairs. In such instances one galaxy can be considered a test object in orbit around the other. The analysis of such a system is complex because both the orientation of the orbit in space and the position of the galaxy in the orbit are unknown. One can, however, resort to the observed properties in a large sample of double galaxies (the difference between the velocities of the two galaxies, their angular separation, and their luminosity) to infer from statistical arguments the probable distribution of orbital elements and M/L ratios appropriate to the galaxies.

Independent analyses by Edwin L. Turner of Princeton, Steven D. Peterson working at Cornell University, Linda Y. Schweizer of the Carnegie Institution and I. D. Karachentsev of the Special Astrophysical Observatory yield mean M/L values in the range between 25 and 100. These values of M/L are an average over a distance equal to the separation of the galaxies in each pair, a distance generally equal to several galaxy diameters, or on the order of 100 kiloparsecs. This result helps to confirm the view that halos of dark matter, with large values of M/L, extend well beyond the optical limits of galaxies.

We can now return to our original question: Does the universe contain enough invisible matter to raise the average density to 5×10^{-30} gram per cubic centimeter, the value needed to close the universe and bring its expansion to a halt? As we have seen, such a density would be reached if the density of nonluminous matter exceeded the density of luminous matter by a factor of about 70. Alternatively what would be needed to close the universe can be expressed in terms of the ratio of total mass to luminosity. That value is roughly 700, compared with 1 for the sun.

Is there any evidence that the M/L value of 700 is approached? Averaged over the visible disks of spiral galaxies, the ratio of total mass (luminous and nonluminous) to luminosity is about 5. For SO and elliptical galaxies the M/L value is higher, on the order of 10. For double galaxies and small groups of galaxies the M/L value increases to between 50 and 100. Analyses of galaxy motions in large clusters indicate M/L values of several hundreds. This increase in mean M/L value with increasing distance from the center of the system was first stressed a decade ago by Einasto, Ants Kaasik, and Enn Saar of the Estonian S.S.R. Academy of Sciences, and also by Ostriker and Peebles and by Amos Yahil of the State University of New York at Stony Brook. So far there is no evidence for the existence of M/L values above the critical one of 700 needed to close the universe. The highest of the derived values, however, comes tantalizingly close. Some physicists consider it significant that the inferred values seem to be converging on the critical one rather than being orders of magnitude either higher or lower.

Investigations encompassing gigantic distances and vast time scales are made more difficult by this new realization that the distribution of light is an unreliable guide to the distribution of mass in the universe. An unknown fraction of the mass in a spiral galaxy is hidden in a nonluminous constituent, and so is an unknown fraction of the mass in clusters of galaxies. One cannot yet state whether regions of the universe that are devoid of galaxies are simply voids of light or are voids of mass as well. To answer this question astronomers will have to be clever in devising novel observing techniques and physicists will have to determine the properties of exotic forms of matter. Only then will it be possible to establish the nature of the ubiquitous dark matter, to determine the full dimensions and mass of galaxies, and to assay the likely fate of the universe.

—June 1983

A NEW LOOK
AT QUASARS

*Recent observations from the Hubble Space Telescope
may reveal the nature and origin of quasars,
the mysterious powerhouse of the cosmos*

Michael Disney

Quasars are the most luminous objects in the universe. They give off hundreds of times as much radiation as a giant galaxy like our own Milky Way, which is itself as luminous as 10 billion suns. Nevertheless, by astrophysical standards quasars are minute objects, no more than a few light-days in diameter, as compared with the tens of thousands of light-years across a typical galaxy. How in heaven can they generate so much energy in such tiny volumes? What are they, and can they be explained by the ordinary laws of physics? To answer these questions, astronomers are training their most advanced instruments—the Hubble Space Telescope in particular—on these celestial superstars.

The first quasar was discovered in 1962, when Cyril Hazard, a young astronomer at the University of Sydney, began to study a powerful source of radio waves in the

Virgo constellation. Hazard could not pinpoint the source, because the radio telescopes of the time were not precise enough, but he realized that the moon would
occult the unknown object when it passed through Virgo. So he and John Bolton,
the director of a newly built radio telescope in Parkes, Australia, pointed the instrument's giant dish toward the radio source and waited for the moon to block it out.
By timing the disappearance and reappearance of the signal, they would be able to
pinpoint the source of radio emissions and identify it with a visible object in the
sky. Unfortunately, by the time the moon arrived the great dish was tipped so far
over that it was running into its safety stops. Apparently unperturbed by the risk,
Bolton sheared off the stops so that the telescope could follow the occultation
downward until the rim of the dish almost touched the ground.

His daring was to be rewarded. From their measurements Hazard was able to
calculate the first accurate position for such a cosmic radio source and then identify
it with a comparatively bright, starlike object in the night sky. The position of that
object—dubbed 3C273—was sent to Maarten Schmidt, an astronomer at the
Mount Palomar Observatory in California, who had the honor of taking its optical
spectrum. After some initial puzzlement Schmidt realized he was looking at the
spectrum of hydrogen shifted redward by the expansion of the universe. The 16
percent redshift meant that 3C273 was about two billion light-years from the
earth. Given the distance and the observed brightness of the object, Schmidt calculated that it had to be emitting several hundred times more light than any galaxy.
The first quasistellar radio source—or quasar—had been discovered.

Spurred by Hazard's and Schmidt's work, astronomers identified many more
quasars in the following years. Observers discovered that the brightness of many
quasars varied wildly; some grew 10 times as bright in just a matter of days. Because no object can turn itself on and off in less time than it takes for light to travel
across it, the astonishing implication was that these highly luminous objects must be
a mere light-week or so across. Some reputable astronomers refused to believe that
the enormous distances and luminosities implied by the redshifts could be so great.
The controversy spilled over to the popular press, where it attracted a younger
generation of scientists, like myself, into astronomy.

Since then astronomers have cataloged thousands of quasars, some with redshifts
as large as 500 percent. They are not difficult to find because unlike stars, and unlike galaxies composed of stars, they emit radiation of all energies from gamma rays
to radio. Ironically, the radio emissions by which they were first discovered turn
out to be, in energetic terms, the least significant portion of their output. For that
reason some astronomers argue that the name "quasar" should be superseded by
QSO, for quasistellar object.

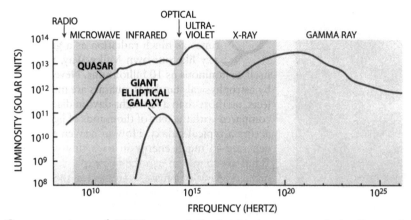

Quasar spectrum of 3C273—one of the brightest quasars and the first to be discovered—is far broader than the spectrum of a typical giant elliptical galaxy. In the optical ranges the quasar is hundreds of times more luminous. Quasars were most numerous when the universe was two to four billion years old. Today quasars are 1,000 times less common. Quasars were also rare in the very early history of the universe, but the exact numbers are uncertain. (Laurie Grace)

There are four big questions facing the quasar astronomer. First, how are quasars related to galaxies and stars? Second, how long does each quasar pour out its enormous energy? In our immediate cosmic neighborhood—within one billion light-years from the earth—there is only one quasar for every million galaxies. But that does not necessarily mean that quasars are much rarer than galaxies; they could be just as common but have much shorter luminous lifetimes. This brings us to the third question: Why were quasars far more numerous in the past? At a redshift of 200 percent—about 10 billion light-years away—the number of quasars jumps 1,000-fold. In the early universe, apparently, quasars were 1,000 times more common than they are today. And last, the most perplexing question: How do quasars generate their prodigious energy?

None of these questions can be easily answered. The typical quasar is so far from the earth that its image on the largest ground-based optical telescope would be 100 million times too small to be resolved. From the outset one school of astronomers felt that quasars had to be sited in galaxies, probably in their nuclei. They gathered evidence to show that all the phenomena observed in quasars were manifested, albeit in a far weaker form, in the nuclei of about 1 percent of the giant galaxies near the Milky Way. A whole zoo of active galactic nuclei were revealed, including radio galaxies, Seyferts, blazars, optically violent variables, superluminal sources, and so on. But astronomers could not tell whether these objects were separate classes of galactic nuclei or representations of the same phenomenon viewed from different angles or at different stages of development. Nor could astronomers explain the exact relation between the active galactic nuclei and quasars. Critics of the theory linking the two types of objects argued that the luminosity of the active nuclei did not even approach that of quasars. And the sheer power of quasars is their most distinctive and mysterious characteristic.

A more direct approach was taken by Jerry Kristian, another astronomer at Mount Palomar, in 1973. He argued that if quasars were inside giant host galaxies, then the images of the closest quasars should show a fuzzy halo of light from the stars in the host galaxy. It would not be an easy observation, because light from the brilliant quasar, scattered by the earth's atmosphere, would swamp the light from the much fainter host. Nevertheless, Kristian was able to demonstrate that the lowest redshift quasars did exhibit this faint, fuzzy halo. His evidence was not very satisfactory, though, because virtually nothing could be discerned about the host galaxies, not even whether they were elliptical or spiral.

■ TROUBLES WITH HUBBLE

When the Hubble Space Telescope was proposed in the mid-1970s, most quasar observers expected it to provide the first clear images of host galaxies, if they really existed. Indeed, finding host galaxies became one of the primary objectives of the telescope. We on the European space telescope team designed the Hubble's Faint Object Camera with quasars very much in mind. For instance, we built in a high-magnification focus and a coronograph specially designed to block off the brilliant light from quasars and thus make the surrounding hosts more visible.

By then astronomers suspected that the only way a quasar could produce so much energy out of such a tiny volume was if the object contained a massive black hole at its core. Such a monster hole, weighing as much as a billion suns, would suck in all the gas and stars in its vicinity. Gas would swirl into the hole at almost the speed of light, generating intense magnetic fields and huge amounts of radiation. Donald Lynden-Bell, then an astronomer at the California Institute of Technology, calculated that a massive black hole could convert up to 40 percent of the infalling matter's rest-mass energy into radiation. Such a process would be 400 times more efficient than the production of thermonuclear energy in stars. For this reason massive black holes became the favored theoretical explanation for quasars. (All the other plausible models would rapidly evolve into black holes anyway.)

One problem with the model, though, was explaining how these monsters could be fed. A black hole of such enormous mass would tend to swallow up all the nearby stars and gas and then go out for lack of fuel. To explore this mystery, the European space telescope team also built a special long-slit spectrograph into the Faint Object Camera. This instrument was designed to measure the rotation speed of material in active galactic nuclei and thus weigh the putative black holes at their cores.

After the much-delayed launch of Hubble in 1990, it was soon discovered that the telescope's main mirror had been incorrectly manufactured. The images were such a travesty that quasar astronomers were devastated. I, for one, felt that 5 to 10 of the most productive years of my astronomical life had been thrown away through unforgivable incompetence. And many others felt likewise. To its credit, however, the National Aeronautics and Space Administration had designed Hubble to be repairable, and astronauts installed new cameras with corrective optics in 1993. Unfortunately, none of the special instruments in the original cameras for observing quasars was recoverable. If we were still going to search for quasar host galaxies, we would have to use the new Wide-Field Planetary Camera, which was not designed for the job. Nevertheless, two teams set out to try: a European team headed by myself and an American team led by astronomer John Bahcall of the Institute for Advanced Study in Princeton, New Jersey.

Observing quasar hosts with the Hubble's new camera was akin to looking into the headlights of an oncoming car in a snowstorm and trying to identify its manufacturer. Astronomers had to take several shots of each object, subtract the high beam—the light from the quasar—and play with the remaining images on their computers. In most cases the final result contained enough detail to make out a galactic structure. Sadly, Jerry Kristian, who pioneered this field, was killed in an ultralight airplane crash in California just before the Hubble results were published.

What did the space telescope reveal? Of the 34 quasars observed, about 75 percent showed the faint, fuzzy halo indicating a host galaxy. The remaining 25 percent showed no such halo, but it is possible that the quasar's dazzling beam is blocking the image in those cases. About half of the host galaxies were elliptical, and half were spiral. The quasars with the strongest radio signals were located primarily in elliptical galaxies, but no other patterns were discernible. Most intriguing, about three quarters of the host galaxies appeared to be colliding with or swallowing other galaxies [see color plate 4].

This finding had already been reported by John Hutchings and his coworkers at Dominion Astrophysical Observatory in Victoria, Canada, who had used a ground-based telescope with adaptive optics to observe quasars. But the Hubble, with its greater resolution, provided much more vivid evidence of the galactic interactions. The images suggest that colliding galaxies supply the fuel for the quasar's energy production. Stars and gas shaken loose by the violence of the impact may be funneling into a massive black hole at the heart of one of the galaxies. The infalling matter then generates the intense radiation.

This process would explain the relative numbers of quasars at different stages in the universe's history. Immediately after the big bang, there were no galaxies and hence no galactic collisions. Even if black holes existed then, there was no mechanism to funnel material toward them and turn them into quasars. Consequently, few quasars are observed at very high redshifts—that is, more than 11 billion years ago. But in the following eons galaxies began to assemble and collide, producing the relatively large number of quasars observed 10 billion light-years from the earth. Finally, the expansion of the universe carried most galaxies away from one another, reducing the number of galactic collisions—and the number of quasars.

Nevertheless, about one quarter of the host galaxies observed by Hubble—such as the spiral galaxy surrounding the quasar PG 0052+251—show no sign that they are colliding with another galaxy. It is possible that a faint companion galaxy is present in these cases, but the quasar's beam is preventing astronomers from seeing it. Or perhaps there is an alternative mechanism that can provide enough fuel to transform a massive black hole into a quasar. What we do know for certain is that the vast majority of galactic interactions do *not* seem to produce quasars; if they did, quasars would be far more common than we observe.

The scarcity of quasars seem to suggest that massive black holes are a rare phenomenon, absent from most galaxies. But this supposition is contradicted by recent evidence gathered by a team of astronomers led by Douglas Richstone of the University of Michigan. Combining observations from Hubble with spectroscopic evidence from ground-based telescopes, the team weighed the nuclei of 27 of the galaxies closest to the Milky Way. In 11 of the galaxies Richstone's group found convincing evidence for the presence of massive dark bodies, most likely black holes.

Furthermore, some of those massive black holes may once have been quasars. In 1994 a group of astronomers led by Holland Ford of Johns Hopkins University used Hubble to look into the heart of M87, a giant elliptical galaxy in the Virgo cluster, about 50 million light-years from the earth. The active nucleus of M87 emits a broad spectrum of radiation, similar to the radiation produced by a quasar but with only a thousandth the intensity. The astronomers discovered that the light from one side of the nucleus was blueshifted (indicating that the source is speeding toward the earth), whereas light from the other side was redshifted (indicating that the source is speeding away). Ford concluded that they were observing a rotating disk of hot gas. What is more, the disk was spinning so rapidly that it could be bound together only by a black hole weighing as much as three billion suns—the same kind of object that is believed to be the quasar's power source. Billions of years ago the nucleuss of M87 may well have been a quasar, too.

■ THE QUASAR QUEST

The recent observations have led astronomers to construct a tentative theory to explain the origin of quasars. According to the theory, most galaxies contain massive black holes capable of generating vast amounts of energy under very special circumstances. The energy production rises dramatically when gas and stars start falling into the black holes at an increased rate, typically about one solar mass a year. This huge infall occurs most often, but not always, as a result of galactic collision or near-misses. Quasars were thus far more prevalent in the epoch of high galaxy density, when the universe was younger and more crowded than it is now.

What can be said of the individual lifetimes of these beasts? Not much for certain. The observed host galaxies show no evidence that the quasars have been radiating long enough to damage them. The hydrogen gas in the host galaxies, for example, has not been substantially ionized, as it might be if quasars were long lived. The observation that so many of the host galaxies are interacting—and the fact that such interactions typically last for one galactic rotation period or less—indicates a quasar lifetime shorter than 100 million years. And if the existence of massive black holes in most galaxies implies a past epoch of quasarlike activity in each case, then the small number of observed quasars—only one for every 1,000 galaxies during their most abundant era—suggests a quasar lifetime of 10 million years or less. If that number is correct, the quasar phenomenon is but a transient phase in the 10-billion-year lifetime of a galaxy. And although the amount of energy generated by each quasar is tremendous, it would account for only about 10 percent of the galaxy's lifetime radiant output.

Obviously more observations are needed to test the theory. The Hubble Space Telescope must be trained on a wider sample of nearby quasars to search for host galaxies. The existing samples of nearby quasars are too small and too narrowly selected for reliable conclusions to be drawn, and the distant host galaxies are too difficult to observe with the current instruments.

Astronomers expect to make new discoveries with the help of two devices recently installed on Hubble: the Near Infrared Camera and Multi-Object Spectrometer (NICMOS), which will allow scientists to peer into the nuclei of galaxies obscured by clouds of dust, and the Space Telescope Imaging Spectrograph (STIS), which has already demonstrated its usefulness by detecting and weighing a black hole in a nearby galaxy in one fortieth the time it would have taken previously.

In 1999 NASA plans to install the Advanced Camera, which will contain a high-resolution coronograph of the kind that was always needed to block the overwhelming quasar light and unmask the host galaxies.

On the theoretical side, we need to understand how and when massive black holes formed in the first place. Did they precede or follow the formation of their host galaxies? And we would like a convincing physical model to explain exactly how such black holes convert infalling matter into all the varieties of quasar radiation, from gamma rays to superluminal radio jets. That may not be easy. Astronomer Carole Mundell of Jodrell Bank Observatory in England once remarked that observing quasars is like observing the exhaust fumes of a car from a great distance and then trying to figure out what is going on under the hood.

—June 1998

GALAXIES IN THE
YOUNG UNIVERSE

*By comparing distant primeval galaxies with older ones
nearby, astronomers hope to determine
how galaxies form and evolve*

F. Duccio Macchetto and Mark Dickinson

Our conception of the universe has changed radically during this past century, as the powers of astronomical observation have steadily improved. Edwin Hubble's pioneering work in the 1920s led to the idea that a "big bang" gave birth to a universe that has been expanding ever since. Later studies revealed that the universe changed in other ways over time as well. Initially it was filled with exceedingly hot, dense, nearly uniform material. Now it is relatively empty. As astronomers look across millions of light-years, the matter they see is collected into a sparse handful of seemingly isolated galaxies. How this transformation occurred, and why the galaxies formed as they did, remains a central question in cosmology today.

229

Considerable effort has gone into examining galaxies nearby—those that are the product of some 10 billion years of evolution. In recent years, however, astronomers have made enormous progress in studying galaxies at cosmological distances—namely, ones that existed when the universe was young. Thanks to the travel time imposed by the finite speed of light, scientists can peer directly into the past by looking out to greater distances. At this time the light cosmologists can see left the most distant galaxies when the universe was less than one fifth its present age. With new instruments and techniques, the hope is to view distant "primeval" galaxies in the process of forming and to trace their evolution to the present day.

Of course such views do not come easily. Young galaxies are so far away that they appear small and faint to even the most powerful telescopes. A galaxy the size of our own Milky Way, observed as it was when the universe was half its current age, would span an angle of mere arc seconds on the sky. Seen through earthbound telescopes, such a galaxy would lose most of its distinguishing structural features; even under excellent conditions cosmologists can often do little more than distinguish remote galaxies from faint stars. In recent years, however, the Hubble Space Telescope—which has extremely sharp vision because it is positioned above the earth's atmosphere—has provided detailed images of the distant universe.

Many Hubble research programs are dedicated to studying young galaxies, but one in particular has taken center stage. In December 1995 Hubble was trained on an unremarkable patch of sky, $\frac{1}{140}$ the apparent size of the full moon, near the Big Dipper. The spot was chosen simply because it afforded a clear view out of our own galaxy and an efficient place to park the telescope. Over the course of 10 full days, Hubble took hundreds of exposures through four filters, covering the spectrum from near-ultraviolet to near-infrared radiation. These images, known as the Hubble Deep Field (HDF) observations, have given us the best view of the distant universe. (In astronomical parlance, "deep" refers to both faintness and distance.)

The HDF images reveal about 3,000 faint galaxies, which take on a bewildering variety of shapes and colors. Many are more than a billion times fainter than what can be seen with the naked eye. The challenge astronomers face is interpreting these two-dimensional pictures of a four-dimensional universe. After all, everything along the line of sight—near and far, young and old—is projected onto the same plane of the sky. The goal is not only to identify primeval galaxies among this mix but to compare their characteristics with those of older galaxies nearby and at all intermediate distances. In doing so, we hope to determine how galaxies form and evolve.

Judging by size or brightness alone, it is not easy to tell how old any one galaxy actually is. Faint objects in our neighborhood and intrinsically bright ones farther away can look very similar. But we do have other ways to determine age. For instance, because the universe is continuously expanding, the distance to a galaxy is proportional to the velocity of its motion away from us. This recession causes a Doppler shift in the light a galaxy emits. As a consequence its characteristic spectral features are displaced toward longer, redder wavelengths. By measuring this so-called redshift, denoted by z, we can determine a galaxy's relative distance and youth.

The timeline of cosmic history is ordered according to redshift: larger values of z represent earlier epochs when the universe was smaller, younger, and more dense. Since any particular redshift z, the universe has expanded by a factor of $(1 + z)$. The relation between z and age is more complex. But generally speaking, at a redshift of z, the universe was at most $\frac{1}{(1 + z)}$ times its present age. Thus, at a redshift of 1, the universe was at most half as old as it is today; at a redshift of 3, it was less than 25 percent as old and perhaps as little as 12.5 percent as old.

Extensive redshift measurements have now determined distances for thousands of faint galaxies out to $z = 1$, the practical limit of current surveys. The Hubble images let us classify the galaxies, compare them with ones nearby, and evaluate their evolutionary state. Many of these galaxies seem to have had a relatively quiescent past: the Hubble pictures reveal bright spiral and elliptical objects, not unlike those nearby, out to redshifts of at least 1. Even at relatively early times these normal-looking galaxies seem to have existed in numbers comparable to the total found in the universe today. We imagine, then, that many galaxies have remained largely the same for billions of years.

Many others, though, have undergone dramatic changes, according to the redshift surveys and the Hubble images. Take the simple exercise of counting galaxies in the sky. There are simply far too many faint ones. At the limits of today's observations, there are at least 10 times as many galaxies as there are in the local universe. The blue colors of these galaxies, and the strong emission features seen in their spectra, suggest that in comparison to galaxies today, they formed stars quite rapidly—an activity that made them brighter and thus more easily seen in the surveys.

Moreover, many of these galaxies have irregular, convoluted morphologies, suggesting that galaxy interactions and perhaps even mergers were common long ago. Although irregular and interacting galaxies can be found closer to home, they seem to have been far more prevalent when the universe was young. At the extreme limits of the HDF the galaxy count is dominated by a vast number of extremely compact objects, barely resolved even by Hubble's sharp eye. Based on these data, astronomers have concluded that the overall star formation rate in the universe has declined dramatically during the latter half of the universe's history and that most action occurred in irregular galaxies.

Cosmologists do not yet understand what physical mechanisms drive this evolution or where these galaxies have gone. Perhaps the rate of galaxy interactions used to be higher only because the universe was smaller and galaxies were closer together. Maybe frequent interactions triggered the star formation researchers now see. Or maybe early galaxies exhausted their gas supplies, stopped forming stars, and faded away to near invisibility. Whatever the case, the findings also tell scientists that the formation of "ordinary" spiral and elliptical galaxies is apparently still out of reach of most redshift surveys; they are found in abundance all the way out to $z = 1$. So to complete this history, astronomers must push the search farther and also probe the nearby universe for remnants of the apparently disappearing "faint blue galaxy" population.

It is not yet possible to ascertain directly distances for the vast majority of galaxies found at the limits of the HDF observations. They do not provide enough light for even the largest telescopes to measure their redshifts. Thus, astronomers employ other techniques to search for galaxies beyond $z = 1$. One method is to rely on distant objects, such as radio sources, as markers or beacons. Certain galaxies generate powerful emissions at radio wavelengths. This radio emission is presumed to originate from an active core, or nucleus, within the galaxy, such as a hidden quasar. Radio galaxies are not common today, but their signature emissions can attract attention from far off in the universe. Indeed, some of the brightest radio sources in the sky are located at vast distances and have redshifts rivaling those of the most remote quasars.

■ RELYING ON RADIO GALAXIES

Powerful radio sources originate most often from elliptical galaxies, which are now generally thought to be quite old. The hope, therefore, has been that the distant galaxies in which radio sources are generated seeded today's elliptical galaxies. On closer inspection, though, distant radio galaxies exhibit highly unusual morphological and spectral traits. New Hubble images display these peculiarities, which include bizarre and complex forms. It seems that a strong radio source can alter a galaxy's appearance and perhaps its evolution as well. Some radio galaxies may well be true primeval galaxies. But at present, because they seem so abnormal, their properties are difficult to interpret. Thus, their pedigree as progenitors of normal modern-day galaxies is suspect.

Fortunately galaxies are gregarious, and where one is found, others are often lurking. Radio galaxies sometimes inhabit clusters, in which they are surrounded by many other faint, more ordinary galaxies. Using powerful new observational tools and techniques, we have sought out and studied these prosaic companions. Such efforts have located rich collections of galaxies around radio sources as distant as $z = 2.3$, when the universe was at most 30 percent of its present age. A few of these distant clusters have been studied in detail using the Hubble, such mighty ground-based telescopes as the Keck 10-meter telescope in Hawaii, and such orbiting x-ray telescopes as ROSAT.

One cluster around the radio source 3C324, at $z = 1.2$, shares many common features with rich clusters nearby. It contains hot gas that shines brightly at x-ray wavelengths. This grouping reveals that some young galaxy clusters were extremely massive—a strong challenge to some theories of cosmic structure formation. Moreover, the cluster around 3C324 contains galaxies remarkably similar to the giant ellipticals that populate closer clusters: they have very red colors and simple, spheroidal forms. Such features indicate that the stars in these distant cluster galaxies were already mature when they emitted the light we observe. Evidently they must have formed much earlier, at some higher redshift, where cosmologists must now extend their search for the stars' birth.

The properties of these galaxies fuel the debate on the age of the universe itself. Recent efforts to determine the rate of cosmic expansion, called the Hubble constant, have suggested that the universe may be younger than was previously thought. Some observations imply that the universe is perhaps less than 10 billion years old, yet astronomers find stars in our own Milky Way they believe to be older than 10 billion years—an impossible contradiction if both the Hubble constant data and the stellar ages are correct! If cosmologists are to believe the elliptical galaxies near 3C324 were already old at a redshift of 1.2, the problems become more pronounced.

■ QUESTING FOR QUASARS

In the search for primeval galaxies some astronomers have turned to quasars, the brightest objects in the universe, as beacons. As the light from a distant quasar travels through space to reach the earth it encounters clouds of gas that imprint characteristic features on the spectrum in the form of absorption lines. Most of these spectral lines are quite weak and are probably produced by tenuous gases unrelated to normal galaxies. Occasionally, however, the lines are broad and deep and totally absorb the quasar radiation at that wavelength. The inferred mass and size of these absorbers suggest they are parts of the disks or halos around galaxies. These strong absorption features can be found easily at redshifts of 3 and beyond, suggesting that there must have been galaxies present in the young universe to produce them.

According to theory, young galaxies should form out of clouds of hydrogen, in which many hot blue stars are constantly being born. As generations of these stars cycle through their short lives, the process of nuclear fusion transforms hydrogen into the heavier elements, which astronomers generally lump together under the name "metals." These early stars then explode as supernovae, ejecting the metals into the surrounding gas clouds. In doing so they also shock and compress the gas clouds, triggering the birth of new generations of stars. Based on this model, the characteristic spectrum for a protogalaxy would be one dominated by blue starlight, with traces of metals. In addition, it might show a strong Lyman-alpha emission line, a feature produced by the plentiful hydrogen that the hot blue stars heat.

Searching for Lyman-alpha emission from young galaxies at high redshift has become a cottage industry, but few good examples have been found. This may be because Lyman-alpha radiation can be easily reabsorbed, especially when dust is present. The mechanisms that produce metals from the hot stars also produce dust, and so young protogalaxies might quench their own Lyman-alpha emission. Without the characteristic Lyman-alpha emission line it can be quite difficult to recognize young galaxies or measure their redshifts. Nevertheless, astronomers have had some success using special filters tuned to detect hydrogen Lyman-alpha emission.

In the field around the distant quasar 0000–263, so named for its coordinates, this technique yielded an exacting find, dubbed with the rather uninspiring name "G2." It was one of the first apparently normal galaxies found at a redshift greater than 3. Spectroscopic measurements subsequently confirmed its distance, and further observations identified several other galaxies in the field at similarly large redshifts; one of these is probably responsible for the Lyman-alpha absorption in the quasar spectrum. Deep Hubble images offered the first clear portrait of galaxies when the universe was somewhere between 10 and 25 percent of its present age. G2 itself appears spheroidal, like a younger blue counterpart to the elliptical found in the 3C324 cluster discussed earlier. Galaxies such as G2 may well be the progenitors of today's elliptical galaxies.

■ A COSMIC BABY BOOM

Most recently scientists have uncovered a treasure trove of galaxies in the early universe, thanks to new search strategies. The most effective method does not capitalize on Lyman-alpha emission but instead takes advantage of a particular color signature that all distant galaxies share. The cause is again hydrogen gas, but the mechanism is different. Hydrogen, which is ubiquitous both within galaxies and in intergalactic space, strongly absorbs all ultraviolet light that is more blue than a certain wavelength. The effect on the starlight a young galaxy emits is dramatic: there is a sharp "cutoff" in its spectrum, producing an unmistakable color signature. Viewed through several filters a distant galaxy should be visible at red or green wavelengths but disappear in the bluest images.

New work shows that these "ultraviolet dropout" galaxies are remarkably common in the deepest astronomical images: they appear everywhere in the sky, and systematic surveys have identified hundreds of them in just the past two years. Their blue colors again suggest that they are forming stars rapidly. Very few of the galaxies, though, exhibit strong Lyman-alpha lines, supporting the notion that these emissions can be easily extinguished by dust. Confirming redshifts without the benefit of Lyman-alpha emission is extremely difficult. Even so, spectra taken with the Keck telescope have now done so for more than 100 galaxies having redshifts between 2 and 3.8, including about 20 (so far) within the HDF.

It is clear that a substantial population of galaxies was already present in the universe when it was only a few billion years old. How these objects relate to galaxies such as the Milky Way is much less certain. Are these the elusive primeval galaxies, the direct ancestors of today's spirals and elliptical? Are they collapsing to form their first generations of stars?

The HDF images show that many of these objects were much smaller than galaxies like our own Milky Way. Many have bright knots and condensations thousands of light-years across—features resembling huge star-forming regions in some nearby galaxies. Many have close companions, suggesting that they are forming from the merging of small galaxies or even subgalactic fragments.

One theory is that such fragments, with sizes around one tenth that of the Milky Way, formed early on and that most of today's galaxies assembled from the merging of these smaller clumps. The mean density of the universe at redshift $z = 3.5$ was 90 times higher than it is now, and the chances for encounters and mergers were correspondingly much larger. Through this process of frequent mergers, and helped by the clumping of matter caused by gravitational attraction, galaxies may have built up gradually until they reached the sizes and masses that are typically found now.

Combining data on nearby galaxies with results from the deep redshift surveys and the ultraviolet dropout techniques, astronomers have sketched the global history of star formation, starting with the first few billion years of the universe's history. The picture that is emerging suggests that the rate of star formation climbed steeply as the universe expanded during the first 20 to 30 percent of its age. It then peaked, perhaps somewhere between redshifts of 2 and 1—an epoch still poorly explored by today's observations. And it has gradually declined again ever since.

Currently, averaging over all galaxies, stars are forming at a rate less than 10 percent of what it was at its peak. The universe has apparently settled into a quiet

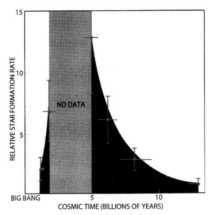

Star formation was swift 12 billion years ago when the universe was young, as shown on this graph based on observations of distant galaxies using the Hubble telescope and earthbound telescopes. About nine billion years ago the star-formation rate—for reasons yet to be discovered—began to decline dramatically. (Nasa P. Madau and J Gitlin STScl)

maturity. This scenario remains incomplete, and new observations will likely revise it, but the fact that it exists demonstrates the astonishing advances made in observational cosmology during the past decade.

■ NEW DIRECTIONS

How will astronomers study the high-redshift universe in the future? One exciting possibility takes advantage of a phenomenon known as gravitational lensing. Large masses, such as dense galaxy clusters, can bend, amplify, and destroy the image of objects beyond them. A faint galaxy behind such a cluster can be magnified and stretched out into a giant arc, revealing morphological details beyond the limits of the best telescopes *[see color plate 5]*.

Gravitational lensing also makes distant galaxies appear much brighter and thus easier to analyze spectroscopically. Using these natural lenses to further boost the resolution of Hubble, astronomers have obtained stunning images of gravitational arcs, some of which have been confirmed to be intrinsically faint galaxies at redshifts comparable to those of the ultraviolet dropout population discussed earlier.

Also, telescopes and instruments now in the making will extend the work described here to larger samples and greater redshifts. A suite of giant telescopes is being built on mountains around the world. In February 1997 U.S. astronauts returned to Hubble and installed two instruments that will enhance the telescope's capabilities and open new windows on the distant universe. And the European Infrared Space Observatory, a satellite probing the longer wavelengths of light where distant, dusty galaxies may emit much of their energy, has recently surveyed the HDF.

The U.S. is preparing its own infrared mission, an advanced telescope called the Space Infrared Telescope Facility. Looking farther ahead, the National Aeronautics and Space Administration has begun considering designs for a Next Generation Space Telescope, a successor to Hubble that could probe the earliest epochs of galaxy history. With these new instruments, and most especially with continued creativity in developing novel techniques for finding and studying objects in the distant universe, astronomers can dig deep enough to characterize the revolution of galaxies and perhaps reveal at least a few of the secrets of their formation.

—May 1997

VI

The Universe

THE EVOLUTION

OF THE UNIVERSE

Some 12 billion years ago the universe emerged from a hot,
dense sea of matter and energy. As the cosmos expanded and
cooled, it spawned galaxies, stars, planets and life

P. James E. Peebles, David N. Schramm,
Edwin L. Turner, and Richard G. Kron

At a particular instant roughly 12 billion years ago all the matter and energy we can observe, concentrated in a region smaller than a dime, began to expand and cool at an incredibly rapid rate. By the time the temperature had dropped to 100 million times that of the sun's core the forces of nature assumed their present properties, and the elementary particles known as quarks roamed freely in a sea of energy. When the universe had expanded an additional 1,000 times, all the matter we can measure filled a region the size of the solar system.

At that time the free quarks became confined in neutrons and protons. After the universe had grown by another factor of 1,000 protons and neutrons combined to form atomic nuclei, including most of the helium and deuterium present today. All of this occurred within the first minute of the expansion. Conditions were still too hot, however, for atomic nuclei to capture electrons. Neutral atoms appeared in abundance only after the expansion had continued for 300,000 years and the universe was 1,000 times smaller than it is now. The neutral atoms then began to coalesce into gas clouds, which later evolved into stars. By the time the universe had expanded to one fifth its present size, the stars had formed groups recognizable as young galaxies.

When the universe was half its present size, nuclear reactions in stars had produced most of the heavy elements from which terrestrial planets were made. Our solar system is relatively young: it formed five billion years ago, when the universe was two thirds its present size. Over time the formation of stars has consumed the supply of gas in galaxies, and hence the population of stars is waning. Fifteen billion years from now stars like our sun will be relatively rare, making the universe a far less hospitable place for observers like us.

Our understanding of the genesis and evolution of the universe is one of the great achievements of 20th-century science. This knowledge comes from decades of innovative experiments and theories. Modern telescopes on the ground and in space detect the light from galaxies billions of light-years away, showing us what the universe looked like when it was young. Particle accelerators probe the basic physics of the high-energy environment of the early universe. Satellites detect the cosmic background radiation left over from the early stages of expansion, providing an image of the universe on the largest scales we can observe.

Our best efforts to explain this wealth of data are embodied in a theory known as the standard cosmological model or the big-bang cosmology. The major claim of the theory is that in the large-scale average, the universe is expanding in a nearly homogenous way from a dense early state. At present there are no fundamental challenges to the big-bang theory, although there are certainly unresolved issues within the theory itself. Astronomers are not sure, for example, how the galaxies were formed, but there is no reason to think the process did not occur within the framework of the big bang. Indeed, the predictions of the theory have survived all tests to date.

Yet the big-bang model goes only so far, and many fundamental mysteries remain. What was the universe like before it was expanding? (No observation we have made allows us to look back beyond the moment at which the expansion began.) What will happen in the distant future, when the last of the stars exhaust the supply of nuclear fuel? No one knows the answers yet.

Our universe may be viewed in many lights—by mystics, theologians, philosophers, or scientists. In science we adopt the plodding route: we accept only what is tested by experiment or observation. Albert Einstein gave us the now well-tested and accepted general theory of relativity, which establishes the relations between mass, energy, space, and time. Einstein showed that a homogenous distribution of matter in space fits nicely with his theory. He assumed without discussion that the universe is static, unchanging in the large-scale average.

In 1922 the Russian theorist Alexander A. Friedmann realized that Einstein's universe is unstable; the slightest perturbation would cause it to expand or contract. At that time Vesto M. Slipher of Lowell Observatory was collecting the first evidence that galaxies are actually moving apart. Then, in 1929, the eminent astronomer Edwin P. Hubble showed that the rate a galaxy is moving away from us is roughly proportional to its distance from us.

The existence of an expanding universe implies that the cosmos has evolved from a dense concentration of matter into the present broadly spread distribution of galaxies. Fred Hoyle, an English cosmologist, was the first to call this process the big bang. Hoyle intended to disparage the theory, but the name was so catchy it gained popularity. It is somewhat misleading, however, to describe the expansion as some type of explosion of matter away from some particular point in space.

That is not the picture at all: in Einstein's universe the concept of space and the distribution of matter are intimately linked; the observed expansion of the system of galaxies reveals the unfolding of space itself. An essential feature of the theory is that the average density in space declines as the universe expands; the distribution of matter forms no observable edge. In an explosion the faster particles move out into empty space, but in the big-bang cosmology particles uniformly fill all space. The expansion of the universe has had little influence on the size of galaxies or even clusters of galaxies that are bound by gravity; space is simply opening up between them. In this sense the expansion is similar to a rising loaf of raisin bread. The dough is analogous to space, and the raisins, to clusters of galaxies. As the dough expands, the raisins move apart. Moreover, the speed with which any two raisins move apart is directly and positively related to the amount of dough separating them.

The evidence for the expansion of the universe has been accumulating for some 60 years. The first important clue is the redshift. A galaxy emits or absorbs some wavelengths of light more strongly than others. If the galaxy is moving away from us, these emission and absorption features are shifted to longer wavelengths—that is, they become redder as the recession velocity increases.

■ HUBBLE'S LAW

Hubble's measurements indicated that the redshift of a distant galaxy is greater than that of one closer to the earth. This relation, now known as Hubble's law, is just what one would expect in a uniformly expanding universe. Hubble's law says the recession velocity of a galaxy is equal to its distance multiplied by a quantity called Hubble's constant. The redshift effect in nearby galaxies is relatively subtle, requiring good instrumentation to detect it. In contrast the redshift of very distant objects—radio galaxies and quasars—is an awesome phenomenon; some appear to be moving away at greater than 90 percent of the speed of light.

Hubble contributed to another crucial part of the picture. He counted the number of visible galaxies in different directions in the sky and found that they appear to be rather uniformly distributed. The value of Hubble's constant seemed to be the same in all directions, a necessary consequence of uniform expansion. Modern surveys confirm the fundamental tenet that the universe is homogenous on large scales. Although maps of the distribution of the nearby galaxies display clumpiness, deeper surveys reveal considerable uniformity.

The Milky Way, for instance, resides in a knot of two dozen galaxies; these in turn are part of a complex of galaxies that protrudes from the so-called local supercluster. The hierarchy of clustering has been traced up to dimensions of about 500 million light-years. The fluctuations in the average density of matter diminish as the scale of the structure being investigated increases. In maps that cover distances that reach close to the observable limit, the average density of matter changes by less than a tenth of a percent.

To test Hubble's law, astronomers need to measure distances to galaxies. One method for gauging distance is to observe the apparent brightness of a galaxy. If one galaxy is four times fainter than an otherwise comparable galaxy, then it can be estimated to be twice as far away. This expectation has now been tested over the whole of the visible range of distances.

Some critics of the theory have pointed out that a galaxy that appears to be smaller and fainter might not actually be more distant. Fortunately there is a direct indication that objects whose redshifts are larger really are more distant. The evidence comes from observations of an effect known as gravitational lensing *[see color plate 5]*. An object as massive and compact as a galaxy can act as a crude lens, producing a distorted, magnified image (or even many images) of any background radiation source that lies behind it. Such an object does so by bending the paths of light rays and other electromagnetic radiation. So if a galaxy sits in the line of sight between the earth and some distant object, it will bend the light rays from the object so that they are observable. During the past decade astronomers have discovered about two dozen gravitational lenses. The object behind the lens is always found to have a higher redshift than the lens itself, confirming the qualitative prediction of Hubble's law.

Hubble's law has great significance not only because it describes the expansion of the universe but also because it can be used to calculate the age of the cosmos. To be precise, the time elapsed since the big bang is a function of the present value of Hubble's constant and its rate of change. Astronomers have determined the approximate rate of the expansion, but no one has yet been able to measure the second value precisely.

Still, one can estimate this quantity from knowledge of the universe's average density. One expects that because gravity exerts a force that opposes expansion, galaxies would tend to move apart more slowly now than they did in the past. The rate of change in expansion is thus related to the gravitational pull of the universe set by its average density. If the density is that of just the visible material in and around galaxies, the age of the universe probably lies between 10 and 15 billion years. (The range allows for the uncertainty in the rate of expansion.)

Yet many researchers believe the density is greater than this minimum value. So-called dark matter would make up the difference. A strongly defended argument holds that the universe is just dense enough that in the remote future the expansion will slow almost to zero. Under this assumption the age of the universe decreases to the range of 7 to 13 billion years.

To improve these estimates, many astronomers are involved in intensive research to measure both the distances to galaxies and the density of the universe. Estimates of the expansion time provide an important test for the big-bang model of the universe. If the theory is correct, everything in the visible universe should be younger than the expansion time computed from Hubble's law.

These two timescales do appear to be in at least rough concordance. For example, the oldest stars in the disk of the Milky Way galaxy are about nine billion years old—an estimate derived from the rate of cooling of white dwarf stars. The stars in the halo of the Milky Way are somewhat older, about 12 billion years—a value derived from the rate of nuclear fuel consumption in the cores of these stars. The ages of the oldest known chemical element are also approximately 12 billion years—a number that comes from radioactive dating techniques. Workers in laboratories have derived these age estimates from atomic and nuclear physics. It is noteworthy that their results agree, at least approximately, with the age that astronomers have derived by measuring cosmic expansion.

Another theory, the steady-state theory, also succeeds in accounting for the expansion and homogeneity of the universe. In 1946 three physicists in England— Hoyle, Hermann Bondi, and Thomas Gold—proposed such a cosmology. In their theory the universe is forever expanding, and matter is created spontaneously to fill the voids. As this material accumulates, they suggested, it forms new stars to

Homogenous distribution of galaxies is apparent in a map that includes objects from 300 million to 1 billion light-years away. The only inhomogeneity, a gap near the centerline, occurs because part of the sky is obscured by the Milky Way. Michael Strauss, of Princeton University, created the map using data from the Infrared Astronomical Satellite. (Johnny Johnson, After Pat McCarthy Carnegie Institutions; NIVR, NASA and SERC)

replace the old. This steady-state hypothesis predicts that ensembles of galaxies close to us should look statistically the same as those far away. The big-bang cosmology makes a different prediction: if galaxies were all formed long ago, distant galaxies should look younger than those nearby because light from them requires a longer time to reach us. Such galaxies should contain more short-lived stars and more gas out of which future generations of stars will form.

■ TESTING THE STEADY-STATE HYPOTHESIS

The test is simple conceptually, but it took decades for astronomers to develop detectors sensitive enough to study distant galaxies in detail. When astronomers examine nearby galaxies that are powerful emitters of radio wavelengths, they see, at optical wavelengths, relatively round systems of stars. Distant radio galaxies, on the other hand, appear to have elongated and sometimes irregular structures. Moreover, in most distant radio galaxies, unlike the ones nearby, the distribution of light tends to be aligned with the pattern of the radio emission.

Likewise, when astronomers study the population of massive, dense clusters of galaxies, they find differences between those that are close and those far away. Distant clusters contain bluish galaxies that show evidence of ongoing star formation. Similar clusters that are nearby contain reddish galaxies in which active star formation ceased long ago. Observations made with the Hubble Space Telescope confirm that at least some of the enhanced star formation in these younger clusters may be the result of collisions between their member galaxies, a process that is much rarer in the present epoch.

So if galaxies are all moving away from one another and are evolving from earlier forms, it seems logical that they were once crowded together in some dense sea of matter and energy. Indeed, in 1927, before much was known about distant galaxies, a Belgian cosmologist and priest, Georges Lemaître, proposed that the expansion of the universe might be traced to an exceedingly dense state he called the primeval "super-atom." It might even be possible, he thought, to detect remnant radiation from the primeval atom. But what would this radiation signature look like?

When the universe was very young and hot, radiation could not travel very far without being absorbed and emitted by some particle. This continuous exchange of energy maintained a state of thermal equilibrium: any particular region was unlikely to be much hotter or cooler than the average. When matter and energy settle to such a state, the result is a so-called thermal spectrum, where the intensity of radiation at each wavelength is a definite function of the temperature. Hence, radiation originating in the hot big bang is recognizable by its spectrum.

In fact this thermal cosmic background radiation has been detected. While working on the development of radar in the 1940s Robert H. Dicke, then at the Massachusetts Institute of Technology, invented the microwave radiometer—a device capable of detecting low levels of radiation. In the 1960s Bell Laboratories used a radiometer in a telescope that would track the early communications satellites Echo-1 and Telstar. The engineer who built this instrument found that it was detecting unexpected radiation. Arno A. Penzias and Robert W. Wilson identified the signal as the cosmic background radiation. It is interesting that Penzias and Wilson were led to this idea by the news that Dicke had suggested that one ought to use a radiometer to search for the cosmic background.

Astronomers have studied this radiation in great detail using the Cosmic Background Explorer (COBE) satellite and a number of rocket-launched, balloon-borne, and ground-based experiments. The cosmic background radiation has two distinctive properties. First, it is nearly the same in all directions. (As the COBE team, led by John Mather of the National Aeronautics and Space Administration Goddard Space Flight Center, showed in 1992, the variation is just 1 part per 100,000.) The interpretation is that the radiation uniformly fills space, as predicted in the big-bang cosmology. Second, the spectrum is very close to that of an object in thermal equilibrium at 2.726 kelvins above absolute zero. To be sure, the cosmic background radiation was produced when the universe was far hotter than 2.726 kelvins, yet researchers anticipated correctly that the apparent temperature of the radiation would be low. In the 1930s Richard C. Tolman of the California Institute of Technology showed that the temperature of the cosmic background would diminish because of the universe's expansion.

The cosmic background radiation provides direct evidence that the universe did expand from a dense, hot state, for this is the condition needed to produce the radiation. In the dense, hot early universe thermonuclear reactions produced elements heavier than hydrogen, including deuterium, helium, and lithium. It is striking that the computed mix of the light elements agrees with the observed abundances. That is, all evidence indicates that the light elements were produced in the hot young universe, whereas the heavier elements appeared later, as products of the thermonuclear reactions that power stars.

The theory for the origin of the light elements emerged from the burst of research that followed the end of World War II. George Gamow and graduate student Ralph A. Alpher of George Washington University and Robert Herman of the Johns Hopkins University Applied Physics Laboratory and others used nuclear physics data from the war effort to predict what kind of nuclear processes might have occurred in the early universe and what elements might have been produced. Alpher and Herman also realized that a remnant of the original expansion would still be detectable in the existing universe.

Despite the fact that significant details of this pioneering work were in error, it forged a link between nuclear physics and cosmology. The workers demonstrated that the early universe could be viewed as a type of thermonuclear reactor. As a result physicists have now precisely calculated the abundances of light elements produced in the big bang and how those quantities have changed because of subsequent events in the interstellar medium and nuclear processes in stars.

■ PUTTING THE PUZZLE TOGETHER

Our grasp of the conditions that prevailed in the early universe does not translate into a full understanding of how galaxies formed. Nevertheless, we do have quite a few pieces of the puzzle. Gravity causes the growth of density fluctuations in the distribution of matter, because it more strongly slows the expansion of denser regions, making them grow still denser. This process is observed in the growth of nearby clusters of galaxies, and the galaxies themselves were probably assembled by the same process on a smaller scale.

The growth of structure in the early universe was prevented by radiation pressure, but that changed when the universe had expanded to about .1 percent of its present size. At that point the temperature was about 3,000 kelvins, cool enough to allow the ions and electrons to combine to form neutral hydrogen and helium. The neutral matter was able to slip through the radiation and to form gas clouds that could collapse into star clusters. Observations show that by the time the universe was one fifth its present size, matter had gathered into gas clouds large enough to be called young galaxies.

A pressing challenge now is to reconcile the apparent uniformity of the early universe with the lumpy distribution of galaxies in the present universe. Astronomers know that the density of the early universe did not vary by much, because they observe only slight irregularities in the cosmic background radiation. So far it has been easy to develop theories that are consistent with the available measurements, but more critical tests are in progress. In particular, different theories for galaxy formation predict quite different fluctuations in the cosmic background radiation on angular scales less than about one degree. Measurements of such tiny fluctuations have not yet been done, but they might be accomplished in the generation of experiments now under way. It will be exciting to learn whether any of the theories of galaxy formation now under consideration survives these tests.

The present-day universe has provided ample opportunity for the development of life as we know it—there are some 100 billion billion stars similar to the sun in the part of the universe we can observe. The big-bang cosmology implies, however, that life is possible only for a bounded span of time: the universe was too hot in the distant past, and it has limited resources for the future. Most galaxies are still producing new stars, but many others have already exhausted their supply of gas. Thirty billion years from now galaxies will be much darker and filled with dead or dying stars, so there will be far fewer planets capable of supporting life as it now exists.

The universe may expand forever, in which case all the galaxies and stars will eventually grow dark and cold. The alternative to this big chill is a big crunch. If the mass of the universe is large enough, gravity will eventually reverse the expansion, and all matter and energy will be reunited. During the next decade, as researchers improve techniques for measuring the mass of the universe, we may learn whether the present expansion is headed toward a big chill or a big crunch.

In the near future we expect new experiments to provide a better understanding of the big bang. New measurements of the expansion rate and the ages of stars are beginning to confirm that the stars are indeed younger than the expanding universe. New telescopes such as the twin 10-meter Keck telescopes in Hawaii and the 2.5-meter Hubble Space Telescope, other new telescopes at the South Pole, and new satellites looking at background radiation as well as new physics experiments searching for "dark matter" may allow us to see how the mass of the universe affects the curvature of space-time, which in turn influences our observations of distant galaxies.

We will also continue to study issues that the big-bang cosmology does not address. We do not know why there was a big bang or what may have existed before.

We do not know whether our universe has siblings—other expanding regions well removed from what we can observe. We do not understand why the fundamental constants of nature have the values they do. Advances in particle physics suggest some interesting ways these questions might be answered; the challenge is to find experimental tests of the ideas.

In following the debate on such matters of cosmology, one should bear in mind that all physical theories are approximations of reality that can fail if pushed too far. Physical science advances by incorporating earlier theories that are experimentally supported into larger, more encompassing frameworks. The big-bang theory is supported by a wealth of evidence: it explains the cosmic background radiation, the abundance of light elements, and the Hubble expansion. Thus, any new cosmology surely will include the big-bang picture. Whatever developments the coming decades may bring, cosmology has moved from a branch of philosophy to a physical science where hypotheses meet the test of observation and experiment.

—*Scientific American Presents,* Spring 1998

PRIMORDIAL DEUTERIUM AND THE BIG BANG

Nuclei of this hydrogen isotope formed in the first moments of the big bang. Their abundance offers clues to the early evolution of the universe and the nature of cosmic dark matter

Craig J. Hogan

The big-bang model of the early universe is extraordinarily simple: it has no structure of any kind on scales larger than individual elementary particles. Even though the behavior it predicts is governed only by general relativity, the Standard Model of elementary particle physics and the energy-distribution rules of basic thermodynamics, it appears to describe the primordial fireball almost perfectly.

Atomic nuclei that formed during the first seconds and minutes of the universe provide additional clues to events in the early universe and to its composition and structure today. The big bang produced a universe made almost entirely of hydrogen and helium. Deuterium, the heavy isotope of hydrogen, was made only at the beginning of the universe; thus, it serves as a particularly important marker. The ratio of deuterium to ordinary hydrogen atoms depends strongly on both the uniformity of matter and the total amount of matter formed in the big bang. During the past few years astronomers have for the first time begun to make reliable, direct measurements of deuterium in ancient gas clouds. Their results promise to provide a precise test of the big-bang cosmogony.

The expansion of the universe appears to have started between 10 and 20 billion years ago. Everything was much closer together and much denser and hotter than it is now. When the universe was only one second old, its temperature was more than 10 billion degrees, 1,000 times hotter than the center of the sun. At that temperature the distinctions between different kinds of matter and energy were not as definite as they are under current conditions: subatomic particles such as neutrons and protons constantly changed back and forth into one another, "cooked" by interactions with plentiful and energetic electrons, positrons, and neutrinos. Neutrons are slightly heavier than protons, however; as things cooled most of the matter settled into the more stable form of protons. As a result, when the temperature fell below 10 billion degrees and the intertransmutation stopped, there were about seven times as many protons as neutrons.

■ OUT OF THE PRIMORDIAL FURNACE

When the universe was a few minutes old (at a temperature of about one billion degrees) the protons and neutrons cooled down enough to stick together into nuclei. Each neutron found a proton partner, creating a pair called a deuteron, and almost all the deuterons in turn stuck together into helium nuclei, which contain two protons and two neutrons. By the time primordial helium had formed, the density of the universe was too low to permit further fusion to form heavier elements in the time available; consequently, almost all the neutrons were incorporated into helium.

Without neutrons to hold them together, protons cannot bind into nuclei because of their electrical repulsion. Because of the limited neutron supply in the primordial fireball, six of every seven protons must therefore remain as isolated hydrogen nuclei. Consequently the big-bang model predicts that about one quarter of the mass of the normal matter of the universe is made of helium and the other three quarters of hydrogen. This simple prediction accords remarkably well with observations. Because hydrogen is the principal fuel of the stars of the universe, its predominance is the basic reason for starlight and sunlight.

During the formation of helium nuclei, perhaps only 1 in 10,000 deuterons remained unpaired. An even smaller fraction fused into nuclei heavier than helium, such as lithium. (All the other familiar elements, such as carbon and oxygen, were produced much later inside stars.) The exact percentages of helium, deuterium, and lithium depend on only one parameter: the ratio of protons and neutrons—particles jointly categorized as baryons—to photons. The value of this ratio, known as η (the Greek letter "eta"), remains essentially constant as the universe expands; because we can measure the number of photons, knowing η tells us how much matter there is. This number is important for understanding the later evolution of the universe, because it can be compared with the actual amount of matter seen in stars and gas in galaxies, as well as the larger amount of unseen dark matter.

For the big bang to make the observed mix of light elements, η must be very small. The universe contains fewer than one baryon per billion photons. The temperature of the cosmic background radiation tells us directly the number of photons left over from the big bang; at present, there are about 411 photons per cubic centimeter of space. Hence, baryons should occur at a density of somewhat less than .4 per cubic meter. Although cosmologists know that η is small, estimates of its exact value currently vary by a factor of almost 10. The most precise and reliable indicators of η are the concentrations of primordial light elements, in particular deuterium. A 5-fold increase in η, for example, would lead to a telltale 13-fold decrease in the amount of deuterium created.

The mere presence of deuterium sets an upper limit on η because the big bang is probably the primary source of deuterium in the universe, and later processing in stars gradually destroys it. One can think of deuterium as a kind of partially spent fuel like charcoal, left over because there was originally not time for all of it to burn completely to ash before the fire cooled. Nucleosynthesis in the big bang lasted only a few minutes, but the nuclear burning in stars lasts for millions or bil-

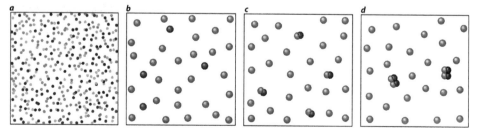

Nucleosynthesis, the formation of atomic nuclei, started instants after the big bang, as the universe cooled, when the fundamental particles, called free quarks (a) condensed into protons and neutrons (b). Protons and neutrons paired off to form deuterons, but because the former outnumber the latter, most of the protons remained alone and became hydrogen nuclei (c). Almost all the deuterons in turn combined to form helium nuclei (d), leaving a tiny remnant to be detected today. (Ian Worpole)

lions of years; as a result any deuterium there is converted to helium or heavier elements. All the deuterium that we find must therefore be a remnant of the big bang—even the 1 molecule in 10,000 of seawater that contains a deuterium atom in place of a hydrogen atom.

■ QUASARS AND GAS CLOUDS

Determining the primordial ratio of deuterium to ordinary hydrogen should be highly informative, but it is not easy, because the universe is not as simple as it used to be. Astronomers can measure deuterium in clouds of atomic hydrogen gas between the stars of our galaxy, but the element's fragility renders the results suspect. We live in a polluted, dissipated, middle-aged galaxy whose gases have undergone a great deal of chemical processing over its 10-billion-year history. Deuterium is very readily destroyed in stars, even in their outer layers and their early prestellar evolution. Stars eject their envelopes when they die, and the gas in our galaxy has been in and out of stars many times. As a result looking at nearby gas clouds can suggest only a lower limit to primordial deuterium abundance.

It would be much better if one could get hold of some truly pristine primordial material that had never undergone chemical evolution. Although we cannot bring such matter into the laboratory, we can look at its composition by its effect on the spectrum of light from distant sources. Bright quasars, the most luminous objects in the universe, are so far away that the light we see now left them when the universe was only one sixth to one quarter of its present size and perhaps only a tenth of its present age. On its way to us the light from these quasars passes through clouds of gas that have not yet condensed into mature galaxies, and the light absorbed by these clouds gives clues to their composition. Some of the clouds that have been detected contain less than one thousandth the proportion of carbon and silicon (both stellar-fusion products) seen in nearby space, a good sign that they retain very nearly the composition they had immediately after the big bang.

There is another advantage to looking so far away. The main component of these clouds, atomic hydrogen, absorbs light at a sharply defined set of ultraviolet wavelengths known as the Lyman series. Each of these absorption lines (so called because of the dark line it leaves in a spectrum) corresponds to the wavelength of a photon exactly energetic enough to excite the electron in a hydrogen atom to a particular energy level. These lines have colors that lie deep in the ultraviolet and cannot usually be seen from the ground because of atmospheric absorption; even the reddest (and most prominent) line, Lyman alpha, appears at a wavelength of 1,215 angstroms. Luckily the expansion of the universe causes a "cosmological redshift" that lengthens the wavelengths of photons that reach the earth to the point where hydrogen absorption lines from sufficiently distant gas clouds reside comfortably within the visible range.

Lyman alpha appears in light from a typical quasar hundreds of times, each time from a different cloud along the line of sight at a different redshift and therefore at a different wavelength. The resulting spectrum is a slice of cosmic history, like a tree-ring sample or a Greenland ice core: these quasar absorption spectra record the history of the conversion of uniform gas from the early big bang into the discrete galaxies we see today over an enormous volume of space. This multiplicity of spectra offers another way to test the primordial character of the absorbing material: the big-bang model predicts that all gas clouds from the early universe should have more or less the same composition. Measuring the abundances of different clouds at vast distances from us and from one another in both time and space will directly test cosmic uniformity.

In some of these clouds we can determine from the quasar spectra both how much ordinary hydrogen there is and how much deuterium. We can separate the signal from deuterium because the added mass in the deuterium nucleus increases the energy required for atomic transitions by about 1 part in 4,000 (twice the ratio of a proton's mass to an electron's mass). As a result the absorption spectrum of deuterium is similar to that of single-nucleon hydrogen, but all the lines show a shift toward the blue end of the spectrum equivalent to that arising from a motion of 82 kilometers per second toward the observer. In spectrographic measurements of a hydrogen cloud, deuterium registers as a faint blueshifted "echo" of the hydrogen.

These spectra also record the velocity and temperature distribution of the atoms. Atoms traveling at different velocities absorb light at slightly different wavelengths because of the Doppler effect, which alters the apparent wavelength of light according to the relative motion of transmitter and receiver. Random thermal motions impel the hydrogen atoms at speeds of about 10 kilometers per second, causing a wavelength shift of 1 part in 30,000; because they are twice as heavy, deuterium atoms at the same temperature move at only about 7 kilometers per second and therefore have a slightly different velocity distribution. A modern spectrograph can resolve these thermal-velocity differences, as well as larger-scale collective flows.

■ WAITING FOR THE LIGHT

Although spectrographs can easily resolve the wavelength differences between ordinary hydrogen and deuterium, splitting the light of a distant quasar into 30,000 colors leaves very little intensity in each color. For more than 20 years these observations proved too difficult. Many of us have spent long nights waiting for photons to drip one by one onto the detectors of the world's largest telescopes, only to find that the weather, instrument problems, and, ultimately, just lack of time had prevented the accumulation of enough light for a convincing result. The technique is now practical only because of improved, more efficient detectors, the 10-meter Keck telescope in Hawaii, and advanced high-resolution, high-throughput spectrographs such as the Keck HIRES.

After many unsuccessful attempts on smaller telescopes, my colleagues Antoinette Songaila and Lennox L. Cowie of the University of Hawaii were allocated their university's first science night on the Keck telescope for this project in November 1993. They trained the telescope on a quasar known as 0014+813, famous among astronomers for its brightness—indeed, it was for some years the brightest single object known in the universe. From earlier studies by Ray J. Weymann of the Observatories of the Carnegie Institution of Washington and Frederic Chaffee, Craig B. Foltz, and Jill Bechtold of the University of Arizona and their collaborators, we knew that a fairly pristine gas cloud lay in front of this quasar.

The first Keck spectrum, obtained in only a few hours, was already of sufficiently high quality to show plausible signs of cosmic deuterium. That spectrum showed the absorption pattern for hydrogen gas moving at various velocities, and it showed an almost perfect echo of the Lyman-alpha line with the characteristic blueshift of deuterium. The amount of absorption in this second signal would be caused by about 2 atoms of deuterium per 10,000 atoms of hydrogen. The result has since been independently confirmed by Robert F. Carswell of the University of Cambridge and his colleagues, using data from the four-meter Mayall telescope at the Kitt Peak National Observatory in Arizona. Subsequent analysis has revealed that the deuterium absorption indeed displays an unusually narrow thermal spread of velocities, as expected.

It is possible that some of the absorption we saw was caused by a chance interposition of a small hydrogen cloud that just happens to be receding from us at 82 fewer kilometers per second than the main cloud we observed. In that case the deuterium abundance would be less than we think. Although the a priori chance of such a coincidence on the first try is small, we ought to regard this estimate as only preliminary. Nevertheless, the effectiveness of the technique is clear. Absorbing clouds in front of many other quasars can be studied with the new technology; we will soon have a statistical sampling of deuterium in primordial material. In fact our group and others have now published measurements and limits for eight different clouds.

One of the most intriguing results is a measurement by David Tytler and Scott Burles of the University of California at San Diego and Xiao-Ming Fan of Columbia

University, who have found a ratio that is apparently almost a factor of 10 lower than our estimate. It remains to be seen whether their result represents the true primordial value. The lower abundance might be a result of deuterium burning in early stars or a sign that the production of deuterium was perhaps not as uniform as the big-bang model predicts.

■ CLUES TO DARK MATTER

If our higher value is correct, the amount of primordial deuterium would fit very well with the standard predictions of the big-bang model for a value of η around two baryons per 10 billion photons. With this value of η the big-bang predictions are also consistent with the amounts of lithium in the oldest stars and estimates of primordial helium seen in nearby metal-poor galaxies. Confirmation of this result would be fabulous news. It would verify that cosmologists understand what happened only one second after the beginning of the expansion of the universe. In addition it would indicate that the history of matter at great distances is like that of nearby matter, as assumed in the simplest possible model of the universe.

This estimate of η fits reasonably well with the number of baryons we actually see in the universe today. The observed density of photons calls for about one atom for every 10 cubic meters of space. This is about the same as the number of atoms counted directly by adding up all the matter in the known gas, stars, planets, and dust, including the quasar absorbers themselves; there is not a huge reservoir of unseen baryons. At the same time, observations suggest that an enormous quantity of dark matter is necessary to explain the gravitational behavior of galaxies and their halos—at least 10 times the mean density of the visible baryons. Thus, our high deuterium abundance indicates that this mass is not made of ordinary atomic matter.

Cosmologists have proposed many candidates for such nonbaryonic forms of dark matter. For example, the big bang predicts that the universe has almost as many neutrinos left over as photons. If each one had even a few billionths as much mass as a proton (equivalent to a few electron volts), neutrinos would contribute to the universe roughly as much mass as all the baryon put together. It is also possible that the early universe created some kind of leftover particle that we have not been able to produce in the laboratory. Either way, the big-bang model, anchored by observation, provides a framework for predicting the astrophysical consequences of such new physical ideas.

—December 1996

Dark Matter in the Universe

As much as 90 percent of the matter in the universe is invisible. Detecting this dark matter will help astronomers better comprehend the universe's destiny

Vera Rubin

Imagine, for a moment, that one night you awaken abruptly from a dream. Coming to consciousness, blinking your eyes against the blackness, you find that, inexplicably, you are standing alone in a vast, pitch-black cavern. Befuddled by this predicament, you wonder: Where am I? What is this space? What are its dimensions?

Groping in the darkness, you stumble upon a book of damp matches. You strike one; it quickly flares, then fizzles out. Again, you try; again, a flash and fizzle. But in that moment you realize that you can glimpse a bit of your surroundings. The next match strike lets you sense faint walls far away. Another flare reveals a strange shadow, suggesting the presence of a big object. Yet another suggests you are moving—or, instead, the room is moving relative to you. With each momentary flare, a bit more is learned.

In some sense this situation recalls our puzzling predicament on the earth. Today, as we have done for centuries, we gaze into the night sky from our planetary platform and wonder where we are in this cavernous cosmos. Flecks of light provide some clues about great objects in space. And what we do discern about their motion and apparent shadows tells us that there is much more that we cannot yet see.

From every photon we collect from the universe's farthest reaches, we struggle to extract information. Astronomy is the study of light that reaches the earth from the heavens. Our task is not only to collect as much light as possible—from ground- and space-based telescopes—but also to use what we can see in the heavens to understand better what we cannot see and yet know must be there.

Based on 50 years of accumulated observations of the motions of galaxies and the expansion of the universe, most astronomers believe that as much as 90 percent of the stuff constituting the universe may be objects or particles that cannot be seen. In other words, most of the universe's matter does not radiate—it provides no glow that we can detect in the electromagnetic spectrum. First posited some 60 years ago by astronomer Fritz Zwicky, this so-called missing matter was believed to reside within clusters of galaxies. Nowadays we prefer to call the missing mass "dark matter," for it is the light, not the matter, that is missing.

Astronomers and physicists offer a variety of explanations for this dark matter. On the one hand, it could merely be ordinary material, such as ultrafaint stars, large or small black holes, cold gas, or dust scattered around the universe—all of which emit or reflect too little radiation for our instruments to detect. It could even be a category of dark objects called MACHOs (Massive Compact Halo Objects) that lurk invisibly in the halos surrounding galaxies and galactic clusters. On the other hand, dark matter could consist of exotic, unfamiliar particles that we have not figured out how to observe. Physicists theorize about the existence of these particles, although experiments have not yet confirmed their presence. A third possibility is that our understanding of gravity needs a major revision—but most physicists do not consider that option seriously.

In some sense our ignorance about dark matter's properties has become inextricably tangled up with other outstanding issues in cosmology—such as how much mass the universe contains, how galaxies formed, and whether or not the universe will expand forever. So important is this dark matter to our understanding of the size, shape, and ultimate fate of the universe that the search for it will very likely dominate astronomy for the next few decades.

■ OBSERVING THE INVISIBLE

Understanding something you cannot see is difficult—but not impossible. Not surprisingly, astronomers currently study dark matter by its effects on the bright matter that we do observe. For instance, when we watch a nearby star wobbling predictably, we infer from calculations that a "dark planet" orbits around it. Applying similar principles to spiral galaxies, we infer dark matter's presence because it accounts for the otherwise inexplicable motions of stars within those galaxies.

When we observe the orbits of stars and clouds of gas as they circle the centers of spiral galaxies, we find that they move too quickly. These unexpectedly high velocities signal the gravitational tug exerted by something more than that galaxy's visible matter. From detailed velocity measurements we conclude that large amounts of invisible matter exert the gravitational force that is holding these stars and gas clouds in high-speed orbits. We deduce that dark matter is spread out around the galaxy, reaching beyond the visible galactic edge and bulging above and below the otherwise flattened, luminous galactic disk. As a rough approximation, try to envision a typical spiral galaxy, such as our Milky Way, as a relatively flat, glowing disk embedded in a spherical halo of invisible material—almost like an extremely diffuse cloud.

Looking at a single galaxy, astronomers see within the galaxy's radius (a distance of about 50,000 light-years) only about one tenth of the total gravitating mass needed to account for how fast individual stars are rotating around the galactic hub.

In trying to discover the amount and distribution of dark matter in a cluster of galaxies, x-ray astronomers have found that galaxies within clusters float immersed in highly diffuse clouds of 100-million-degree gas—gas that is rich in energy yet difficult to detect. Observers have learned to use the x-ray-emitting gas's temperature and extent in much the same way that optical astronomers use the velocities of stars in a single galaxy. In both cases the data provide clues to the nature and location of the unseen matter.

In a cluster of galaxies the extent of the x-ray-emitting region and temperature of the gas enable us to estimate the amount of gravitating mass within the cluster's radius, which measures almost 100 million light-years. In a typical case, when we add together the luminous matter and the x-ray-emitting hot gas, we are able to sense roughly 20 to 30 percent of the cluster's total gravitating mass. The remainder, which is dark matter, remains undetected by present instruments.

Subtler ways to detect invisible matter have recently emerged. One clever method involves spotting rings or arcs around clusters of galaxies. These "Einstein rings" arise from an effect known as gravitational lensing, which occurs when gravity from a massive object bends light passing by. For instance, when a cluster of galaxies blocks our view of another galaxy behind it, the cluster's gravity warps the more distant galaxy's light, creating rings or arcs, depending on the geometry involved. Interestingly, the nearer cluster acts as nature's telescope, bending light into our detectors—light that would otherwise have traveled elsewhere in the universe. Someday we may exploit these natural telescopes to view the universe's most distant objects.

Using computer models, we can calculate the mass of the intervening cluster, estimating the amount of invisible matter that must be present to produce the observed geometric deflection. Such calculations confirm that clusters contain far more mass than the luminous matter suggests.

Even compact dark objects in our own galaxy can gravitationally lens light. When a foreground object eclipses a background star, the light from the background star is distorted into a tiny ring, whose brightness far exceeds the star's usual brightness. Consequently we observe an increase, then a decrease, in the background star's brightness. Careful analysis of the light's variations can tease out the mass of the dark foreground lensing object.

■ WHERE IS DARK MATTER?

Several teams search nightly for nearby lensing events, caused by invisible MACHOs in our own Milky Way's halo. The search for them covers millions of stars in the Magellanic Clouds and the Andromeda galaxy. Ultimately the search will limit the amount of dark matter present in our galaxy's halo.

Given the strong evidence that spiral and elliptical galaxies lie embedded in large dark-matter halos, astronomers now wonder about the location, amount, and distribution of the invisible material.

To answer those questions, researchers compare and contrast observations from specific nearby galaxies. For instance, we learn from the motions of the Magellanic Clouds, two satellite galaxies gloriously visible in the Southern Hemisphere, that they orbit within the Milky Way galaxy's halo and that the halo continues beyond the clouds, spanning a distance of almost 300,000 light-years. In fact motions of our galaxy's most distant satellite objects suggest that its halo may extend twice as far—to 600,000 light-years.

Because our nearest neighboring spiral galaxy, Andromeda, lies a mere two million light-years away, we now realize that our galaxy's halo may indeed span a significant fraction of the distance to Andromeda and its halo. We have also determined that clusters of galaxies lie embedded in even larger systems of dark matter.

At the farthest distances for which we can deduce the masses of galaxies, dark matter appears to dwarf luminous matter by a factor of at least 10, possibly as much as 100 *[see color plate 6]*.

Overall we believe dark matter associates loosely with bright matter, because the two often appear together. Yet, admittedly, this conclusion may stem from biased observations, because bright matter typically enables us to find dark matter.

By meticulously studying the shapes and motions of galaxies over decades, astronomers have realized that individual galaxies are actively evolving, largely because of the mutual gravitational pull of galactic neighbors. Within individual galaxies stars remain enormously far apart relative to their diameters, thus little affecting one another gravitationally. For example, the separation between the sun and its nearest neighbor, Proxima Centauri, is so great that 30 million suns could fit between the two. In contrast galaxies lie close together, relative to their diameters—nearly all have neighbors within a few diameters. So galaxies do alter one another gravitationally, with dark matter's added gravity a major contributor to these interactions.

As we watch many galaxies—some growing, shrinking, transforming, or colliding—we realize that these galactic motions would be inexplicable without taking dark matter into account. Right in our own galactic neighborhood, for instance, such interactions are under way. The Magellanic Clouds, our second nearest neighboring galaxies, pass through our galaxy's plane every billion years. As they do, they mark their paths with tidal tails of gas and, possibly, stars. Indeed, on every passage they lose energy and spiral inward. In less than 10 billion years they will fragment and merge into the Milky Way.

Recently astronomers identified a still nearer neighboring galaxy, the Sagittarius dwarf, which lies on the far side of the Milky Way, close to its outer edge. (Viewed from the earth, this dwarf galaxy appears in the constellation Sagittarius.) As it turns out, gravity from our galaxy is pulling apart this dwarf galaxy, which will cease to exist as a separate entity after several orbits. Our galaxy itself may be made up of dozens of such previous acquisitions.

Similarly the nearby galaxy M31 and the Milky Way are now hurtling toward each other at the brisk clip of 130 kilometers (81 miles) per second. As eager spectators, we must watch its encounter for a few decades to know if M31 will strike our galaxy or merely slide by. If they do collide, we will lose: the Milky Way will merge into the more massive M31. Computer models predict that in about four billion years the galactic pair will become one spheroidal galaxy. Of course by then our sun will have burned out—so others in the universe will have to enjoy the pyrotechnics.

In many ways our galaxy, like all large galaxies, behaves as no gentle neighbor. It gobbles up nearby companions and grinds them into building blocks for its own growth. Just as the Earth's continents slide beneath our feet, so too does our galaxy evolve around us. By studying the spinning, twisting, and turning motions and structures of many galaxies as they hurtle through space, astronomers can figure out the gravitational forces required to sustain their motions—and the amount of invisible matter they must contain.

How much dark matter does the universe contain? The destiny of the universe hinges on one still-unknown parameter: the total mass of the universe. If we live in a high-density, or "closed," universe, then mutual gravitational attraction will ultimately halt the universe's expansion, causing it to contract—culminating in a big crunch, followed perhaps by reexpansion. If, on the other hand, we live in a low-density, or "open," universe, then the universe will expand forever.

Observations thus far suggest that the universe—or at least the region we can observe—is open, forever expanding. When we add up all the luminous matter we can detect, plus all the dark matter that we infer from observations, the total still comes to only a fraction—perhaps 20 percent—of the density needed to stop the universe from expanding forever.

I would be content to end the story there, except that cosmologists often dream of, and model, a universe with "critical" density—meaning one that is finely balanced between high and low density. In such a universe the density is just right. There is enough matter to slow the universe's continuous expansion, so that it eventually coasts nearly to a halt. Yet this model does not describe the universe we actually measure. As an observer, I recognize that more matter may someday be detected, but this does not present sufficient reason for me to adopt a cosmological model that observations do not yet require.

Another complicating factor to take into account is that totally dark systems may exist—that is, there may be agglomerations of dark matter into which luminous matter has never penetrated. At present we simply do not know if such totally dark systems exist, because we have no observational data either to confirm or to deny their presence.

■ WHAT IS DARK MATTER?

Whatever dark matter turns out to be, we know for certain that the universe contains large amounts of it. For every gram of glowing material we can detect, there may be tens of grams of dark matter out there. Currently the astronomical jury is still out as to exactly what constitutes dark matter. In fact one could say we are still at an early stage of exploration. Many candidates exist to account for the invisible mass, some relatively ordinary, others rather exotic.

Nevertheless, there is a framework in which we must work. Nucleosynthesis, which seeks to explain the origin of elements after the big bang, sets a limit to the number of baryons—particles of ordinary, run-of-the-mill matter—that can exist in the universe. This limit arises out of the Standard Model of the early universe, which has one free parameter—the ratio of the number of baryons to the number of photons.

From the temperature of the cosmic microwave background—which has been measured—the number of photons is now known. Therefore, to determine the number of baryons, we must observe stars and galaxies to learn the cosmic abundance of light nuclei, the only elements formed immediately after the big bang.

Without exceeding the limits of nucleosynthesis, we can construct an acceptable model of a low-density, open universe. In that model we take approximately equal amounts of baryons and exotic matter (nonbaryonic particles), but in quantities that add up to only 20 percent of the matter needed to close the universe. This model universe matches all our actual observations. On the other hand, a slightly different model of an open universe in which all matter is baryonic would also satisfy observations. Unfortunately this alternative model contains too many baryons, violating the limits of nucleosynthesis. Thus, any acceptable low-density universe has mysterious properties: most of the universe's baryons would remain invisible, their nature unknown, and in most models much of the universe's matter is exotic.

■ EXOTIC PARTICLES

Theorists have posited a virtual smorgasbord of objects to account for dark matter, although many of them have fallen prey to observational constraints. As leading possible candidates for baryonic dark matter, there are black holes (large and small), brown dwarfs (stars too cold and faint to radiate), sun-size MACHOs, cold gas, dark galaxies, and dark clusters, to name only a few.

The range of particles that could constitute nonbaryonic dark matter is limited only slightly by theorists' imaginations. The particles include photinos, neutrinos, gravitinos, axions, and magnetic monopoles, among many others. Of these, researchers have detected only neutrinos—and whether neutrinos have any mass remains unknown. Experiments are under way to detect other exotic particles. If they exist, and if one has a mass in the correct range, then that particle might pervade the universe and constitute dark matter. But these are very large "ifs."

To a great extent, the details of the evolution of galaxies and clusters depend on properties of dark matter. Without knowing those properties it is difficult to explain how galaxies evolved into the structures observed today. As knowledge of the early universe deepens, I remain optimistic that we will soon known much more about both galaxy formation and dark matter.

What we fail to see with our eyes, or detectors, we can occasionally see with our minds, aided by computer graphics. Computers now play a key role in the search for dark matter. Historically, astronomers have focused on observations; now the field has evolved into an experimental science. Today's astronomical experimenters sit neither at lab benches nor at telescopes but at computer terminals. They scrutinize cosmic simulations in which tens of thousands of points, representing stars, gas, and dark matter, interact gravitationally over a galaxy's lifetime. A cosmologist can tweak a simulation by adjusting the parameters of dark matter and then watch what happens as virtual galaxies evolve in isolation or in a more realistic, crowded universe.

Computer models can thus predict galactic behavior. For instance, when two galaxies suffer a close encounter, violently merging or passing briefly in the night, they sometimes spin off long tidal tails. Yet from the models we now know these tails appear only when the dark matter of each galaxy's halo is 3 to 10 times greater than its luminous matter. Heavier halos produce stubbier tails. This realization through modeling has helped observational astronomers to interpret what they see and to understand more about the dark matter they cannot see. For the first time in the history of cosmology computer simulations actually guide observations.

New tools, no less than new ways of thinking, give us insight into the structure of the heavens. Less than 400 years ago Galileo put a small lens at one end of a cardboard tube and a big brain at the other end. In so doing he learned that the faint stripe across the sky, called the Milky Way, in fact comprised billions of single stars and stellar clusters. Suddenly a human being understood what a galaxy is. Perhaps in the coming century another—as yet unborn—big brain will put her eye to a clever new instrument and definitively answer, What is dark matter?

—*Scientific American Presents,* Spring 1998

THE INFLATIONARY UNIVERSE

A new theory of cosmology suggests that the observable universe is embedded in a much larger region of space that had an extraordinary growth spurt a fraction of a second after the primordial big bang

Alan H. Guth and Paul J. Steinhardt

In the past few years certain flaws in the standard big-bang theory of cosmology have led to the development of a new model of the very early history of the universe. The model, known as the inflationary universe, agrees precisely with the generally accepted description of the observed universe for all times after the first 10^{-30} second. For this first fraction of a second, however, the story is dramatically different. According to the inflationary model, the universe had a brief period of extraordinarily rapid inflation, or expansion, during which its diameter increased by a factor perhaps 10^{50} times larger than had been thought. In the course of this stupendous growth spurt all the matter and energy in the universe could have been created from virtually nothing. The inflationary process also has important implications for the present universe. If the new model is correct, the observed universe is only a very small fraction of the entire universe.

The inflationary model has many features in common with the standard big-bang model. In both models the universe began between 10 and 15 billion years ago as a primeval fireball of extreme density and temperature, and it has been expanding and cooling ever since. This picture has been successful in explaining many aspects of the observed universe, including the redshifting of the light from distant galaxies, the cosmic microwave background radiation, and the primordial abundances of the lightest elements. All these predictions have to do only with events that presumably took place after the first second, when the two models coincide.

Until about five years ago there were few serious attempts to describe the universe during its first second. The temperature in the period is thought to have been higher than 10 billion degrees Kelvin, and little was known about the properties of matter under such conditions. Relying on recent developments in the physics of elementary particles, however, cosmologists are now attempting to understand the history of the universe back to 10^{-45} second after its beginning. (At even earlier times the energy density would have been so great that Einstein's general theory of relativity would have to be replaced by a quantum theory of gravity, which so far does not exist.) When the standard big-bang model is extended to these earlier times, various problems arise. First, it becomes clear that the model requires a number of stringent, unexplained assumptions about the initial conditions of the universe. In addition most of the new theories of elementary particles imply that the standard model would lead to a tremendous overproduction of the exotic particles called magnetic monopoles (each of which corresponds to an isolated north or south magnetic pole).

The inflationary universe was invented to overcome these problems. The equations that describe the period of inflation have a very attractive feature: from almost any initial conditions the universe evolves to precisely the state that had to be assumed as the initial one in the standard model. Moreover, the predicted density of magnetic monopoles becomes small enough to be consistent with observations. In the context of the recent developments in elementary-particle theory the inflationary model seems to be a natural solution to many of the problems of the standard big-bang picture.

The standard big-bang model is based on several assumptions. First, it is assumed that the fundamental laws of physics do not change with time and that the effects of gravitation are correctly described by Einstein's general theory of relativity. It is also assumed that the early universe was filled with an almost perfectly uniform, expanding, intensely hot gas of elementary particles in thermal equilibrium. The gas filled all of space, and the gas and space expanded together at the same rate. When they are averaged over large regions, the densities of matter and energy have remained nearly uniform from place to place as the universe has evolved. It is further assumed that any changes in the state of the matter and the radiation have been so smooth that they have had a negligible effect on the thermodynamic history of the universe. The violation of the last assumption is a key to the inflationary universe model.

The big-bang model leads to three important, experimentally testable predictions. First, the model predicts that as the universe expands, galaxies recede from one another with a velocity proportional to the distance between them. In the 1920s Edwin P. Hubble inferred just such an expansion law from his study of the redshifts of distant galaxies. Second, the big-bang model predicts that there should be a background of microwave radiation bathing the universe as a remnant of the intense heat of its origin. The universe became transparent to this radiation several hundred thousand years after the big bang. Ever since then the matter has been clumping into stars, galaxies, and the like, but the radiation has simply continued to expand and redshift, and in effect to cool. In 1964 Arno A. Penzias and Robert W. Wilson of the Bell Telephone Laboratories discovered a background of microwave radiation received uniformly from all directions with an effective temperature of about three degrees K. Third, the model leads to successful predictions of the formation of light atomic nuclei from protons and neutrons during the first minutes after the big bang. Successful predictions can be obtained in this way for the abundance of helium 4, deuterium, helium 3, and lithium 7. (Heavier nuclei are thought to have been produced much later in the interior of stars.)

Unlike the successes of the big-bang model, all of which pertain to events a second or more after the big bang, the problems all concern times when the universe was much less than a second old. One set of problems has to do with the special conditions the model requires as the universe emerged from the big bang.

The first problem is the difficulty of explaining the large-scale uniformity of the observed universe. The large-scale uniformity is most evident in the microwave background radiation, which is known to be uniform in temperature to about 1 part in 10,000. In the standard model the universe evolves much too quickly to allow this uniformity to be achieved by the usual processes whereby a system approaches thermal equilibrium. The reason is that no information or physical process can propagate faster than a light signal. At any given time there is a maximum distance, known as the horizon distance, that a light signal could have traveled since the beginning of the universe. In the standard model the sources of the microwave background radiation observed from opposite directions in the sky were separated from each other by more than 90 times the horizon distance when the radiation was emitted. Since the regions could not have communicated, it is difficult to see how they could have evolved conditions so nearly identical.

The puzzle of explaining why the universe appears to be uniform over distances that are large compared with the horizon distance is known as the horizon problem. It is not a genuine inconsistency of the standard model; if the uniformity is assumed in the initial conditions, the universe will evolve uniformly. The problem is that one of the most salient features of the observed universe—its large-scale uniformity—cannot be explained by the standard model; it must be assumed as an initial condition.

Even with the assumption of large-scale uniformity, the standard big-bang model requires yet another assumption to explain the nonuniformity observed on smaller scales. To account for the clumping of matter into galaxies, clusters of galaxies, superclusters of clusters, and so on, a spectrum of primordial inhomogeneities must be assumed as part of the initial conditions. The fact that the spectrum of inhomogeneities has no explanation is a drawback in itself, but the problem becomes even more pronounced when the model is extended back to 10^{-45} second after the big bang. The incipient clumps of matter develop rapidly with time as a result of their gravitational self-attraction, and so a model that begins at a very early time must begin with very small inhomogeneities. To begin at 10^{-45} second the matter must start in a peculiar state of extraordinary but not quite perfect uniformity. A normal gas in thermal equilibrium would be far too inhomogenous, owing to the random motion of particles. This peculiarity of the initial state of matter required by the standard model is called the smoothness problem.

Another subtle problem of the standard model concerns the energy density of the universe. According to general relativity, the space of the universe can in principle be curved, and the nature of the curvature depends on the energy density. If the energy density exceeds a certain critical value, which depends on the expansion rate, the universe is said to be closed: space curves back on itself to form a finite volume with no boundary. (A familiar analogy is the surface of a sphere, which is finite in area and has no boundary.) If the energy density is less than the critical density, the universe is open: space curves but does not turn back on itself, and the volume is infinite. If the energy density is just equal to the critical density, the universe is flat: space is described by the familiar Euclidean geometry (again with infinite volume).

The ratio of the energy density of the universe to the critical density is a quantity cosmologists designate by the Greek letter Ω (omega). The value $\Omega = 1$ (corresponding to a flat universe) represents a state of unstable equilibrium. If Ω was ever exactly equal to 1, it would remain exactly equal to 1 forever. If Ω differed slightly from 1 an instant after the big bang, however, the deviation from 1 would grow rapidly with time. Given this instability, it is surprising that Ω is measured today as being between .1 and 2. (Cosmologists are still not sure whether the universe is open, closed, or flat.) In order for Ω to be in this rather narrow range today, its value a second after the big bang had to equal 1 to within one part in 10^{15}. The standard model offers no explanation of why Ω began to close to 1 but merely assumes the fact as an initial condition. This shortcoming of the standard model, called the flatness problem, was first pointed out in 1979 by Robert H. Dicke and P. James E. Peebles of Princeton University.

The successes and drawbacks of the big-bang model we have considered so far involve cosmology, astrophysics, and nuclear physics. As the big-bang model is traced backward in time, however, one reaches an epoch for which these branches of physics are no longer adequate. In this epoch all matter is decomposed into its elementary-particle constituents. In an attempt to understand this epoch cosmologists have made use of recent progress in the theory of elementary particles. Indeed, one of the important developments of the past decade has been the fusing of interests in particle physics, astrophysics, and cosmology. The result for the big-bang model appears to be at least one more success and at least one more failure.

Perhaps the most important development in the theory of elementary particles over the past decade has been the notion of grand unified theories, the prototype of which was proposed in 1974 by Howard M. Georgi and Sheldon Lee Glashow of Harvard University. The theories are difficult to verify experimentally because their most distinctive predictions apply to energies far higher than those that can be reached with particle accelerators. Nevertheless, the theories have some experimental support, and they unify the understanding of elementary-particle interactions so elegantly that many physicists find them extremely attractive.

The basic idea of a grand unified theory is that what were perceived to be three independent forces—the strong, the weak, and the electromagnetic—are actually parts of a single unified force. In the theory a symmetry relates one force to another. Since experimentally the forces are very different in strength and character, the theory is constructed so that the symmetry is spontaneously broken in the present universe.

A spontaneously broken symmetry is one that is present in the underlying theory describing a system but is hidden in the equilibrium state of the system. For example, a liquid described by physical laws that are rotationally symmetric is itself rotationally symmetric: the distribution of molecules looks the same no matter how the liquid is turned. When the liquid freezes into a crystal, however, the atoms arrange themselves along crystallographic axes and the rotational symmetry is broken. One would expect that if the temperature of a system in a broken-symmetry state were raised, it could undergo a kind of phase transition to a state in which the symmetry is restored, just as a crystal can melt into a liquid. Grand unified theories predict such a transition at a critical temperature of roughly 10^{27} degrees.

One novel property of the grand unified theories has to do with the particles called baryons, a class whose most important members are the proton and the neutron. In all physical processes observed up to now the number of baryons minus the number of antibaryons does not change; in the language of particle physics the total baryon number of the system is said to be conserved. A consequence of such a conservation law is that the proton must be absolutely stable; because it is the lightest baryon, it cannot decay into another particle without changing the total baryon number. Experimentally the lifetime of the proton is known to exceed 10^{31} years.

Grand unified theories imply that baryon number is not exactly conserved. At low temperature, in the broken-symmetry phase, the conservation law is an excellent approximation, and the observed limit on the proton lifetime is consistent with at least many versions of grand unified theories. At high temperature, however, processes that change the baryon number of a system of particles are expected to be quite common.

One direct result of combining the big-bang model with grand unified theories is the successful prediction of the asymmetry of matter and antimatter in the universe. It is thought that all the stars, galaxies, and dust observed in the universe are in the form of matter rather than antimatter; their nuclear particles are baryons rather than antibaryons. It follows that the total baryon number of the observed universe is about 10^{78}. Before the advent of grand unified theories, when baryon number was thought to be conserved, this net baryon number had to be postulated as yet another initial condition of the universe. When grand unified theories and the big-bang picture are combined, however, the observed excess of matter over antimatter can be produced naturally by elementary-particle interactions at temperatures just below the critical temperature of the phase transition. Calculations in the grand unified theories depend on too many arbitrary parameters for a quantitative prediction, but the observed matter-antimatter asymmetry can be produced with a reasonable choice of values for the parameters.

A serious problem that results from combining grand unified theories with the big-bang picture is that a large number of defects are generally formed during the transition from the symmetric phase to the broken-symmetry phase. The defects are created when regions of symmetric phase undergo a transition to different broken-symmetry states. In an analogous situation, when a liquid crystallizes, different regions may begin to crystallize with different orientations of the crystallographic axes. The domains of different crystal orientation grow and coalesce, and it is energetically favorable for them to smooth the misalignment along their boundaries. The smoothing is often imperfect, however, and localized defects remain.

In the grand unified theories there are serious cosmological problems associated with pointlike defects, which correspond to magnetic monopoles, and surfacelike defects, called domain walls. Both are expected to be extremely stable and extremely massive. (The monopole can be shown to be about 10^{16} times as heavy as the proton.) A domain of correlated broken-symmetry phase cannot be much larger than the horizon distance at that time, and so the minimum number of defects created during the transition can be estimated. The result is that there would be so many defects after the transition that their mass would dominate the energy density of the universe and thereby speed up its subsequent evolution. The microwave background radiation would reach its present temperature of three degrees Kelvin only 30,000 years after the big bang instead of 10 billion years, and all the successful predictions of the big-bang model would be lost. Thus any successful union of grand unified theories and the big-bang picture must incorporate some mechanism to drastically suppress the production of magnetic monopoles and domain walls.

The inflationary-universe model appears to provide a satisfactory solution to these problems. Before the model can be described, however, we must first explain a few more of the details of symmetry breaking and phase transitions in grand unified theories.

All modern particle theories, including the grand unified theories, are examples of quantum field theories. The best-known field theory is the one that descries electromagnetism. According to the classical (nonquantum) theory of electromagnetism developed by James Clerk Maxwell in the 1860s, electric and magnetic fields have a well-defined value at every point in space, and their variation with time is described by a definite set of equations. Maxwell's theory was modified early in the 20th century in order to achieve consistency with the quantum theory. In the classical theory it is possible to increase the energy of an electromagnetic field by any amount, but in the quantum theory the increases in energy can come only in discrete lumps, the quanta, which in this case are called photons. The photons have both wavelike and particlelike properties, but in the lexicon of modern physics they are usually called particles. In general the formulation of a quantum field theory begins with a classical theory of fields, and it becomes a theory of particles when the rules of the quantum theory are applied.

As we have already mentioned, an essential ingredient of grand unified theories is the phenomenon of spontaneous symmetry breaking. The detailed mechanism of spontaneous symmetry breaking in grand unified theories is simpler in many ways than the analogous mechanism in crystals. In a grand unified theory spontaneous symmetry breaking is accomplished by including in the formulation of the theory a special set of fields known as Higgs fields (after Peter W. Higgs of the University of Edinburgh). The symmetry is unbroken when all the Higgs fields have a value of zero, but it is spontaneously broken whenever at least one of the Higgs fields acquires a nonzero value. Furthermore, it is possible to formulate the theory in such a way that a Higgs field has a nonzero value in the state of lowest energy density, which in this context is known as the true vacuum. At temperatures greater than about 10^{27} degrees thermal fluctuations drive the equilibrium value of the Higgs field to zero, resulting in a transition to the symmetric phase.

We have now assembled enough background information to describe the inflationary model of the universe, beginning with the form in which it was first proposed by one of us (Guth) in 1980. Any cosmological model must begin with some assumptions about the initial conditions, but for the inflationary model the initial conditions can be rather arbitrary. One must assume, however, that the early universe included at least some regions of gas that were hot compared with the critical temperature of the phase transition and that were also expanding. In such a hot region the Higgs field would have a value of zero. As the expansion caused the temperature to fall it would become thermodynamically favorable for the Higgs field to acquire a nonzero value, bringing the system to its broken-symmetry phase.

For some values of the unknown parameters of the grand unified theories this phase transition would occur very slowly compared with the cooling rate. As a result the system could cool to well below 10^{27} degrees with the value of the Higgs field remaining at zero. This phenomenon, known as supercooling, is quite common in condensed-matter physics; water, for example, can be supercooled to more than 20 degrees below its freezing point, and glasses are formed by rapidly supercooling a liquid to a temperature well below its freezing point.

As the region of gas continued to supercool, it would approach a peculiar state of matter known as a false vacuum. This state of matter has never been observed, but it has properties that are unambiguously predicted by quantum field theory. The temperature, and hence the thermal component of the energy density, would rapidly decrease and the energy density of the state would be concentrated entirely in the Higgs field. A zero value for the Higgs field implies a large energy density for the false vacuum. In the classical form of the theory such a state would be absolutely stable, even though it would not be the state of lowest energy density. States with a lower energy density would be separated from the false vacuum by an intervening energy barrier, and there would be no energy available to take the Higgs field over the barrier.

In the quantum version of the model the false vacuum is not absolutely stable. Under the rules of the quantum theory all the fields would be continually fluctuating. As was first described by Sidney R. Coleman of Harvard, a quantum fluctuation would occasionally cause the Higgs field in a small region of space to "tunnel" through the energy barrier, nucleating a "bubble" of the broken-symmetry phase. The bubble would then start to grow at a speed that would rapidly approach the speed of light, converting the false vacuum into the broken-symmetry phase. The rate at which bubbles form depends sensitively on the unknown parameters of the grand unified theory; in the inflationary model it is assumed that the rate would be extremely low.

The most peculiar property of the false vacuum is probably its pressure, which is both large and negative. To understand why, consider again the process by which a bubble of true vacuum would grow into a region of false vacuum. The growth is favored energetically because the true vacuum has a lower energy density than the false vacuum. The growth also indicates, however, that the pressure of the true vacuum must be higher than the pressure of the false vacuum, forcing the bubble wall to grow outward. Because the pressure of the true vacuum is zero, the pressure of the false vacuum must be negative. A more detailed argument shows that the pressure of the false vacuum is equal to the negative value of its energy density (when the two quantities are measured in the same units).

The negative pressure would not result in mechanical forces within the false vacuum, because mechanical forces arise only from differences in pressure. Nevertheless, there would be gravitational effects. Under ordinary circumstances the expansion of the region of gas would be slowed by the mutual gravitational attraction of the matter within it. In Newtonian physics this attraction is proportional to the mass density, which in relativistic theories is equal to the energy density divided by the square of the speed of light. According to general relativity, the pressure also contributes to the attraction; to be specific, the gravitational force is proportional to the energy density plus three times the pressure. For the false vacuum the contribution made by the pressure would overwhelm the energy density contribution and would have the opposite sign. Hence the bizarre notion of negative pressure leads to the even more bizarre effect of a gravitational force that is effectively repulsive. As a result the expansion of the region would be accelerated and the region would grow exponentially, doubling in diameter during each interval of about 10^{-34} second.

This period of accelerated expansion is called the inflationary era, and it is the key element of the inflationary model of the universe. According to the model, the inflationary era continued for 10^{-32} second or longer, and during this period the diameter of the universe increased by a factor of 10^{50} or more. It is assumed that after this colossal expansion the transition to the broken-symmetry phase finally took place. The energy density of the false vacuum was then released, resulting in a tremendous amount of particle production. The region was reheated to a temperature of almost 10^{27} degrees. (In the language of thermodynamics the energy released is called the latent heat; it is analogous to the energy released when water freezes.) From this point on the region would continue to expand and cool at the rate described by the standard big-bang model. A volume the size of the observable universe would lie well within such a region.

The horizon problem is avoided in a straightforward way. In the inflationary model the observed universe evolves from a region that is much smaller in diameter (by a factor of 10^{50} or more) than the corresponding region in the standard model. Before inflation begins the region is much smaller than the horizon distance, and it has time to homogenize and reach thermal equilibrium. This small homogenous region is then inflated to become large enough to encompass the observed universe. Thus the sources of the microwave background radiation arriving today from all directions in the sky were once in close contact; they had time to reach a common temperature before the inflationary era began.

The flatness problem is also evaded in a simple and natural way. The equations describing the evolution of the universe during the inflationary era are different from those for the standard model, and it turns out that the ratio Ω is driven rapidly toward 1, no matter what value it had before inflation. This behavior is most easily understood by recalling that a value of $\Omega = 1$ corresponds to a space that is geometrically flat. The rapid expansion causes the space to become flatter just as the surface of a balloon becomes flatter when it is inflated. The mechanism driving Ω toward 1 is so effective that one is led to an almost rigorous prediction: the value of Ω today should be very accurately equal to 1. Many astronomers (although not all) think a value of 1 is consistent with current observations, but a more reliable determination of Ω would provide a crucial test of the inflationary model.

In the form in which the inflationary model was originally proposed it had a crucial flaw: under the circumstances described the phase transition itself would create inhomogeneities much more extreme than those observed today. As we have already described, the phase transition would take place by the random nucleation of bubbles of the new phase. It can be shown that the bubbles would always remain in finite clusters disconnected from one another, and that each cluster would be dominated by a single largest bubble. Almost all the energy in the cluster would be initially concentrated in the surface of the largest bubble, and there is no apparent mechanism to redistribute energy uniformly. Such a configuration bears no resemblance to the observed universe.

For almost two years after the invention of the inflationary-universe model it remained a tantalizing but clearly imperfect solution to a number of important cosmological problems. Near the end of 1981, however, a new approach was developed by A. D. Linde of the P. N. Lebedev Physical Institute in Moscow and independently by Andreas Albrecht and one of us (Steinhardt) of the University of Pennsylvania. This approach, known as the new inflationary universe, avoids all the problems of the original model while maintaining all its successes.

The key to the new approach is to consider a special form of the energy-density function that describes the Higgs field. Quantum field theories with energy-density functions of this type were first studied by Coleman, working in collaboration with Erick J. Weinberg of Columbia University. In contrast to the more typical case there is no energy barrier separating the false vacuum from the true vacuum; instead the false vacuum lies at the top of a rather flat plateau. In the context of grand unified theories such an energy-density function is achieved by a special choice of parameters. As we shall explain below, this energy-density function leads to a special type of phase transition that is sometimes called a slow-rollover transition.

The scenario begins just as it does in the original inflationary model. Again one must assume the early universe had regions that are hotter than about 10^{27} degrees and were also expanding. In these regions thermal fluctuations would drive the equilibrium value of the Higgs fields to zero and the symmetry would be unbroken. As the temperature fell it would become thermodynamically favorable for the system to undergo a phase transition in which at least one of the Higgs fields acquired a nonzero value, resulting in a broken-symmetry phase. As in the previous case, however, the rate of this phase transition would be extremely low compared with the rate of cooling. The system would supercool to a negligible temperature with the Higgs field remaining at zero, and the resulting state would again be considered a false vacuum.

The important difference in the new approach is the way in which the phase transition would take place. Quantum fluctuations or small residual thermal fluctuations would cause the Higgs field to deviate from zero. In the absence of an energy barrier the value of the Higgs field would begin to increase steadily; the rate of increase would be much like that of a ball rolling down a hill of the same shape as the curve of the energy-density function, under the influence of a frictional drag force. Since the energy-density curve is almost flat near the point where the Higgs field vanishes, the early stage of the evolution would be very slow. As long as the Higgs field remained close to zero, the energy density would be almost the same as it is in the false vacuum. As in the original scenario, the region would undergo accelerated expansion, doubling in diameter every 10^{-34} second or so. Now, however, the expansion, would cease to accelerate when the value of the Higgs field reached the steeper part of the curve. By computing the time required for the Higgs field to evolve, the amount of inflation can be determined. An expansion factor of 10^{50} or more is quite plausible, but the actual factor depends on the details of the particle theory one adopts.

So far the description of the phase transition has been slightly oversimplified. There are actually many different broken-symmetry states, just as there are many possible orientations for the axes of a crystal. There are a number of Higgs fields, and the various broken-symmetry states are distinguished by the combination of Higgs fields that acquire nonzero values. Since the fluctuations that drive the Higgs fields from zero are random, different regions of the primordial universe would be driven toward different broken-symmetry states, each region forming a domain with an initial radius of roughly the horizon distance. At the start of the phase transition the horizon distance would be about 10^{-24} centimeter. Once the domain formed, with the Higgs fields deviating slightly from zero in a definite combination, it would evolve toward one of the stable broken-symmetry states and would inflate by a factor of 10^{50} or more. The size of the domain after inflation would then be greater than 10^{26} centimeters. The entire observable universe, which at that time would be only about 10 centimeters across, would be able to fit deep inside a single domain.

In the course of this enormous inflation any density of particles that might have been present initially would be diluted to virtually zero. The energy content of the region would then consist entirely of the energy stored in the Higgs field. How could this energy be released? Once the Higgs field evolved away from the flat part of the energy-density curve, it would start to oscillate rapidly about the true-vacuum value. Drawing on the relation between particles and fields implied by quantum field theory, this situation can also be described as a state with a high density of Higgs particles. The Higgs particles would be unstable, however: they would rapidly decay to lighter particles, which would interact with one another and possibly undergo subsequent decays. The system would quickly become a hot gas of elementary particles in thermal equilibrium, just as was assumed in the initial conditions for the standard model. The reheating temperature is calculable and is typically a factor of between 2 and 10 below the critical temperature of the phase transition. From this point on the scenario coincides with that of the standard big-bang model, and so all the successes of the standard model are retained.

Note that the crucial flaw of the original inflationary model is deftly avoided. Roughly speaking, the isolated bubbles that were discussed in the original model are replaced here by the domains. The domains of the slow-rollover transition would be surrounded by other domains rather than by false vacuum, and they would tend not to be spherical. The term "bubble" is therefore avoided. The key difference is that in the new inflationary model each domain inflates in the course of its formation, producing a vast essentially homogenous region within which the observable universe can fit.

Since the reheating temperature is near the critical temperature of the grand-unified-theory phase transition, the matter-antimatter asymmetry could be produced by particle interactions just after the phase transition. The production mechanism is the same as the one predicted by grand unified theories for the standard big-bang model. In contrast to the standard model, however, the inflationary model does not allow the possibility of assuming the observed net baryon number of the universe as an initial condition; the subsequent inflation would dilute any initial baryon-number density to an imperceptible level. Thus the viability of the inflationary model depends crucially on the viability of particle theories, such as the grand unified theories, in which baryon number is not conserved.

One can now grasp the solutions to the cosmological problems discussed above. The horizon and flatness problems are resolved by the same mechanisms as in the original inflationary-universe model. In the new inflationary scenario the problem of monopoles and domain walls can also be solved. Such defects would form along the boundaries separating domains, but the domains would have been inflated to such an enormous size that the defects would lie far beyond any observable distance. (A few defects might be generated by thermal effects after the transition, but they are expected to be negligible in number.)

Thus with a few simple ideas the improved inflationary model of the universe leads to a successful resolution of several major problems that plague the standard big-bang picture: the horizon, flatness, magnetic-monopole, and domain-wall problems. Unfortunately the necessary slow-rollover transition requires the fine-tuning of parameters; calculations yield reasonable predictions only if the parameters are assigned values in a narrow range. Most theorists (including both of us) regard such fine-tuning as implausible. The consequences of the scenario are so successful, however, that we are encouraged to go on in the hope we may discover realistic versions of grand unified theories in which such a slow-rollover transition occurs without fine-tuning.

The successes already discussed offer persuasive evidence in favor of the new inflationary model. Moreover, it was recently discovered that the model may also resolve an additional cosmological problem not even considered at the time the model was developed: the smoothness problem. The generation of density inhomogeneities in the new inflationary universe was addressed in the summer of 1982 at the Nuffield Workshop on the Very Early Universe by a number of theorists, including James M. Bardeen of the University of Washington, Stephen W. Hawking of the University of Cambridge, So-Young Pi of Boston University, Michael S. Turner of the University of Chicago, A. A. Starobinsky of the L. D. Landau Institute of Theoretical Physics in Moscow, and the two of us. It was found that the new inflationary model, unlike any previous cosmological model, leads to a definite prediction for the spectrum of inhomogeneities. Basically the process of inflation first smoothes out any primordial inhomogeneities that might have been present in the initial conditions. Then in the course of the phase transition inhomogeneities are generated by the quantum fluctuations of the Higgs field in a way that is completely determined by the underlying physics. The inhomogeneities are created on a very small scale of length, where quantum phenomena are important, and they are then enlarged to an astronomical scale by the process of inflation.

The predicted shape for the spectrum of inhomogeneities is essentially scale-invariant; that is, the magnitude of the inhomogeneities is approximately equal on all length scales of astrophysical significance. This prediction is comparatively insensitive to the details of the underlying grand unified theory. It turns out that a spectrum of precisely this shape was proposed in the early 1970s as a phenomenological model for galaxy formation by Edward R. Harrison of the University of Massachusetts at Amherst and Yakov B. Zel'dovich of the Institute of Physical Problems in Moscow, working independently. The details of galaxy formation are complex and are still not well understood, but many cosmologists think a scale-invariant spectrum of inhomogeneities is precisely what is needed to explain how the present structure of galaxies and galactic clusters evolved.

The new inflationary model also predicts the magnitude of the density inhomogeneities, but the prediction is quite sensitive to the details of the underlying particle theory. Unfortunately the magnitude that results from the simplest grand unified theory is far too large to be consistent with the observed uniformity of the cosmic microwave background. This inconsistency represents a problem, but it is not yet known whether the simplest grand unified theory is the correct one. In particular the simplest grand unified theory predicts a lifetime for the proton that appears to be lower than present experimental limits. On the other hand, one can construct more complicated grand unified theories that result in density inhomogeneities of the desired magnitude. Many investigators imagine that with the development of the correct particle theory the new inflationary model will add the resolution of the smoothness problem to its list of successes.

One promising line of research involves a class of quantum field theories with a new kind of symmetry called supersymmetry. Supersymmetry relates the properties of particles with integer angular momentum to those of particles with half-integer angular momentum; it thereby highly constrains the form of the theory. Many theorists think supersymmetry might be necessary to construct a consistent quantum theory of gravity and to eventually unify gravity with the strong, the weak, and the electromagnetic forces. A tantalizing property of models incorporating supersymmetry is that many of them give slow-rollover phase transitions without any fine-tuning of parameters. The search is on to find a supersymmetry model that is realistic as far as particle physics is concerned and that also gives rise to inflation and to the correct magnitude for the density inhomogeneities.

In short the inflationary model of the universe is an economical theory that accounts for many features of the observable universe lacking an explanation in the standard big-bang model. The beauty of the inflationary model is that the evolution of the universe becomes almost independent of the details of the initial conditions, about which little if anything is known. It follows, however, that if the inflationary model is correct, it will be difficult for anyone to ever discover observable consequences of the conditions existing before the inflationary phase transition. Similarly the vast distance scale created by inflation would make it essentially impossible to observe the structure of the universe as a whole. Nevertheless, one can still discuss these issues, and a number of remarkable scenarios seem possible.

The simplest possibility for the very early universe is that it actually began with a big bang, expanded rather uniformly until it cooled to the critical temperature of the phase transition, and then proceeded according to the inflationary scenario. Extrapolating the big-bang model back to zero time brings the universe to a cosmological singularity, a condition of infinite temperature and density in which the known laws of physics do not apply. The instant of creation remains unexplained. A second possibility is that the universe began (again without explanation) in a random, chaotic state. The matter and temperature distributions would be nonuniform, with some parts expanding and other parts contracting. In this scenario certain small regions that were hot and expanding would undergo inflation, evolving into huge regions easily capable of encompassing the observable universe. Outside these regions there would remain chaos, gradually creeping into the regions that had inflated.

Recently there has been some serious speculation that the actual creation of the universe is describable by physical laws. In this view the universe would originate as a quantum fluctuation, starting from absolutely nothing. The idea was first proposed by Edward P. Tryon of Hunter College of the City University of New York in 1973 and it was put forward again in the context of the inflationary model by Alexander Vilenkin of Tufts University in 1982. In this context "nothing" might refer to empty space, but Vilenkin uses it to describe a state devoid of space, time, and matter. Quantum fluctuations of the structure of space-time can be discussed only in the context of quantum gravity, and so these ideas must be considered highly speculative until a working theory of quantum gravity is formulated. Nevertheless, it is fascinating to contemplate that physical laws may determine not only the deevolution of a given state of the universe but also the initial conditions of the observable universe.

As for the structure of the universe as a whole, the inflationary model allows for several possibilities. (In all cases the observable universe is a very small fraction of the universe as a whole; the edge of our domain is likely to be 10^{35} or more light-years away.) The first possibility is that the domains meet one another and fill all space. The domains are then separated by domain walls, and in the interior of each wall is the symmetric phase of the grand unified theory. Protons or neutrons passing through such a wall would decay instantly. Domain walls would tend to straighten with time. After 10^{35} years or more smaller domains (possibly even our own) would disappear and larger domains would grow.

Alternatively some versions of grand unified theories do not allow for the formation of sharp domain walls. In these theories it is possible for different broken-symmetry states in two neighboring domains to merge smoothly into each other. At the interface of two domains one would find discontinuities in the density and velocity of matter, and one would also find an occasional magnetic monopole.

A quite different possibility would result if the energy density of the Higgs fields were described by a curve. As in the other two cases, regions of space would super-cool into the false-vacuum state and undergo accelerated expansion. As in the original inflationary model, the false-vacuum state would decay by the mechanism of random bubble formation: quantum fluctuations would cause at least one of the Higgs fields in a small region of space to tunnel through the energy barrier, to the value marked A in the illustration. In contrast to the original inflationary scenario, the Higgs field would then evolve very slowly (because of the flatness of the curve) to its true-vacuum value. The accelerated expansion would continue, and the single bubble would become large enough to encompass the observed universe. If the rate of bubble formation were low, bubble collisions would be rare. The fraction of space filled with bubbles would become closer to 1 as the system evolved, but space would be expanding so fast that the volume remaining in the false-vacuum state would increase with time. Bubble universes would continue to form forever, and there would be no way of knowing how much time had elapsed before our bubble was formed. This picture is much like the old steady-state cosmological model on the very large scale, and yet the interior of each bubble would evolve according to the big-bang model, improved by inflation.

From a historical point of view probably the most revolutionary aspect of the inflationary model is the notion that all the matter and energy in the observable universe may have emerged from almost nothing. This claim stands in marked contrast to centuries of scientific tradition in which it was believed that something cannot come from nothing. The tradition, dating back at least as far as the Greek philosopher Parmenides in the fifth century B.C., has manifested itself in modern times in the formulation of a number of conservation laws, which state that certain physical quantities cannot be changed by any physical process. A decade or so ago the list of quantities thought to be conserved included energy, linear momentum, angular momentum, electric charge, and baryon number.

Since the observed universe apparently has a huge baryon number and a huge energy, the idea of creation from nothing has seemed totally untenable to all but a few theorists. (The other conservation laws mentioned above present no such problems: the total electric charge and the angular momentum of the observed universe have values consistent with zero, whereas the total linear momentum depends on the velocity of the observer and so cannot be defined in absolute terms.) With the advent of grand unified theories, however, it now appears quite plausible that baryon number is not conserved. Hence only the conservation of energy needs further consideration.

The total energy of any system can be divided into a gravitational part and a nongravitational part. The gravitational part (that is, the energy of the gravitational field itself) is negligible under laboratory conditions, but cosmologically it can be quite important. The nongravitational part is not by itself conserved; in the standard big-bang model it decreases drastically as the early universe expands, and the rate of energy loss is proportional to the pressure of the hot gas. During the era of inflation, on the other hand, the region of interest is filled with a false vacuum that has a large negative pressure. In this case the nongravitational energy increases drastically. Essentially all the nongravitational energy of the universe is created as the false vacuum undergoes its accelerated expansion. This energy is released when the phase transition takes place, and it eventually evolves to become stars, planets, human beings, and so forth. Accordingly the inflationary model offers what is apparently the first plausible scientific explanation for the creation of essentially all the matter and energy in the observable universe.

Under these circumstances the gravitational part of the energy is somewhat ill defined, but crudely speaking one can say that the gravitational energy is negative, and that it precisely cancels the nongravitational energy. The total energy is then zero and is consistent with the evolution of the universe from nothing.

If grand unified theories are correct in their prediction that baryon number is not conserved, there is no known conservation law that prevents the observed universe from evolving out of nothing. The inflationary model of the universe provides a possible mechanism by which the observed universe could have evolved from an infinitesimal region. It is then tempting to go one step farther and speculate that the entire universe evolved from literally nothing.

—May 1984

THE SELF-REPRODUCING

INFLATIONARY UNIVERSE

Recent versions of the inflationary scenario describe
the universe as a self-generating fractal that
sprouts other inflationary universes

Andrei Linde

If my colleagues and I are right, we may soon be saying good-bye to the idea that our universe was a single fireball created in the big bang. We are exploring a new theory based on a 15-year-old notion that the universe went through a stage of inflation. During that time, the theory holds, the cosmos became exponentially large within an infinitesimal fraction of a second. At the end of this period the universe continued its evolution according to the big-bang model. As workers refined this inflationary scenario, they uncovered some surprising consequences. One of them constitutes a fundamental change in how the cosmos is seen. Recent versions of inflationary theory assert that instead of being an expanding ball of fire the universe is a huge, growing fractal. It consists of many inflating balls that produce new balls, which in turn produce more balls, ad infinitum.

Cosmologists did not arbitrarily invent this rather peculiar vision of the universe. Several workers, first in Russia and later in the U.S., proposed the inflationary hypothesis that is the basis of its foundation. We did so to solve some of the complications left by the old big-bang idea. In its standard form the big-bang theory maintains that the universe was born about 15 billion years ago from a cosmological singularity—a state in which the temperature and density are infinitely high. Of course one cannot really speak in physical terms about these quantities as being infinite. One usually assumes that the current laws of physics did not apply then. They took hold only after the density of the universe dropped below the so-called Planck density, which equals about 10^{94} grams per cubic centimeter.

As the universe expanded, it gradually cooled. Remnants of the primordial cosmic fire still surround us in the form of the microwave background radiation. This radiation indicates that the temperature of the universe has dropped to 2.7 kelvins. The 1965 discovery of this background radiation by Arno A. Penzias and Robert W. Wilson of Bell Laboratories proved to be the crucial evidence in establishing the big-bang theory as the preeminent theory of cosmology. The big-bang theory also explained the abundances of hydrogen, helium, and other elements in the universe.

As investigators developed the theory, they uncovered complicated problems. For example, the standard big-bang theory, coupled with the modern theory of elementary particles, predicts the existence of many superheavy particles carrying magnetic charge—that is, objects that have only one magnetic pole. These magnetic monopoles would have a typical mass 10^{16} times that of the proton, or about .00001 milligram. According to the standard big-bang theory, monopoles should have emerged very early in the evolution of the universe and should now be as abundant as protons. In that case the mean density of matter in the universe would be about 15 orders of magnitude greater than its present value, which is about 10^{-29} gram per cubic centimeter.

■ QUESTIONING STANDARD THEORY

This and other puzzles forced physicists to look more attentively at the basic assumptions underlying the standard cosmological theory. And we found many to be highly suspicious. I will review six of the most difficult. The first, and main, problem is the very existence of the big bang. One may wonder: What came before? If space-time did not exist then, how could everything appear from nothing? What arose first: the universe or the laws determining its evolution? Explaining this initial singularity—where and when it all began—still remains the most intractable problem of modern cosmology.

A second trouble spot is the flatness of space. General relativity suggests that space may be very curved, with a typical radius on the order of the Planck length, or 10^{-33} centimeter. We see, however, that our universe is just about flat on a scale of 10^{28} centimeters, the radius of the observable part of the universe. This result of our observation differs from theoretical expectations by more than 60 orders of magnitude.

A similar discrepancy between theory and observations concerns the size of the universe, a third problem. Cosmological examinations show that our part of the universe contains at least 10^{88} elementary particles. But why is the universe so big? If one takes a universe of a typical initial size given by the Planck length and a typical initial density equal to the Planck density, then, using the standard big-bang theory, one can calculate how many elementary particles such a universe might encompass. The answer is rather unexpected: the entire universe should only be large enough to accommodate just 1 elementary particle—or at most 10 of them. It would be unable to house even a single reader of *Scientific American,* who consists of about 10^{29} elementary particles. Obviously, something is wrong with this theory.

The fourth problem deals with the timing of the expansion. In its standard form the big-bang theory assumes that all parts of the universe began expanding simultaneously. But how could all the different parts of the universe synchronize the beginning of their expansion? Who gave the command?

Fifth, there is the question about the distribution of matter in the universe. On the very large scale, matter has spread out with remarkable uniformity. Across more than 10 billion light-years, its distribution departs from perfect homogeneity by less than 1 part in 10,000. For a long time nobody had any idea why the universe was so homogenous. But those who do not have ideas sometimes have principles. One of the cornerstones of the standard cosmology was the "cosmological principle," which asserts that the universe must be homogenous. This assumption, however, does not help much, because the universe incorporates important deviations from homogeneity, namely, stars, galaxies, and other agglomerations of matter. Hence, we must explain why the universe is so uniform on large scales and at the same time suggest some mechanism that produces galaxies.

Finally, there is what I call the uniqueness problem. Albert Einstein captured its essence when he said, "What really interests me is whether God had any choice in the creation of the world." Indeed, slight changes in the physical constants of nature could have made the universe unfold in a completely different manner. For example, many popular theories of elementary particles assume that space-time originally had considerably more than four dimensions (three spatial and one temporal).

In order to square theoretical calculations with the physical world in which we live, these models state that the extra dimensions have been "compactified," or shrunk to a small size and tucked away. But one may wonder why compactification stopped with four dimensions, not two or five.

Moreover, the manner in which the other dimensions become rolled up is significant, for it determines the values of the constants of nature and the masses of particles. In some theories compactification can occur in billions of different ways. A few years ago it would have seemed rather meaningless to ask why space-time has four dimensions, why the gravitational constant is so small, or why the proton is almost 2,000 times heavier than the electron. Now developments in elementary-particle physics make answering these questions crucial to understanding the construction of our world.

All these problems (and others I have not mentioned) are extremely perplexing. That is why it is encouraging that many of these puzzles can be resolved in the context of the theory of the self-reproducing, inflationary universe.

The basic features of the inflationary scenario are rooted in the physics of elementary particles. So I would like to take you on a brief excursion into this realm— in particular, to the unified theory of weak and electromagnetic interactions. Both these forces exert themselves through particles. Photons mediate the electromagnetic force; the W and Z particles are responsible for the weak force. But whereas photons are massless, the W and Z particles are extremely heavy. To unify the weak and electromagnetic interactions despite the obvious differences between photons and the W and Z particles, physicists introduced what are called scalar fields.

Although scalar fields are not the stuff of everyday life, a familiar analogue exists. That is the electrostatic potential—the voltage in a circuit is an example. Electrical fields appear only if this potential is uneven, as it is between the poles of a battery or if the potential changes in time. If the entire universe had the same electrostatic potential—say, 110 volts—then nobody would notice it; the potential would seem to be just another vacuum state. Similarly a constant scalar field looks like a vacuum: we do not see it even if we are surrounded by it.

These scalar fields fill the universe and mark their presence by affecting properties of elementary particles. If a scalar field interacts with the W and Z particles, they become heavy. Particles that do not interact with the scalar field, such as photons, remain light.

To describe elementary-particle physics, therefore, physicists begin with a theory in which all particles initially are light and in which no fundamental difference between weak and electromagnetic interactions exists. This difference arises only later, when the universe expands and becomes filled by various scalar fields. The process by which the fundamental forces separate is called symmetry breaking. The particular value of the scalar field that appears in the universe is determined by the position of the minimum of its potential energy.

■ SCALAR FIELDS

Scalar fields play a crucial role in cosmology as well as in particle physics. They provide the mechanism that generates the rapid inflation of the universe. Indeed, according to general relativity, the universe expands at a rate (approximately) proportional to the square root of its density. If the universe were filled by ordinary matter, then the density would rapidly decrease as the universe expanded. Thus, the expansion of the universe would rapidly slow down as density decreased. But because of the equivalence of mass and energy established by Einstein, the potential energy of the scalar field also contributes to the expansion. In certain cases this energy decreases much more slowly than does the density of ordinary matter.

The persistence of this energy may lead to a stage of extremely rapid expansion, or inflation, of the universe. This possibility emerges even if one considers the very simplest version of the theory of a scalar field. In this version the potential energy reaches a minimum at the point where the scalar field vanishes. In this case, the larger the scalar field, the greater the potential energy. According to Einstein's theory of gravity, the energy of the scalar field must have caused the universe to expand very rapidly. The expansion slowed down when the scalar field reached the minimum of its potential energy.

One way to imagine the situation is to picture a ball rolling down the side of a large bowl. The bottom of the bowl represents the energy minimum. The position of the ball corresponds to the value of the scalar field. Of course the equations describing the motion of the scalar field in an expanding universe are somewhat more complicated than the equations for the ball in an empty bowl. They contain an extra term corresponding to friction, or viscosity. This friction is akin to having molasses in the bowl. The viscosity of this liquid depends on the energy of the field: the higher the ball in the bowl is, the thicker the liquid will be. Therefore, if the field initially was very large, the energy dropped extremely slowly.

The sluggishness of the energy drop in the scalar field has a crucial implication in the expansion rate. The decline was so gradual that the potential energy of the scalar field remained almost constant as the universe expanded. This behavior contrasts sharply with that of ordinary matter, whose density rapidly decreases in an expanding universe. Thanks to the large energy of the scalar field, the universe continued to expand at a speed much greater than that predicted by preinflation cosmological theories. The size of the universe in this regime grew exponentially.

This stage of self-sustained, exponentially rapid inflation did not last long. Its duration could have been as short as 10^{-35} second. Once the energy of the field declined, the viscosity nearly disappeared and inflation ended. Like the ball as it reaches the bottom of the bowl, the scalar field began to oscillate near the minimum of its potential energy. As the scalar field oscillated it lost energy, giving it up in the form of elementary particles. These particles interacted with one another and eventually settled down to some equilibrium temperature. From this time on the standard big-bang theory can describe the evolution of the universe.

The main difference between inflationary theory and the old cosmology becomes clear when one calculates the size of the universe at the end of inflation. Even if the universe at the beginning of inflation was as small as 10^{-33} centimeter, after 10^{-35} second of inflation this domain acquires an unbelievable size. According to some inflationary models, this size in centimeters can equal $10^{10^{12}}$—that is, a 1 followed by a trillion zeros. These numbers depend on the models used, but in most versions this size is many orders of magnitude greater than the size of the observable universe, or 10^{28} centimeters.

This tremendous spurt immediately solves most of the problems of the old cosmological theory. Our universe appears smooth and uniform because all inhomogeneities were stretched $10^{10^{12}}$ times. The density of primordial monopoles and other undesirable "defects" becomes exponentially diluted. (Recently we have found that monopoles may inflate themselves and thus effectively push themselves out of the observable universe.) The universe has become so large that we can now see just a tiny fraction of it. That is why, just like a small area on a surface of a huge inflated balloon, our part looks flat. That is why we do not need to insist that all parts of the universe began expanding simultaneously. One domain of a smallest possible size of 10^{-33} centimeter is more than enough to produce everything we see now.

■ AN INFLATIONARY UNIVERSE

Inflationary theory did not always look so conceptually simple. Attempts to obtain the stage of exponential expansion of the universe have a long history. Unfortunately, because of political barriers, this history is only partially known to American readers.

The first realistic version of the inflationary theory came in 1979 from Alexei A. Starobinsky of the L. D. Landau Institute of Theoretical Physics in Moscow. The Starobinsky model created a sensation among Russian astrophysicists, and for two years it remained the main topic of discussion at all conferences on cosmology in the Soviet Union. His model, however, was rather complicated (it was based on the theory of anomalies in quantum gravity) and did not say much about how inflation could actually start.

In 1981 Alan H. Guth of the Massachusetts Institute of Technology suggested that the hot universe at some intermediate stage could expand exponentially. His model derived from a theory that interpreted the development of the early universe as a series of phase transitions. This theory was proposed in 1972 by David A. Kirzhnits and me at the P. N. Lebedev Physics Institute in Moscow. According to this idea, as the universe expanded and cooled, it condensed into different forms. Water vapor undergoes such phase transitions. As it becomes cooler the vapor condenses into water, which, if cooling continues, becomes ice.

Guth's idea called for inflation to occur when the universe was in an unstable, supercooled state. Supercooling is common during phase transitions; for example, water under the right circumstances remains liquid below zero degrees Celsius. Of course supercooled water eventually freezes. That event would correspond to the end of the inflationary period. The idea to use supercooling for solving many problems of the big-bang theory was very attractive. Unfortunately, as Guth himself pointed out, the postinflation universe of his scenario becomes extremely inhomogenous. After investigating his model for a year, he finally renounced it in a paper he coauthored with Erick J. Weinberg of Columbia University.

In 1982 I introduced the so-called new inflationary-universe scenario, which Andreas Albrecht and Paul J. Steinhardt of the University of Pennsylvania also later discovered. This scenario shrugged off the main problems of Guth's model. But it was still rather complicated and not very realistic.

Only a year later did I realize that inflation is a naturally emerging feature in many theories of elementary particles, including the simplest model of the scalar field discussed earlier. There is no need for quantum gravity effects, phase transitions, supercooling, or even the standard assumption that the universe originally was hot. One just considers all possible kinds and values of scalar fields in the early universe and then checks to see if any of them leads to inflation. Those places where inflation does not occur remain small. Those domains where inflation takes place become exponentially large and dominate the total volume of the universe. Because the scalar fields can take arbitrary values in the early universe, I called this scenario chaotic inflation.

In many ways chaotic inflation is so simple that it is hard to understand why the idea was not discovered sooner. I think the reason was purely psychological. The glorious successes of the big-bang theory hypnotized cosmologists. We assumed that the entire universe was created at the same moment, that initially it was hot, and that the scalar field from the beginning resided close to the minimum of its potential energy. Once we began relaxing these assumptions, we immediately found that inflation is not an exotic phenomenon invoked by theorists for solving their problems. It is a general regime that occurs in a wide class of theories of elementary particles.

That a rapid stretching of the universe can simultaneously resolve many difficult cosmological problems may seem too good to be true. Indeed, if all inhomogeneities were stretched away, how did galaxies form? The answer is that while removing previously existing inhomogeneities, inflation at the same time made new ones.

These inhomogeneities arise from quantum effects. According to quantum mechanics, empty space is not entirely empty. The vacuum is filled with small quantum fluctuations. These fluctuations can be regarded as waves, or undulations in physical fields. The waves have all possible wavelengths and move in all directions. We cannot detect these waves, because they live only briefly and are microscopic.

In the inflationary universe the vacuum structure becomes even more complicated. Inflation rapidly stretches the waves. Once their wavelengths become sufficiently large, the undulations begin to "feel" the curvature of the universe. At this moment they stop moving because of the viscosity of the scalar field (recall that the equations describing the field contain a friction term).

The first fluctuations to freeze are those that have large wavelengths. As the universe continues to expand, new fluctuations become stretched and freeze on top of other frozen waves. At this stage one cannot call these waves quantum fluctuations anymore. Most of them have extremely large wavelengths. Because these waves do not move and do not disappear, they enhance the value of the scalar field in some areas and depress it in others, thus creating inhomogeneities. These disturbances in the scalar field cause the density perturbations in the universe that are crucial for the subsequent formation of galaxies.

■   TESTING INFLATIONARY THEORY

In addition to explaining many features of our world, inflationary theory makes several important and stable predictions. First, density perturbations produced during inflation affect the distribution of matter in the universe. They may also accompany gravitational waves. Both density perturbations and gravitational waves make their imprint on the microwave background radiation. They render the temperature of this radiation slightly different in various places in the sky. This nonuniformity was found in 1992 by the Cosmic Background Explorer (COBE) satellite, a finding later confirmed by several other experiments.

Although the COBE results agree with the predictions of inflation, it would be premature to claim that COBE has confirmed inflationary theory. But it is certainly true that the results obtained by the satellite at their current level of precision could have definitively disproved most inflationary models, and it did not happen. At present no other theory can simultaneously explain why the universe is so homogenous and still predict the "ripples in space" discovered by COBE.

Inflation also predicts that the universe should be nearly flat. Flatness of the universe can be experimentally verified because the density of a flat universe is related in a simple way to the speed of its expansion. So far observational data are consistent with this prediction. A few years ago it seemed that if someone were to show that the universe is open rather than flat, then inflationary theory would fall apart. Recently, however, several models of an open inflationary universe have been found. The only consistent description of a large homogeneous open universe that we currently know is based on inflationary theory. Thus, even if the universe is open, inflation is still the best theory to describe it. One may argue that the only way to disprove the theory of inflation is to propose a better theory.

One should remember that inflationary models are based on the theory of elementary particles, and this theory is not completely established. Some versions (most notably, superstring theory) do not automatically lead to inflation. Pulling inflation out of the superstring model may require radically new ideas. We should certainly continue the search for alternative cosmological theories. Many cosmologists, however, believe inflation, or something very similar to it, is absolutely essential for constructing a consistent cosmological theory. The inflationary theory itself changes as particle-physics theory rapidly evolves. The list of new models includes extended inflation, natural inflation, hybrid inflation, and many others. Each model has unique features that can be tested through observation or experiment. Most, however, are based on the idea of chaotic inflation.

Here we come to the most interesting part of our story, to the theory of an eternally existing, self-reproducing inflationary universe. This theory is rather general, but it looks especially promising and leads to the most dramatic consequences in the context of the chaotic inflation scenario.

As I already mentioned, one can visualize quantum fluctuations of the scalar field in an inflationary universe as waves. They first moved in all possible directions and then froze on top of one another. Each frozen wave slightly increased the scalar field in some parts of the universe and decreased it in others.

Now consider those places of the universe where these newly frozen waves persistently increased the scalar field. Such regions are extremely rare, but still they do exist. And they can be extremely important. Those rare domains of the universe where the field jumps high enough begin exponentially expanding with ever-increasing speed. The higher the scalar field jumps, the faster the universe expands. Very soon those rare domains will acquire a much greater volume than other domains.

From this theory it follows that if the universe contains at least one inflationary domain of a sufficiently large size, it begins unceasingly producing new inflationary domains. Inflation in each particular point may end quickly, but many other places will continue to expand. The total volume of all these domains will grow without end. In essence, one inflationary universe sprouts other inflationary bubbles, which in turn produce other inflationary bubbles.

This process, which I have called eternal inflation, keeps going as a chain reaction, producing a fractal-like pattern of universes. In this scenario the universe as a whole is immortal. Each particular part of the universe may stem from a singularity somewhere in the past, and it may end up in a singularity somewhere in the future. There is, however, no end for the evolution of the entire universe.

The situation with the very beginning is less certain. There is a chance that all parts of the universe were created simultaneously in an initial big-bang singularity. The necessity of this assumption, however, is no longer obvious.

Furthermore, the total number of inflationary bubbles on our "cosmic tree" grows exponentially in time. Therefore, most bubbles (including our own part of the universe) grow indefinitely far away from the trunk of this tree. Although this scenario makes the existence of the initial big bang almost irrelevant, for all practical purposes one can consider the moment of formation of each inflationary bubble as a new "big bang." From this perspective inflation is not a part of the big-bang

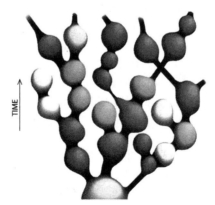

The self-reproducing cosmos appears as an extended branching of inflationary bubbles. Changes in shade represent "mutations" in the laws of physics from parent universes. The properties of space in each bubble do not depend on the time when the bubble formed. In this sense the universe as a whole may be stationary, even though the interior of each bubble can be described by the big-bang theory. (Jared Schneidman Design)

theory, as we thought 15 years ago. On the contrary, the big bang is a part of the inflationary model.

In thinking about the process of self-reproduction of the universe, one cannot avoid drawing analogies, however superficial they may be. One may wonder: Is not this process similar to what happens with all of us? Some time ago we were born. Eventually we will die, and the entire world of our thoughts, feelings, and memories will disappear. But there were those who lived before us, there will be those who will live after, and humanity as a whole, if it is clever enough, may live for a long time.

Inflationary theory suggests that a similar process may occur with the universe. One can draw some optimism from knowing that even if our civilization dies, there will be other places in the universe where life will emerge again and again, in all its possible forms.

■ A NEW COSMOLOGY

Could matters become even more curious? The answer is yes. Until now we have considered the simplest inflationary model with only one scalar field, which has only one minimum of its potential energy. Meanwhile realistic models of elementary particles propound many kinds of scalar fields. For example, in the unified theories of weak, strong, and electromagnetic interactions, at least two other scalar fields exist. The potential energy of these scalar fields may have several different minima. This condition means that the same theory may have different "vacuum states," corresponding to different types of symmetry breaking between fundamental interactions and, as a result, to different laws of low-energy physics. (Interactions of particles at extremely large energies do not depend on symmetry breaking.)

Such complexities in the scalar field mean that after inflation the universe may become divided into exponentially large domains that have different laws of low-energy physics. Note that this division occurs even if the entire universe originally began in the same state, corresponding to one particular minimum of potential energy. Indeed, large quantum fluctuations can cause scalar fields to jump out of their minima. That is, they jiggle some of the balls out of their bowls and into other ones. Each bowl corresponds to alternative laws of particle interactions. In some inflationary models quantum fluctuations are so strong that even the number of dimensions of space and time can change.

If this model is correct, then physics alone cannot provide a complete explanation for all properties of our allotment of the universe. The same physical theory may yield large parts of the universe that have diverse properties. According to this scenario, we find ourselves inside a four-dimensional domain with our kind of physical laws, not because domains with different dimensionality and with alternative properties are impossible or improbable but simply because our kind of life cannot exist in other domains.

Does this mean that understanding all the properties of our region of the universe will require, besides a knowledge of physics, a deep investigation of our own nature, perhaps even including the nature of our consciousness? This conclusion would certainly be one of the most unexpected that one could draw from the recent developments in inflationary cosmology.

The evolution of inflationary theory has given rise to a completely new cosmological paradigm, which differs considerably from the old big-bang theory and even from the first version of the inflationary scenario. In it the universe appears to be both chaotic and homogenous, expanding and stationary. Our cosmic home grows, fluctuates, and eternally reproduces itself in all possible forms, as if adjusting itself for all possible types of life.

Some parts of the new theory, we hope, will stay with us for years to come. Many others will have to be considerably modified to fit with new observational data and with the ever-changing theory of elementary particles. It seems, however, that the past 15 years of development of cosmology have irreversibly changed our understanding of the structure and fate of our universe and of our own place in it.

—*Scientific American Presents,* Spring 1998

BLACK HOLES AND THE INFORMATION PARADOX

*What happens to the information in matter destroyed by
a black hole? Searching for that answer, physicists are
groping toward a quantum theory of gravity*

Leonard Susskind

Somewhere in outer space Professor Windbag's time capsule has been sabotaged by
his archrival, Professor Goulash. The capsule contains the only copy of a vital mathematical formula, to be used by future generations. But Goulash's diabolical
scheme to plant a hydrogen bomb on board the capsule has succeeded. Bang! The
formula is vaporized into a cloud of electrons, nucleons, photons, and an occasional
neutrino. Windbag is distraught. He has no record of the formula and cannot remember its derivation.

Later, in court, Windbag charges that Goulash has sinned irrevocably: "What that
fool has done is irreversible. Why, the fiend has destroyed my formula and must pay.
Off with his tenure!"

"Nonsense," says an unflustered Goulash. "Information can never be destroyed. It's just your laziness, Windbag. Although it's true that I've scrambled things a bit, all you have to do is go and find each particle in the debris and reverse its motion. The laws of nature are time-symmetric, so on reversing everything your stupid formula will be reassembled. That proves, beyond a shadow of a doubt, that I could never have destroyed your precious information." Goulash wins the case.

Windbag's revenge is equally diabolical. While Goulash is out of town his computer is burglarized, along with all his files, including his culinary recipes. Just to make sure that Goulash will never again enjoy his famous Matelote d'anguilles with truffles, Windbag launches the computer into outer space and straight into a nearby black hole.

At Windbag's trial Goulash is beside himself. "You've gone too far this time, Windbag. There's no way to get my files out. They're inside the black hole, and if I go in to get them I'm doomed to be crushed. You've truly destroyed information, and you'll pay."

"Objection, Your Honor!" Windbag jumps up. "Everyone knows that black holes eventually evaporate. Wait long enough, and the black hole will radiate away all its mass and turn into outgoing photons and other particles. True, it may take 10^{70} years, but it's the principle that counts. It's really no different from the bomb. All Goulash has to do is reverse the paths of the debris, and his computer will come flying back out of the black hole."

"Not so!" cries Goulash. "This is different. My recipe was lost behind the black hole's boundary, its horizon. Once something crosses the horizon, it can never get back out without exceeding the speed of light. And Einstein taught us that nothing can ever do that. There is no way the evaporation products, which come from outside the horizon, can contain my lost recipes even in scrambled form. He's guilty, Your Honor."

Her Honor is confused. "We need some expert witnesses. Professor Hawking, what do you say?"

Stephen W. Hawking of the University of Cambridge comes to the stand. "Goulash is right. In most situations information is scrambled and in a practical sense is lost. For example, if a new deck of cards is tossed in the air, the original order of the cards vanishes. But in principle, if we know the exact details of how the cards are thrown, the original order can be reconstructed. This is called microreversibility. But in my 1976 paper I showed that the principle of microreversibility, which has always held in classical and quantum physics, is violated by black holes. Because information cannot escape from behind the horizon, black holes are a fundamental new source of irreversibility in nature. Windbag really did destroy information."

Her Honor turns to Windbag: "What do you have to say to that?" Windbag calls on Professor Gerard 't Hooft of Utrecht University.

"Hawking is wrong," 't Hooft begins. "I believe black holes must not lead to violation of the usual laws of quantum mechanics. Otherwise the theory would be out of control. You cannot undermine microscopic reversibility without destroying energy conservation. If Hawking were right, the universe would heat up to a temperature of 10^{31} degrees in a tiny fraction of a second. Because this has not happened, there must be some way out of this problem."

Twenty more famous theoretical physicists are called to the stand. All that becomes clear is that they cannot agree.

■ THE INFORMATION PARADOX

Windbag and Goulash are, of course, fictitious. Not so Hawking and 't Hooft, nor the controversy of what happens to information that falls into a black hole. Hawking's claim that a black hole consumes information has drawn attention to a potentially serious conflict between quantum mechanics and the general theory of relativity. The problem is known as the information paradox.

When something falls into a black hole, one cannot expect it ever to come flying back out. The information coded in the properties of its constituent atoms is, according to Hawking, impossible to retrieve. Albert Einstein once rejected quantum mechanics with the protest: "God does not play dice." But Hawking states that "God not only plays dice. He sometimes throws the dice where they cannot be seen"—into a black hole.

The problem, 't Hooft points out, is that if the information is truly lost, quantum mechanics breaks down. Despite its famed indeterminacy, quantum mechanics controls the behavior of particles in a very specific way: it is reversible. When one particle interacts with another, it may be absorbed or reflected or may even break up into other particles. But one can always reconstruct the initial configurations of the particles from the final products.

If this rule is broken by black holes, energy must be created or destroyed, threatening one of the most essential underpinnings of physics. The conservation of energy is ensured by the mathematical structure of quantum mechanics, which also guarantees reversibility; losing one means losing the other. As Thomas Banks, Michael Peskin, and I showed in 1980 at Stanford University, information loss in a black hole leads to enormous amounts of energy being generated. For such reasons 't Hooft and I believe the information that falls into a black hole must somehow become available to the outside world.

Some physicists feel the question of what happens in a black hole is academic or even theological, like counting angels on pinheads. But it is not so at all: at stake are the future rules of physics. Processes inside a black hole are merely extreme examples of interactions between elementary particles. At the energies that particles can acquire in today's largest accelerators (about 10^{12} electron volts), the gravitational attraction between them is negligible. But if the particles have a "Planck energy" of about 10^{28} electron volts, so much energy—and therefore mass—becomes concentrated in a tiny volume that gravitational forces outweigh all others. The resulting collisions involve quantum mechanics and the general theory of relativity in equal measure.

It is to Planckian accelerators that we would nominally look for guidance in building future theories of physics. Alas, Shmuel Nussinov of Tel Aviv University concludes that such an accelerator would have to be at least as big as the entire known universe.

Nevertheless, the physics at Planck energies may be revealed by the known properties of matter. Elementary particles have a variety of attributes that lead physicists to suspect they are not so elementary after all: they must actually have a good deal of undiscovered internal machinery, which is determined by the physics at Planck energies. We will recognize the right confluence of general relativity and quantum physics—or quantum gravity—by its ability to explain the measurable properties of electrons, photons, quarks, or neutrinos.

Very little is known with absolute certainty about collisions at energies beyond the Planck scale, but there is a good educated guess. Head-on collisions at these energies involve so much mass concentrated in a tiny volume that a black hole will form and subsequently evaporate. So figuring out whether black holes violate the rules of quantum mechanics or not is essential to unraveling the ultimate structure of particles.

A black hole is born when so much mass or energy gathers in a small volume that gravitational forces overwhelm all others and everything collapses under its own weight. The material squeezes into an unimaginably small region called a singularity, the density inside of which is essentially infinite. But it is not the singularity itself that will interest us.

Surrounding the singularity is an imaginary surface called the horizon. For a black hole with the mass of a galaxy the horizon is 10^{11} kilometers from the center—as far as the outermost reaches of the solar system are from the sun. For a black hole of solar mass the horizon is roughly a kilometer away; for a black hole with the mass of a small mountain the horizon is 10^{-13} centimeter away, roughly the size of a proton.

The horizon separates space into two regions that we can think of as the interior and exterior of the black hole. Suppose that Goulash, who is scouting for his computer near the black hole, shoots a particle away from the center. If he is not too close and the particle has a high velocity, then it may overcome the gravitational pull of the black hole and fly away. It will be most likely to escape if it is shot with the maximum velocity—that of light. If, however, Goulash is too close to the singularity, the gravitational force will be so great that even a light ray will be sucked in. The horizon is the place with the (virtual) warning sign: POINT OF NO RETURN. No particle or signal of any kind can cross it from the inside to the outside.

■ AT THE HORIZON

An analogy inspired by William G. Unruh of the University of British Columbia, one of the pioneers in black-hole quantum mechanics, helps to explain the relevance of the horizon. Imagine a river that gets swifter downstream. Among the fish that live in it, the fastest swimmers are the "lightfish." But at some point the river flows at the fish's maximum speed; clearly any lightfish that drifts past this point can never get back up. It is doomed to be crushed on the rocks below Singularity Falls, located farther downstream. To the unsuspecting lightfish, though, passing the point of no return is a nonevent. No currents or shock waves warn it of the crossing.

What happens to Goulash, who in a careless moment gets too close to the black hole's horizon? Like the freely drifting fish, he senses nothing special: no great forces, no jerks or flashing lights. He checks his pulse with his wristwatch— normal. His breathing rate—normal. To him the horizon is just like any other place.

But Windbag, watching Goulash from a spaceship safely outside the horizon, sees Goulash acting in a bizarre way. Windbag has lowered to the horizon a cable equipped with a camcorder and other probes, to better keep an eye on Goulash. As Goulash falls toward the black hole, his speed increases until it approaches that of light. Einstein found that if two persons are moving fast relative to each other, each sees the other's clock slow down; in addition, a clock that is near a massive object will run slowly compared with one in empty space. Windbag sees a strangely lethargic Goulash. As he falls, the latter shakes his fist at Windbag. But he appears to be moving ever more slowly; at the horizon, Windbag sees Goulash's motions slow to a halt. Although Goulash falls through the horizon, Windbag never quite sees him get there.

In fact, not only does Goulash seem to slow down, but his body looks as if it is being squashed into a thin layer. Einstein also showed that if two persons move fast with respect to each other, each will see the other as being flattened in the direction of motion. More strangely, Windbag should also see all the material that ever fell

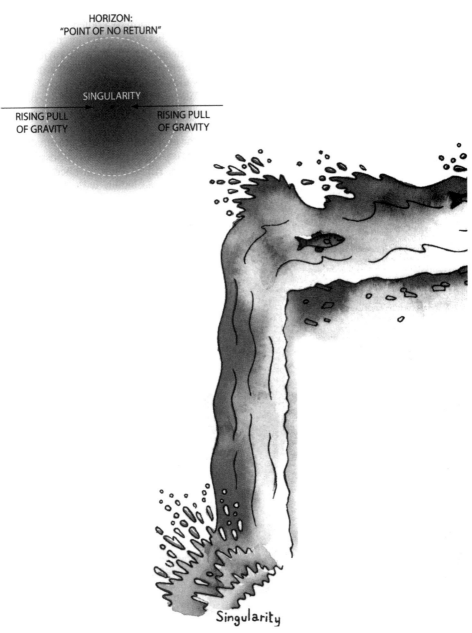

HORIZON:
"POINT OF NO RETURN"

SINGULARITY

RISING PULL
OF GRAVITY

RISING PULL
OF GRAVITY

Singularity

Invisible horizon is represented in this analogy as a line in a river. To the left of it the water flows faster than "lightfish" can swim. So if a lightfish happens to drift beyond this line, it can never get back upstream; it is doomed to be crushed in the falls. But the fish notices nothing special at the line. Likewise, a light ray or person who is inside the horizon can never get back out; the object inevitably falls into the singularity at the black hole's center, but without noticing anything special about the horizon. (Bryan Christie, left. Yan Nascimbene, right)

into the black hole, including the original matter that made it up—and Goulash's computer—similarly flattened and frozen at the horizon. With respect to an outside observer, all of that matter suffers a relativistic time dilation. To Windbag, the black hole consists of an immense junkyard of flattened matter at its horizon. But Goulash sees nothing unusual until much later, when he reaches the singularity, there to be crushed by ferocious forces.

Black-hole theorists have discovered over the years that from the outside, the properties of a black hole can be described in terms of a mathematical membrane above the horizon. This layer has many physical qualities, such as electrical conductivity and viscosity. Perhaps the most surprising of its properties was postulated in the early 1970s by Hawking, Unruh, and Jacob D. Bekenstein of the Hebrew University in Israel. They found that as a consequence of quantum mechanics, a black hole—in particular, its horizon—behaves as though it contains heat. The horizon is a layer of hot material of some kind.

The temperature of the horizon depends on just where it is measured. Suppose one of the probes that Windbag has attached to his cable is a thermometer. Far from the horizon he finds that the temperature is inversely proportional to the black hole's mass. For a black hole of solar mass this "Hawking temperature" is about 10^{-8} degree—far colder than intergalactic space. As Windbag's thermometer approaches the horizon, however, it registers higher temperatures. At a distance of a centimeter, it measures about a thousandth of a degree; at a nuclear diameter it records 10 billion degrees. The temperature ultimately becomes so high that no imaginable thermometer can measure it.

Hot objects also possess an intrinsic disorder called entropy, which is related to the amount of information a system can hold. Think of a crystal lattice with N sites; each site can house one atom or none at all. Thus, every site holds one "bit" of information, corresponding to whether an atom is there or not; the total lattice has N such bits and can contain N units of information. Because there are two choices for each site and N ways of combining these choices, the total system can be in any one of 2^N states (each of which corresponds to a different pattern of atoms). The entropy (or disorder) is defined as the logarithm of the number of possible states. It is roughly equal to N—the same number that quantifies the capacity of the system for holding information.

Bekenstein found that the entropy of a black hole is proportional to the area of its horizon. The precise formula, derived by Hawking, predicts an entropy of 3.2×10^{64} per square centimeter of horizon area. Whatever physical system carries the bits of information at the horizon must be extremely small and densely distributed: their linear dimensions have to be $\frac{1}{10^{20}}$ the size of a proton's. They must also be very special for Goulash to completely miss them as he passes through.

The discovery of entropy and other thermodynamic properties of black holes led Hawking to a very interesting conclusion. Like other hot bodies, a black hole must radiate energy and particles into the surrounding space. The radiation comes from the region of the horizon and does not violate the rule that nothing can escape from within. But it causes the black hole to lose energy and mass. In the course of time an isolated black hole radiates away all its mass and vanishes.

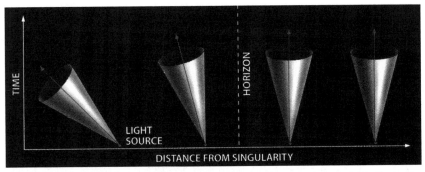

Light cones describe the path of light rays emanating from a point. Outside the horizon the light cones point upward—that is, forward in time. But inside, the light cones tip so that light falls straight into the black hole's center. (Bryan Christie)

All of the above, though peculiar, has been known to relativists for some decades. The true controversies arise when, following Hawking, we seek the fate of the information that fell into the black hole during and after its formation. In particular, can it be carried way by the evaporation products—albeit in a very scrambled form—or is it lost forever behind the horizon?

Goulash, who followed his computer into the black hole, would insist that its contents passed behind the horizon, where they were lost to the outside world; this in a nutshell is Hawking's argument. The opposing point of view might be described by Windbag: "I saw the computer fall toward the horizon, but I never saw it fall through. The temperature and radiation grew so intense I lost track of it. I believe the computer was vaporized; later, its energy and mass came back out in the form of thermal radiation. The consistency of quantum mechanics requires that this evaporating energy also carried away all the information in the computer." This is the position that 't Hooft and I take.

■ BLACK-HOLE COMPLEMENTARITY

Is it possible that Goulash and Windbag are in a sense both correct? Can it be that Windbag's observations are indeed consistent with the hypothesis that Goulash and his computer are thermalized and radiated back into space before ever reaching the horizon, even though Goulash discovers nothing unusual until long after, when he encounters the singularity? The idea that these are not contradictory but complementary scenarios was first put forward as the principle of black-hole complementarity by Lárus Thorlacius, John Uglum, and me at Stanford. Very similar ideas are also found in 't Hooft's work. Black-hole complementarity is a new principle of relativity. In the special theory of relativity we find that although different observers disagree about the lengths of time and space intervals, events take place at definite space-time locations. Black-hole complementarity does away with even that.

How this principle actually comes into play is clearer when applied to the structure of subatomic particles. Suppose that Windbag, whose cable is also equipped with a powerful microscope, watches an atom fall toward the horizon. At first he sees the atom as a nucleus surrounded by a cloud of negative charge. The electrons in the cloud move so rapidly they form a blur. But as the atom gets closer to the black hole, its internal motions seem to slow down, and the electrons become visible. The protons and neutrons in the nucleus still move so fast that its structure is obscure. But a little later the electrons freeze, and the protons and neutrons start to show up. Later yet, the quarks making up these particles are revealed. (Goulash, who falls with the atom, sees no changes.)

Many physicists believe elementary particles are made of even smaller constituents. Although there is no definitive theory for this machinery, one candidate stands out as being the most promising—namely, string theory. In this theory an elementary particle does not resemble a point; rather it is like a tiny rubber band that can vibrate in many modes. The fundamental mode has the lowest frequency; then there are higher harmonics, which can be superimposed on top of one another. There are an infinite number of such modes, each of which corresponds to a different elementary particle.

Here another analogy helps. One cannot see the wings of a hovering hummingbird, because its wings flutter too fast. But in a photograph taken with a fast shutter speed, one can see the wings—so the bird looks bigger. If a hummer falls into the black hole, Windbag will see its wings take form as the bird approaches the horizon and the vibrations appear to slow down; it seems to grow. Now suppose that the wings have feathers that flap even faster. Soon these, too, would come into view, adding further to the apparent size of the bird. Windbag sees the hummer enlarge continuously. But Goulash, who falls with the bird, sees no such strange growth.

Like the hummingbird's wings, the string's oscillations are usually too rapid to detect. A string is a minute object, $\frac{1}{10^{20}}$ the seize of a proton. But as it falls into a black hole, its vibrations slow down and more of them become visible. Mathematical studies done at Stanford by Amanda Peet, Thorlacius, Arthur Mezhlumian, and me have demonstrated the behavior of a string as its higher modes freeze out. The string spreads and grows, just as if it were being bombarded by particles and radiation in a very hot environment. In a relatively short time the string and all the information that it carries are smeared over the entire horizon.

This picture applies to all the material that ever fell into the black hole—because according to string theory, everything is ultimately made of strings. Each elementary string spreads and overlaps all the others until a dense tangle covers the hori-

Cascade of vibrations on a string slow down and become visible if the string falls into a black hole. Strings are small enough to encode all the information that ever fell into a black hole and offer a way out of the information paradox. (Bryan Christie)

zon. Each minute segment of string, measuring 10^{-33} centimeter across, functions as a bit. Thus strings provide a means for the black hole's surface to hold the immense amount of information that fell in during its birth and thereafter.

■ STRING THEORY

It seems, then, that the horizon is made of all the substance in the black hole, resolved into a giant tangle of strings. The information, as far as an outside observer is concerned, never actually fell into the black hole; it stopped at the horizon and was later radiated back out. String theory offers a concrete realization of black-hole complementarity and therefore a way out of the information paradox. To outside observers—that is, us—information is never lost. Most important, it appears that the bits at the horizon are minute segments of string.

Tracing the evolution of a black hole from beginning to end is far beyond the current techniques available to string theorists. But some exciting new results are giving quantitative flesh to these ghostly ideas. Mathematically the most tractable black holes are the "extremal" black holes. Whereas black holes that have no electrical charge evaporate until all their mass is radiated away, black holes with electrical or (in theory) magnetic charge cannot do that; their evaporation ceases when the gravitational attraction equals the electrostatic or magnetostatic repulsion of whatever is inside the black hole. The remaining stable object is called an extremal black hole.

Following earlier suggestions of mine, Ashoke Sen of the Tata Institute of Fundamental Research (TIFR) showed in 1995 that for certain extremal black holes with electrical charge, the number of bits predicted by string theory exactly accounts for the entropy as measured by the area of the horizon. This agreement was the first powerful evidence that black holes are consistent with quantum-mechanical strings.

Sen's black holes were, however, microscopic. More recently Andrew Strominger of the University of California at Santa Barbara, Cumrun Vafa of Harvard University, and, slightly later, Curtis G. Callan and Juan Maldacena of Princeton University extended this analysis to black holes with both electrical and magnetic charge. Unlike Sen's tiny black holes, these new black holes can be large enough to allow Goulash to fall through unharmed. Again, the theorists find complete consistency.

Two groups have done an even more exciting new calculation of Hawking radiation: Sumit R. Das of TIFR, with Samir Mathur of the Massachusetts Institute of Technology; and Avinash Dhar, Gautam Mandal, and Spenta R. Wadia, also at TIFR. The researchers studied the process by which an extremal black hole with some excess energy or mass radiates off this flab. String theory fully accounted for the Hawking radiation that was produced. Just as quantum mechanics describes the radiation of an atom by showing how an electron jumps from a high-energy "excited" state to a low-energy "ground" state, quantum strings seem to account for the spectrum of radiation from an excited black hole.

Quantum mechanics, I believe, will in all likelihood turn out to be consistent with the theory of gravitation; these two great streams of physics are merging into a quantum theory of gravity based on string theory. The information paradox, which appears to be well on its way to being resolved, has played an extraordinary role in this ongoing revolution in physics. And although Goulash would never admit it, Windbag will probably turn out to be right: the recipe for Matelote d'anguilles is not forever lost to the world.

—April 1997

The Expansion Rate
and Size of
the Universe

The age, evolution, and fate of the universe depend on just how fast it is expanding. By measuring the size of the universe using a variety of new techniques, astronomers have recently improved estimates of the expansion rate

Wendy L. Freedman

Our Milky Way and all other galaxies are moving away from one another as a result of the big bang, the fiery birth of the universe. As we near the end of the millennium it is interesting to reflect that during the 20th century, cosmologists discovered this expansion, detected the microwave background radiation from the original explosion, deduced the origin of chemical elements in the universe, and mapped the large-scale structure and motion of galaxies. Despite these advances, elementary questions remain. When did the colossal expansion begin? Will the universe expand forever, or will gravity eventually halt its expansion and cause it to collapse back on itself?

For decades, cosmologists have attempted to answer such questions by measuring the universe's size scale and expansion rate. To accomplish this task, astronomers must determine both how fast galaxies are moving and how far away they are. Techniques for measuring the velocities of galaxies are well established, but estimating the distances to galaxies has proved far more difficult [see color plate 7]. During the past decade several independent groups of astronomers have developed better methods for measuring the distances to galaxies, leading to completely new estimates of the expansion rate. Recently the superb resolution of the Hubble Space Telescope has extended and strengthened the calibration of the extragalactic distance scale, leading to new estimates of the expansion rate.

At present several lines of evidence point toward a high expansion rate, implying that the universe is relatively young, perhaps only 10 billion years old. The evidence also suggests that the expansion of the universe may continue indefinitely. Still, many astronomers and cosmologists do not yet consider the evidence definitive. We actively debate the merits of our techniques.

An accurate measurement of the expansion rate is essential not only for determining the age of the universe and its fate but also for constraining theories of cosmology and models of galaxy formation. Furthermore, the expansion rate is important for estimating fundamental quantities, from the density of the lightest elements (such as hydrogen and helium) to the amount of nonluminous matter in galaxies, as well as clusters of galaxies. Because we need accurate distance measurements to calculate the luminosity, mass, and size of astronomical objects, the issue of the cosmological distance scale, or the expansion rate, affects the entire field of extragalactic astronomy.

Astronomers began measuring the expansion rate of the universe some 70 years ago. In 1929 the eminent astronomer Edwin P. Hubble of the Carnegie Institution's observatories made the remarkable observation that the velocity of a galaxy's recession is proportional to its distance. His observations provided the first evidence that the entire universe is expanding.

■ THE HUBBLE CONSTANT

Hubble was the first to determine the expansion rate. Later this quantity became known as the Hubble constant: the recession velocity of the galaxy divided by its distance. A very rough estimate of the Hubble constant is 100 kilometers per second per megaparsec. (Astronomers commonly represent distances in terms of megaparsecs, where one megaparsec is the distance light travels in 3.26 million years.) Thus, a typical galaxy at a distance of 50 megaparsecs moves away at about 5,000 kilometers (3,000 miles) per second. A galaxy at 500 megaparsecs therefore moves at about 50,000 kilometers per second, or more than 100 million miles per hour!

For seven decades astronomers have hotly debated the precise value of the expansion rate. Hubble originally obtained a value of 500 kilometers per second per megaparsec (km/s/Mpc). After Hubble's death in 1953, his protégé Allan R. Sandage, also at Carnegie, continued to map the expansion of the universe. As Sandage and others made more accurate and extensive observations, they revised Hubble's original value downward into the range of 50 to 100 km/s/Mpc, thereby indicating a universe far older and larger than suggested by the earliest measurements.

During the past two decades new estimates of the Hubble constant have continued to fall within this same range, but preferentially toward the two extremes. Notably, Sandage and his longtime collaborator Gustav A. Tammann of the University of Basel have argued for a value of 50 km/s/Mpc, whereas the late Gérard de Vaucouleurs of the University of Texas advocated a value of 100 km/s/Mpc. The controversy has created an unsatisfactory situation in which scientists have been free to choose any value of the Hubble constant between the two extremes.

In principle determining the Hubble constant is simple, requiring only a measurement of velocity and distance. Measuring a galaxy's velocity is straightforward: astronomers disperse light from a galaxy and record its spectrum. A galaxy's spectrum has discrete spectral lines, which occur at characteristic wavelengths caused by emission or absorption of elements in the gas and stars making up the galaxy. For a galaxy receding from the earth, these spectral lines shift to longer wavelengths by an amount proportional to the velocity—an effect known as redshift.

If the measurement of the Hubble constant is so simple in principle, then why has it remained one of the outstanding problems in cosmology for almost 70 years? In practice measuring the Hubble constant is extraordinarily difficult, primarily for two reasons. First, although we can measure their velocities accurately, galaxies interact gravitationally with their neighbors. In so doing their velocities become perturbed, inducing "peculiar" motions that are superimposed onto the general expansion of the universe. Second, establishing an accurate distance scale has turned out to be much more difficult than anticipated. Consequently, an accurate measure of the Hubble constant requires us not only to establish an accurate extragalactic distance scale but also to do this already difficult task at distances great enough that peculiar motions of galaxies are small compared with the overall expansion, or

Hubble flow. To determine the distance to a galaxy, astronomers must choose from a variety of complicated methods. Each has its advantages, but none is perfect.

■ MEASURING DISTANCES TO GALAXIES

Astronomers can most accurately measure distances to nearby galaxies by monitoring a type of star commonly known as a Cepheid variable. Over time the star changes in brightness in a periodic and distinctive way. During the first part of the cycle its luminosity increases very rapidly, whereas during the remainder of the cycle the luminosity of the Cepheid decreases slowly. On average Cepheid variables are about 10,000 times brighter than the sun.

Remarkably, the distance to a Cepheid can be calculated from its period (the length of its cycle) and its average apparent brightness (its luminosity as observed from the earth). In 1908 Henrietta S. Leavitt of Harvard College Observatory discovered that the period of a Cepheid correlates closely with its brightness. She found that the longer the period, the brighter the star. This relation arises from the fact that a Cepheid's brightness is proportional to its surface area. Large, bright Cepheids pulsate over a long period just as, for example, large bells resonate at a low frequency (or longer period).

By observing a Cepheid's variations in luminosity over time, astronomers can obtain its period and average apparent luminosity, thereby calculating its absolute luminosity (that is, the apparent brightness the star would have if it were a standard distance of 10 parsecs away). Furthermore, they know that the apparent luminosity decreases as the distance it travels increases—because the apparent luminosity falls off in proportion to the square of the distance to an object. Therefore, we can compute the distance to the Cepheid from the ratio of its absolute brightness to its apparent brightness.

During the 1920s Hubble used Cepheid variables to establish that other galaxies existed far beyond the Milky Way. By measuring apparent brightnesses and periods of faint, starlike images that he discovered on photographs of objects such as the Andromeda Nebula (also known as M31), the Triangulum Nebula (M33), and NGC 6882, he could show that these objects were located more than several hundred thousand light-years from the sun, well outside the Milky Way. From the 1930s to the 1960s Hubble, Sandage, and others struggled to find Cepheids in nearby galaxies. They succeeded in measuring the distances to about a dozen galaxies. About half these galaxies are useful for the derivation of the Hubble constant.

One of the difficulties with the Cepheid method is that dust between stars diminishes apparent luminosity. Dust particles absorb, scatter, and redden light from all types of stars. Another complication is that it is hard to establish how Cepheids of different chemical element abundances differ in brightness. The effects of both dust and element abundances are most severe for blue and ultraviolet light. Astronomers must either observe Cepheids at infrared wavelengths, where the effects are less significant, or observe them at many different optical wavelengths so that they can assess the effects and correct for them.

During the 1980s my collaborator (and husband) Barry F. Madore of the California Institute of Technology and I remeasured the distances to the nearest galaxies using charge-coupled devices (CCDs) and the large reflecting telescopes at many sites, including Mauna Kea in Hawaii, Las Campanas in Chile, and Mount Palomar in California. As a result we determined the distances to nearby galaxies with much greater accuracy than has been done before.

These new CCD observations proved critical to correct for the effects of dust and to improve previous photographic photometry. In some cases we revised distances to nearby galaxies downward by a factor of two. Were it feasible, we would use Cepheids directly to measure distances associated with the universe's expansion. Unfortunately so far we cannot detect Cepheids in galaxies sufficiently far away so that we know they are part of a "pure" Hubble expansion of the universe.

Nevertheless, astronomers have developed several other methods for measuring relative distances between galaxies on vast scales, well beyond Cepheid range. Because we must use the Cepheid distance scale to calibrate these techniques, they are considered secondary distance indicators.

During the past decade astronomers have made great strides developing techniques to measure such relative distances. These methods include observing and measuring a special category of supernovae: catastrophic explosions signaling the death of certain low-mass stars. Sandage and his collaborators are now determining the Hubble constant by studying such supernovae based on the calibration of Cepheids. Other secondary distance-determining methods include measuring the brightnesses and rotations of velocities of entire spiral galaxies, the fluctuations (or graininess) in the light of elliptical galaxies, and the analysis and measurement of the expansion properties of another category of younger, more massive supernovae. The key to measuring the Hubble constant using these techniques is to determine the distance to selected galaxies using Cepheids; their distances can, in turn, be used to calibrate the relative extragalactic distance scale by applying secondary methods.

Yet scientists have not reached a consensus about which, if any, secondary indicators are reliable. As the saying goes, "The devil is in the details." Astronomers disagree on how to apply these methods, whether they should be adjusted for various effects that might bias the results, and what the true uncertainties are. Differences in the choice of secondary methods lie at the root of most current debates about the Hubble constant.

■ ESTABLISHING A DISTANCE SCALE

One technique for measuring great distances, the Tully-Fisher relation, relies on a correlation between a galaxy's brightness and its rotation rate. High-luminosity galaxies typically have more mass than low-luminosity galaxies, and so bright galaxies rotate slower than dim galaxies. Several groups have tested the Tully-Fisher method and shown that the relation does not appear to depend on environment; it remains the same in the dense and outer parts of rich clusters and for relatively isolated galaxies. The Tully-Fisher relation can be used to estimate distances as far away as 300 million light-years. A disadvantage is that astronomers lack a detailed theoretical understanding of the Tully-Fisher relation.

Another distance indicator that has great potential is a particular kind of supernova known as type Ia. Type Ia supernovae, astronomers believe, occur in double-star systems in which one of the stars is a very dense object known as a white dwarf. When a companion star transfers its mass to a white dwarf, it triggers an explosion. Because supernovae release tremendous amounts of radiation, astronomers should be able to see supernovae as far away as five billion light-years—that is a distance spanning a radius of half the visible universe.

Type Ia supernovae make good distance indicators because, at the peak of their brightness, they all produce roughly the same amount of light. Using this information, astronomers can infer their distance.

If supernovae are also observed in galaxies for which Cepheid distances can be measured, then the brightnesses of supernovae can be used to infer distances. In practice, however, the brightnesses of supernovae are not all the same; there is a range of brightnesses that must be taken into account. A difficulty is that supernovae are very rare events, so the chance of seeing one nearby is very small. Unfortunately a current limitation of this method is that about half of all supernovae observed in galaxies close enough to have Cepheid distances were observed decades ago, and these measurements are of low quality.

An interesting method, developed by John L. Tonry of the Massachusetts Institute of Technology and his colleagues, exploits the fact that nearby galaxies appear grainy, whereas remote galaxies are more uniform in their surface-brightness distribution. The graininess decreases with distance because the task of resolving individual stars becomes increasingly difficult. Hence, the distance to a galaxy can be gauged by how much the apparent brightness of the galaxy fluctuates over its surface. This method cannot currently extend as far as the Tully-Fisher relation or supernovae, but it and other methods offer an important, independent way to test and compare relative distances. These comparisons yield excellent agreement, representing one of the most important advancements in recent years.

For decades astronomers have recognized that the solution to the impasse on the extragalactic distance scale would require observations made at very high spatial resolution. The Hubble telescope can now resolve Cepheids at distances 10 times farther (and therefore in a volume 1,000 times larger) than we can do from the ground. A primary motivation for building an orbiting optical telescope was to enable the discovery of Cepheids in remoter galaxies and to measure accurately the Hubble constant.

More than a decade ago several colleagues and I were awarded time on the Hubble telescope to undertake this project. This program involves 26 astronomers, led by me, Jeremy R. Mould of Mount Stromlo and Siding Springs Observatory, and Robert C. Kennicutt of Steward Observatory. Our effort involves measuring Cepheid distances to about 20 galaxies, enough to calibrate a wide range of secondary distance methods. We aim to compare and contrast results from many techniques and to assess the true uncertainties in the measurement of the Hubble constant.

Though still incomplete, new Cepheid distances to a dozen galaxies have been measured as part of this project. Preliminary results yield a value of the Hubble constant of about 70 km/s/Mpc with an uncertainty of about 15 percent. This value is based on a number of methods, including the Tully-Fisher relation, type Ia supernovae, type II supernovae, surface-brightness fluctuations, and Cepheid measurements to galaxies in the nearby Virgo and Fornax clusters.

Sandage and his collaborators have reported a value of 59 km/s/Mpc, based on type Ia supernovae. Other groups (including our own) have found a value in the middle-60s range, based on the same type Ia supernovae. Nevertheless, these current disagreements are much smaller than the earlier discrepancies of a factor of two, which have existed until now. This progress is encouraging.

Two other methods for determining the Hubble constant spark considerable interest because they do not involve the Cepheid distance scale and can be used to measure distances on vast cosmological scales. The first of these alternative methods relies on an effect called gravitational lensing: if light from some distant source travels near a galaxy on its way to the earth, the light can be deflected as a result of gravity, according to Einstein's general theory of relativity. The light may take many different paths around the galaxy, some shorter, some longer, and consequently arrives at the earth at different times. If the brightness of the source varies in some distinctive way, the signal will be seen first in the light that takes the shortest path and will be observed again, some time later, in the light that traverses the longest path. The difference in the arrival times reveals the difference in length between the two light paths. By applying a theoretical model of the mass distribution of the galaxy, astronomers can calculate a value for the Hubble constant.

The second method uses a phenomenon known as the Sunyaev-Zel'dovich (SZ) effect. When photons from the microwave background travel through galaxy clusters, they can gain energy as they scatter off the hot plasma (x-ray) electrons found in the clusters. The net result of the scattering is a decrease in the microwave background toward the position of the cluster. By comparing the microwave and x-ray distributions, a distance to the cluster can be inferred. To determine the distance, however, astronomers must also know the average density of the electrons, as well as their distribution and temperature, and have an accurate measure of the decrement in the temperature of the microwave background. By calculating the distance to the cluster and measuring its recessional velocity, astronomers can then obtain the Hubble constant.

The SZ method and the gravitational-lensing technique are promising. Yet, to date, few objects are available with the required characteristics. Hence, these methods have not yet been tested rigorously. Fortunately, impressive progress is being made in both these areas with large, new surveys. Current applications of these methods result in values of the Hubble constant in the range of 40 to 80 km/s/Mpc.

The debate continues regarding the best method for determining distances to remote galaxies. Consequently astronomers hold many conflicting opinions about what the best current estimate is for the Hubble constant.

■ HOW OLD IS THE UNIVERSE?

The value of the Hubble constant has many implications for the age, evolution, and fate of the universe. A low value for the Hubble constant implies an old age for the universe, whereas a high value suggests a young age. For example, a value of 100 km/s/Mpc indicates the universe is about 6.5 to 8.5 billion years old (depending on the amount of matter in the universe and the corresponding deceleration caused by that matter). A value of 50 km/s/Mpc suggests, however, an age of 13 to 16.5 billion years.

And what of the ultimate fate of the universe? If the average density of matter in the universe is low, as current observations indicate, the standard cosmological model predicts that the universe will expand forever.

Nevertheless, theory and observation suggest that the universe contains more mass than what can be attributed to luminous matter. A very active area of cosmological research is the search for this additional "dark" matter in the universe. To answer the question about the fate of the universe unambiguously, cosmologists require not only a knowledge of the Hubble constant and the average mass density of the universe but also an independent measure of the age of the universe. These three quantities are needed to specify uniquely the geometry and the evolution of the universe.

If the Hubble constant turns out to be high, it would have profound implications for our understanding of the evolution of galaxies and the universe. A Hubble constant of 70 km/s/Mpc yields an age estimate of 9 to 12 billion years (allowing for uncertainty in the value of the average density of the universe). A high-density universe corresponds to an age of about nine billion years. A low-density universe corresponds to an age of about 12 billion years for this same value of the Hubble constant.

These estimates are all shorter than what theoretical models suggest for the age of old stellar systems known as globular clusters. Globular clusters are believed to be among the first objects to form in our galaxy, and their age is estimated to be between 13 and 17 billion years. Obviously the ages of the globular clusters cannot be older than the age of the universe itself.

Age estimates for globular clusters are often cited as a reason to prefer a low value for the Hubble constant and therefore an older age of the universe. Some astronomers argue, however, that the theoretical model of globular clusters on which these estimates depend may not be complete and may be based on inaccurate assumptions. For instance, the models rely on knowing precise ratios of certain elements in globular clusters, particularly oxygen and iron. Moreover, accurate ages require accurate measures of luminosities of globular cluster stars, which in turn require accurate measurements of the distances to the globular clusters.

Recent measurements from the Hipparcos satellite suggest that the distances to globular clusters might have to be increased slightly. The resulting effect of this change, if confirmed, would be to lower the globular cluster ages, perhaps to 11 or 12 billion years. Given the current uncertainties in the measurements of both the Hubble constant and the models and distances for globular clusters, these new results may indicate that no serious discrepancy exists between the age of the universe, based on expansion, and the age of globular clusters.

In any case these subtle inconsistencies highlight the importance of accurate distance measurements, not only for studying galaxies and determining the Hubble constant but also for understanding globular clusters and their ages.

A high value for the Hubble constant raises another potentially serious problem: it disagrees with standard theories of how galaxies are formed and distributed in space. For example, the theories predict how much time is required for large-scale clustering, which has been observed in the distribution of galaxies, to occur. If the Hubble constant is large (that is, the universe is young), the models cannot reproduce the observed distribution of galaxies.

Scientists are excited about results in the next decade. The recently installed NICMOS infrared camera on the Hubble telescope will allow us to refine the Cepheid distances measured so far. Large, ground-based telescope surveys will increase the number of galaxies for which we can measure relative distances beyond the reach of Cepheids.

Promising space missions loom on the horizon, such as the National Aeronautics and Space Administration's Microwave Anisotropy Probe (MAP) and the European Space Agency's Planck Surveyor. These two experiments will permit detailed mapping of small fluctuations in the cosmic-microwave background. If current cosmological theories prove correct, these measurements will robustly determine the density of matter in the universe and independently constrain the Hubble constant.

Although the history of science suggests that ours is not the last generation to wrestle with these questions, the next decade promises much excitement. There are many reasons to be optimistic that the current disagreement over values of the cosmological parameters governing the evolution of the universe will soon be resolved.

—*Scientific American Presents,* Spring 1998

VII

Technology

Early Results from the Hubble Space Telescope

Although hampered by optical and mechanical flaws,
Hubble has relayed a plentitude of eye-opening images
and revealing spectral portraits of cosmic objects

Eric J. Chaisson

On April 25, 1990, the Hubble Space Telescope was deployed from the bay of the space shuttle Discovery, marking the beginning of a new era in optical astronomy. Earth-bound optical telescopes, ranging from Galileo's primitive spyglasses to the brand-new Keck telescope, have always been hindered by the earth's restless, distorting atmosphere. From its vantage 610 kilometers above the surface, Hubble was designed to observe the cosmos in unprecedented clarity.

As nearly everyone knows, the telescope has not functioned as intended. A number of mechanical and design failings—most notably a misshapen main mirror—have degraded the telescope's capabilities. These difficulties dismayed many astronomers and attracted critical commentary from the media. But thanks to several improvised procedural changes and innovative computer image correction techniques, Hubble can match the sensitivity, and exceed the resolving power, of the finest ground-based telescopes. It can also detect ultraviolet rays (radiation having wavelengths slightly shorter than those of visible light), which do not penetrate the earth's atmosphere.

During its first two years of operation the space telescope has served up dramatic views of storms on Saturn, of the birth and death of stars, and of enigmatic objects lurking in the hearts of galaxies—perhaps giant black holes. Individually none of Hubble's discoveries yet qualifies as revolutionary. But taken together they are sending astronomers scrambling to rewrite their textbooks.

The 11,500-kilogram Hubble Space Telescope (six times the weight of a full-size automobile) is the most complex and sensitive civilian observatory ever launched into space. Its 2.4-meter-diameter primary mirror is the smoothest and cleanest one in existence. An advanced guidance system keeps the telescope locked on its targets even as it whips around the earth once every 96 minutes. Five instruments analyze the light that it collects.

The Faint Object Camera offers exceptional sensitivity and resolution, whereas the Wide Field and Planetary Camera provides a broader view. Two spectrographs (the Faint Object Spectrograph and the Goddard High-Resolution Spectrograph) similarly share the duty of splitting light into its component wavelengths to reveal the dynamics and physical makeup of the object observed. A photometer determines the exact brightness of sources. In addition, Hubble's guidance sensors perform astrometry, the precise measurement of the angular positions of stars.

Immediately after Hubble's launch operators at the National Aeronautics and Space Administration Goddard Space Flight Center and at the Space Telescope Science Institute began an extensive series of systems checks and calibrations. The first test images revealed an inherent focusing problem, technically known as spherical aberration. A close examination of the images revealed that the telescope's main mirror had been ground to the wrong shape: it is two microns flatter at the edges than stipulated by design (a micron is one millionth of a meter). Small though the error may seem, it is a gross mistake by the standards of modern precision optics.

The shape of the mirror makes it impossible to focus all the light collected by Hubble to a single point. Hubble's designers intended that the telescope should be able to concentrate 70 percent of the light of a point source—a distant star, for example—into a spot .1 arc second across (an arc second is a tiny angle, equal to $\frac{1}{1,800}$ the apparent diameter of the moon). Actually only 15 percent of the light falls into this central image; the other 85 percent spills over into an unwanted halo several arc seconds in diameter.

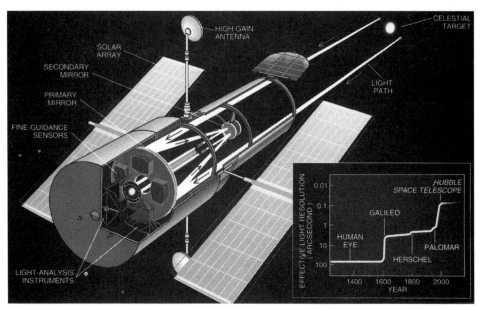

Hubble Space Telescope follows essentially the same design as do modern reflecting telescopes on the earth. The 2.4-meter-diameter primary mirror collects light, which is distributed among five analytic instruments. The telescope's focusing problem results from the primary mirror's incorrect curvature. Hubble improves on the resolution of ground-based telescopes by nearly the same extent that Galileo's telescope improved on the resolution of the human eye (inset). (Ian Worpole)

Various other difficulties have surfaced. Twice each orbit, when Hubble passes in and out of the earth's shadow, the sudden temperature change causes the telescope's large solar cell panels to flap up and down about 30 centimeters every 10 seconds. The resulting jitter can disrupt the telescope's pointing system and cause additional blurring of astronomical images. Two of Hubble's six gyroscopes have failed, and a third works only intermittently; the telescope needs at least three gyroscopes to perform its normal science operations. Faulty electrical contacts threaten to shut down the High-Resolution Spectrograph.

NASA hopes to address some of these problems in 1994, when astronauts are scheduled to visit Hubble. They will attempt to replace the telescope's solar panels and two of the gyroscopes. The astronauts may also try to install a package of corrective optics and an upgraded Wide Field and Planetary Camera if the new devices are completed by then.

In the meantime scientists have quickly learned how to wring as much performance from the space telescope as possible. Because Hubble's mirror was ground to fine precision and because its error is well understood, computer enhancement can restore many images to their intended sharpness. The resulting astronomical views have eloquently refuted some early pessimism about the telescope's scientific capabilities. Regrettably, attaining such resolution often involves discarding the smeared halos that appear around celestial targets, literally throwing away most of the light captured by Hubble.

The greatest blow to Hubble's scientific mission therefore has been not a loss of resolution but a loss of sensitivity. Hubble was designed to be able to detect objects a billion times fainter than those visible to the human eye. At present the telescope is limited to observing objects roughly 20 times brighter than intended. Hubble cannot detect some particularly elusive targets, such as extremely distant galaxies and quasars or possible planets around nearby stars. Astronomers have had to postpone many of their potentially most significant observations until the telescope is fixed.

Although designed to home in on some of the most remote cosmic objects, Hubble has proved well suited to studying objects within the solar system. For example, it has captured stunning views of the giant planets, Jupiter and Saturn. NASA's two Voyager space probes closely scrutinized Jupiter in 1979 and Saturn in 1980 and 1981. The space telescope can routinely produce images of Jupiter and Saturn comparable in detail to those obtained by the Voyagers only a few days before their closest approaches to the two planets.

When Hubble examined Jupiter it found a world remarkably changed from the one visited by Voyagers 1 and 2. New cloud bands have come and gone, many spots (cyclonelike storms, some of them thousands of kilometers across) have appeared, and turbulent structure has emerged on the edge of the planet's huge South Equatorial Belt. The famed Great Red Spot, a seemingly perpetual hurricane twice the diameter of the earth, has turned a dull, brownish orange.

Hubble's early images of Saturn dramatically illustrate the telescope's resolving power. William A. Baum of the Lowell Observatory and Shawn P. Ewald of the Space Telescope Science Institute assembled a color image of Saturn by instructing Hubble to capture three one-second exposures of the planet, one in red light, one in green light, and one in blue light. Imaging experts in the institute's Astronomy Visualization Laboratory prepared a computer-enhanced color image that captures the detailed structure in Saturn's cloud bands and rings. The image contains the first high-resolution views of the planet's north polar region.

In the fall of 1990 Hubble took more than 100 additional images of Saturn in order to track a 50,000-kilometer-wide storm of ammonia ice crystals, termed the Great White Spot. These observations are especially valuable because the spot appears only once every 60 years or so.

The distinct advantage of Hubble for planetary astronomers is that it can provide image clarity comparable to that from a space probe whenever an observer calls for it. Philip B. James of the University of Toledo and several colleagues will monitor the atmosphere of Mars for several years to study that planet's weather patterns and to try to understand the events that trigger periodic, planetwide dust storms. Such research is an important prerequisite for any manned expedition to the red planet. James A. Westphal of the California Institute of Technology plans a similar systematic study of Jupiter's potent weather systems.

Hubble has also turned its gaze on tiny Pluto, orbiting at the dim outer reaches of the solar system nearly five billion kilometers from the sun. Pluto and its relatively large moon Charon orbit a mere 19,000 kilometers from one another, so the pair appear as hardly more than a lopsided blob when viewed from the earth. Hubble's Faint Object Camera yielded, for the first time, clear, separate images of Pluto and Charon.

Detailed analysis of the brightness variations of the two bodies will provide data about the changing structures of their thin, methane atmospheres. Rudolf Albrecht of the Space Telescope European Coordinating Facility, who directed the Hubble observations of Pluto, hopes computer enhancement may even reveal some surface markings. Precise measurements of the orbits of Pluto and Charon about each other will enable researchers to measure accurately their individual masses and densities. That information will provide clues regarding the compositions and origins of these enigmatic objects.

Looking beyond the solar system, Hubble turned to the beautiful Orion nebula, a patch of glowing ionized gas (atoms stripped of some of their electrons) visible to the naked eye as the middle "star" in the sword of Orion. Huge clouds of this kind are the places where stars are born. The Orion nebula lies 1,500 light-years distant, making it the closest bright star-forming region. Astronomers have examined it extensively over the years and felt they knew its structure rather well—at least until the Hubble images came in [see color plate 8].

The Wide Field and Planetary Camera took three 10-minute exposures of the nebula, which were later assembled into a color image of the object. The result reveals structures as small as .1 arc second (or about six billion kilometers, roughly the radius of the solar system, at the distance of the Orion nebula). Previously unseen wispy arcs, filaments, and sheets of ionized gas streak across the image. C. Robert O'Dell of Rice University, who was in charge of the Orion nebula study, also noticed numerous luminous knots in the nebula. These knots consist of gas ionized by ultraviolet rays from hot, young stars embedded in the nebula. Some researchers tempered enthusiasm about the image, however, worrying that computer image enhancement procedures could have produced spurious features. The Orion nebula images epitomize the mixture of thrill and frustration astronomers experience when they work with Hubble.

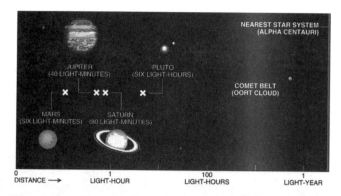

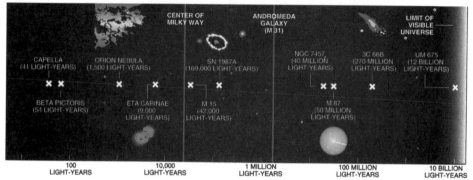

*Astronomical vistas available to Hubble span a tremendous range of distances. Pluto
lies an average of six light-hours from the earth (one light-hour equals roughly one billion kilometers). In comparison the quasar UM 675 sits some 12 billion light-years
away. Distances are shown on a logarithmic scale. (Ian Worpole)*

One highly intriguing discovery by the telescope concerns a later stage in the
birth of stars, when surrounding planets may begin to form. In 1983 the Infrared
Astronomical Satellite revealed that the nearby star Beta Pictoris (54 light-years
away) is encircled by a thin disk of gas and dust at least 80 billion kilometers across,
or nearly 10 times the size of Pluto's orbit. Such disks are thought to be the raw
material from which planetary systems coalesce. To learn more about the disk, Albert Boggess of the Goddard Space Flight Center and his colleagues used the High-
Resolution Spectrograph to examine ultraviolet radiation from Beta Pictoris.

Boggess finds that some of the circumstellar gas is falling in toward the star.
Moreover, the star's spectrum varies in appearance, probably because as the disk revolves different patches of radiation-absorbing matter pass between Beta Pictoris
and the earth. The rapidity of the variation—some spectral changes occur in less
than a month—suggests that the disk is rather clumpy. These findings hint that new
material is being released into the disk, possibly gas evaporating from cometlike objects orbiting the star.

The Hubble data do not tell whether fully formed planets are present around Beta Pictoris. Most astronomers will not be convinced of the discovery of extrasolar planets until a camera takes a picture of one. In this area Hubble's optical defect has dealt a crushing blow. The foggy halo of scattered starlight from the misshapen mirror would obliterate the faint light reflected from planets orbiting even the nearest stars. A dedicated search for extrasolar planets, one of the key projects intended for Hubble, will have to wait until the telescope's optical system is corrected.

Beta Pictoris is a fairly sedate, sunlike star. In contrast the highest-priority target in Hubble's first round of observations was the unstable, rapidly evolving star Eta Carinae, located 9,000 light-years away in the southern sky. Eta Carinae may be the most massive and energetic star in the Milky Way, about 100 times as massive and 4 million times as luminous as the sun. It flared up in 1843, briefly becoming the second brightest star in the sky. Subsequent observations showed that a small nebula, called the Homunculus, had started forming around the star.

The best ground-based observations show the Homunculus as a small, fuzzy oblong nebula. Hubble unveiled a much more complicated picture of the object. The Homunculus has a peculiar, peanut-shaped form; two opposing, tightly focused jets gush from its middle. The cloud appears clumpy and has a sharp outer edge, suggesting that it is actually a thin, dusty shell of matter rather than a filled volume. Presumably it consists of material either ejected from or swept up by Eta Carinae as a result of its outburst.

One of the jets terminates in a U-shaped feature. This structure is probably a bow shock, analogous to the wake around the bow of a moving ship, that formed when the jet penetrated the slow-moving interstellar matter around the star. A baffling series of parallel lines of luminous gas, resembling the rungs of a ladder, protrudes to one side of Eta Carinae. Perhaps the rungs are light-year-long standing waves, like the sound waves inside an organ pipe. Or they may be ripples that develop as matter rapidly flows along the jet's bow shock. Once again, the possible existence of artifacts from the computer enhancement process complicates interpretation of the image.

While Hubble has deepened the mystery of Eta Carinae, it has answered a long-standing question about a more distant denizen of the galaxy, the globular cluster M 15. Globular clusters are dense spherical swarms of up to one million stars. Such a tremendous concentration of stars, many astronomers reason, should be conducive to the formation of a black hole, a collapsed object whose gravity is so strong that even light cannot escape. Other researchers disagree, suggesting that the rapid motions of stars, especially binary stars, near the center could help buoy the core and prevent a catastrophic collapse.

One likely black-hole candidate was the bright cluster M 15, located 42,000 light-years way in the constellation Pegasus. Although the hole itself would be invisible, vast quantities of radiation ought to blaze from its immediate environs, where matter under extreme stress and tidal friction grows tremendously hot before vanishing forever. Therefore, a hole should produce a characteristic bright point of light at the center of the cluster. Until now, no telescope could resolve the core in sufficient detail to disclose such a "brightness spike." Hubble can discern details in M 15 as small as about .02 light-year, close to the theoretical diameter of a black hole having 1,000 times the mass of the sun.

Tod R. Lauer of Kitt Peak National Observatory searched for such a brightness spike in M 15. When he failed to find it, he and his colleagues turned to another technique. Hubble was able to resolve the bright red giant stars in the outskirts of M 15, enabling the researchers to subtract them out of the image. What remained was a core region containing thousands of faint stars spread out over a surprisingly large radius of about .4 light-year, 10 times that predicted by the black-hole models. The Hubble results strongly suggest that M 15 does not harbor a black hole at its center. This finding comes as a relief to those of us made uncomfortable by the popular tendency to invoke an invisible black hole to explain every powerful object in the universe. Astronomers have been especially eager to obtain Hubble views of the remnant of SN 1987A, the bright supernova that appeared in 1987 in the Large Magellanic Cloud, a satellite galaxy of the Milky Way. On August 23, 1990, Hubble's Faint Object Camera transmitted a 28-minute exposure of SN 1987A. The image that appeared on our computer screens at the Space Telescope Science Institute showed a remarkable luminescent ring of matter 1.4 light-years across surrounding the supernova remnant. Most of us in the room were astounded by the existence of the ring. Our colleague Nino Panagia at the institute was not; he had actually expected the formation of a peculiar outer structure.

The intriguing aspect about the supernova ring is that it looks elliptical, not round. Its shape implies that the feature is not a three-dimensional shell (shells exist around many planetary nebulae; because of perspective effects, they often resemble circular rings). It appears to be a genuine circular torus of material inclined 43 degrees to our line of sight, giving it an elliptical appearance. Such a formation could not have been produced by the supernova itself. Rather the ring must be a ghostly relic of the expelled outer layers of the progenitor red giant star.

Panagia thinks that thousands of years before the explosion a gentle stellar wind carried off the star's outer envelope, mostly in the equatorial direction. A subsequent, faster wind compressed the material into a gaseous ring. Ultraviolet radiation from the supernova explosion heated and ionized the gas, causing it to glow. Within a few decades the ring should disintegrate when it is disrupted by debris from the supernova that is now moving outward at an average velocity of 10,000 kilometers per second. Hubble will continue to monitor the protean structure of the remnant of SN 1987A.

Hubble observations of the supernova have led to a vastly improved knowledge of the distance to the Large Magellanic Cloud. The images show very precisely the angular size of the ring. The International Ultraviolet Explorer satellite monitored the timing of when the near and far edges of the ring first began to glow; this information, when combined with the well-known speed of light, yields the ring's true diameter. Simple trigonometry then reveals that the distance to the supernova, and hence to the surrounding galaxy, is 169,000 light-years. This estimate is accurate to within 5 percent, more than three times better than earlier measurements. Such information will be important for calibrating the distance scale to other, more remote cosmic objects.

Unfortunately, Hubble's general ability to measure galactic distances has been severely compromised. Its optical flaw makes it impossible for the telescope to distinguish individual Cepheid variable stars in faraway galaxies. These stars are of tremendous interest because their brightness fluctuates in a regular manner: the period of variation is related to their absolute luminosity. Observations of Cepheids can therefore provide an unambiguous measurement of the distance of a galaxy. The study of Cepheids will be an important task for Hubble when it is repaired in 1994.

Many people mistakenly think that because of Hubble's impaired light-gathering ability, it cannot study distant celestial objects. Nothing could be farther from the truth. The space telescope has made substantive observations of objects lying nearly at the limits of the visible universe. It has also enabled optical astronomers to study at an unprecedented level of resolution the cores of galaxies beyond our local galaxy cluster; the result has been many remarkable and often unexpected findings.

One of Hubble's earliest targets outside the Milky Way was NGC 7457, an elliptical galaxy about 40 million light-years away. The galaxy was chosen as a seemingly normal test subject. But when Lauer and his colleagues used the Wide Field and Planetary Camera to explore the central regions of NGC 7457, they were surprised by what they saw. A significant fraction of the galaxy's light arises from a pointlike source, no more than 10 light-years across, lying at the very heart of the nucleus. Stars there must be packed together at least 30,000 times as tightly as the stars in the sun's galactic neighborhood, hundreds of times the stellar density that astronomers theoretically expected.

The central brightness spike may denote the location where vast quantities of material—perhaps entire stars—spiral into a black hole having millions of times the mass of the sun. Alternatively the bright region might be something less exotic but still unanticipated, such as an exceptionally rich cluster of stars. Researchers working with Hubble's spectrographs will soon attempt to measure the orbital velocity of gas and stars in the core, thereby indicating the total amount of matter present there and helping to determine the true nature of the central object.

Hubble has also produced corroborating evidence that a huge black hole may indeed exist where astronomers did expect to find one: in the giant elliptical galaxy M 87, located in the Virgo galaxy cluster, roughly 50 million light-years away. M 87 strongly emits radio waves and x-rays. A gigantic jet of ionized gas (thousands of times the length of the jets from Eta Carinae) points outward from the galaxy's center. Theorists speculated that a massive black hole might be the central engine driving all this activity.

Lauer, working with Sandra M. Faber of the University of California at Santa Cruz, C. Roger Lynds of the National Optical Astronomy Observatories, and several other collaborators, called on the Wide Field and Planetary Camera to help settle the question. They found that, as with NGC 7457, stars crowd together in the central region of M 87 hundreds of times more tightly than they would in a normal galaxy. If a black hole is in fact responsible for M 87's dense and overly bright core, it must have a mass a few billion times that of the sun.

F. Duccio Macchetto of the European Space Agency is interested in the violent processes occurring in even more distant active galaxies. He turned the Faint Object Camera to collect ultraviolet radiation from the galaxy 3C 66B, which sits 270 million light-years from the earth. Hubble revealed the details of 3C 66B's extraordinary jet of glowing plasma. The jet extends 10,000 light-years from the center of the galaxy, twice the length of M 87's gaseous protrusion.

Macchetto exploited computer image processing to make the jet more visible by subtracting out the image of the host galaxy (this is possible because the jet is brightest at ultraviolet wavelengths, at which the galaxy is faint). He and his coworkers then corrected for Hubble's spherical aberration. They were thus able to glimpse gaseous filaments, bright knots, and odd kinks in the jet material, details never before seen through an optical telescope. The jet exhibits an odd, braided structure consisting of two plasma strands situated 500 light-years apart.

The observed features closely correspond to those identified by their radio emission. But visible radiation from the jet is produced by high-speed electrons that lose their energy far more rapidly than do the relatively sluggish electrons that produce the radio jet. The most energetic, light-emitting electrons exist in regions that have been recently disrupted. In comparison to radio astronomy studies Hubble's data trace out much more recent behavior of the potent forces that produced the huge jet.

Further Hubble observations of active galaxies and of their more energetic cousins, quasars and Seyfert galaxies, will show how energy travels outward along the jets. Hubble studies will also help define the role of magnetic fields in channeling matter (mostly electrons moving at nearly the speed of light) from the cores of galaxies into intergalactic space. A better knowledge of active galaxies will help determine whether black holes can account for their prodigious energy output or if astronomers need to come up with new, perhaps even more exotic, explanations for these celestial powerhouses.

Hubble is also helping test and refine the big-bang theory, which forms the foundation of modern cosmology. The theory states that the present universe, including all matter and all space, exploded outward from a single point roughly 15 billion years ago. If the big-bang is correct, most of the helium in the universe was created in the moments after the birth of the universe. Since then, however, some additional helium has been synthesized in the interiors of stars via nuclear fusion.

Margaret E. Burbidge of the University of California at San Diego sought to measure the abundance of helium near the quasar UM 675, which lies some 12 billion light-years from the earth. Because of its great distance, humans now see UM 675 as it was 12 billion years ago, when the universe was one fifth its present age. The big-bang theory predicts that UM 675 should contain nearly as much helium as do modern, nearby objects. If the theory is wrong, the helium abundance in UM 675 might be close to zero. Hubble's Faint Object Spectrograph showed a clear signature of helium, lending credence to the big bang.

In another experiment Jeffrey Linsky of the University of Colorado and his colleagues attempted to measure the cosmic abundance of deuterium, a heavy version of the element hydrogen. Theory implies that the amount of deuterium created immediately after the big bang reflects the overall density of the universe. Linsky and his coworkers used Hubble's High-Resolution Spectrograph to observe the spectrum of the bright star Capella, situated 41 light-years away in the constellation Auriga. The exact shape of the spectrum indicates the amount of radiation absorbed by hydrogen and by deuterium atoms between Capella and the earth.

Linsky's analysis of the spectrum suggests that the universe contains only one tenth the amount of ordinary matter necessary to halt the present expansion. Many cosmologists speculate that exotic, undiscovered particles may add considerably to the total mass of the universe. If not, the Hubble results imply that the universe is infinite and that its expansion will continue for the rest of eternity.

Despite a plethora of optical and mechanical shortfalls, Hubble is proving a powerful scientific research tool. The results from its first two years of operation offer only a small taste of what the telescope should be capable of if it receives an optical fix in 1994. After centuries of being condemned to watching stars twinkle and dance in the earth's turbulent atmosphere, optical astronomers have finally entered a new age of space-based research. The Hubble Space Telescope, I hope, is only the first step.

—June 1992

THE INTERNATIONAL SPACE STATION

A Work in Progress

Tim Beardsley

The construction site in space that is for the next six years the International Space Station is nothing if not ambitious. Writers have an array of superlatives they can choose from to describe the program: it is by far the most complex in-orbit project ever attempted and arguably one of the biggest engineering endeavors of any kind. More than 100 separate elements weighing 455,000 kilograms (over a million pounds) on the earth will be linked together during the assembly operation, making it the most massive thing in orbit: it will have the equivalent of two 747 jetliners' worth of laboratory and living space. The job will need 45 flights by U.S. shuttles and Russian rockets, and over 50 more launches will take up supplies, crew, and fuel to maintain the station in its orbit. Contributions come from 16 countries, making it the most cosmopolitan space program. Hooking the pieces together will take at least 1,700 hours of space walks, many more than have been made during the entire history of space exploration to date. Robotic arms and hands will be required, and free-flying robotic "eyes" might be employed for inspection flights.

But one remarkable aspect of the project received little attention during the hoopla surrounding the successful launch and mating of the first two components in 1997. With construction work on the station well under way in its orbit 400 kilometers (250 miles) up, the final configuration of the edifice is not yet settled. Indeed, it could look very different from current artists' impressions.

In large part the changes are the result of pressure that Congress has put on the National Aeronautics and Space Administration to reduce the program's near-total reliance on Russia as a provider of essential station components and rocket launches. Concern has focused especially on the Russian Service Module, which is scheduled to provide living quarters, life support, propulsion, navigation, and communications for the station during the early years of assembly. The Service Module will, if all goes well, be the next major component in orbit after the Zarya tug and Unity Connecting Module that are now flying.

But all has not been going well with construction of the Service Module at the Khrunichev State Research and Production Space Center in Moscow. Originally scheduled for completion in April 1998, the module has been a victim of Russia's financial crisis. Work on the module, which was originally to be part of a Russian space station, started as long ago as 1985, long before Russia joined the International Space Station. Russia's failure to finish the component in time is the main reason the start of station assembly was delayed from 1997 until late 1998. Without the propulsion provided by the Service Module, the station as originally envisaged would be incapable of staying in orbit for more than 500 days. Friction with the sparse air molecules in low-earth orbit would gradually cause it to lose altitude.

NASA has had to employ creative accounting techniques to justify sending the Russian Space Agency ever-mounting sums to complete the module. Last year it gave the Russians an extra $60 million (the official explanation was that these funds would purchase additional stowage space and experiment time for the U.S. during the construction phase). But NASA has acknowledged that over the next four years it will most likely have to send a further $600 million to ensure the completion of other modules. Many Russian space workers have not been paid for months.

■   THE PRICE OF PROGRESS

This $660-million contribution is in addition to $728 million NASA has already paid the Russians between 1994 and 1998 for space station work and the joint flights on the Russian space station Mir, according to the Congressional Research Service. Although having Russia in the program was originally intended to save money, NASA now admits that it has actually added about $1 billion to the station's cost. NASA has had to work hard to secure from the Russians an agreement that they will shut down the Mir space station this summer, despite opposition from Russian nationalists. Keeping Mir alive could drain Russian resources from the international station, NASA fears.

Not that cost overruns are restricted to Russia. NASA figures indicate that U.S. construction costs are running 30 percent over projections, and an independent commission headed by Jay Chabrow, a former TRW executive, estimated that the overrun will reach 42 percent. NASA has irked scientists who had planned to run experiments on the station by transferring some $460 million from science accounts to help meet U.S. construction costs. The station's expense, including the cost of shuttle flights, is now likely to exceed $40 billion, and it has become "an albatross around the agency's neck," in the view of space policy expert Marcia S. Smith of the Congressional Research Service. The General Accounting Office puts the total cost of the program at $95.6 billion.

All these estimates assume nothing major goes wrong during assembly. The British magazine *New Scientist* has decided, on the basis of a statistical analysis of risks, that there is a 73.6 percent chance of at least one catastrophic failure that would result in the loss of station hardware during one of the U.S. or Russian assembly launches.

While the costs of keeping Russia as a partner have been growing, its planned contributions have declined. Russian officials have announced a "core program" on the space station that no longer includes a science power platform, two research laboratories, and a life-support module. Russia is discussing constructing one laboratory with Ukraine—but "we don't see much design and development work" on the life-support module, says W. Michael Hawes, Sr., senior engineer for the space station. Hawes says the changing design has now made the Russian life-support module redundant. The status of other Russian components is unclear. Perhaps more worrying, Russia is unlikely to be able to supply the seven Progress and two Soyuz refueling and crew-rotation flights each year that it had undertaken to do: congressional overseers now think five such flights each year is more realistic.

To satisfy Congress's demands for a backup plan, NASA has quietly been changing the assembly sequence and designing and modifying hardware to reduce its vulnerability. The first of these late-arriving additions is a $156-million Interim Control Module, which is now nearing completion at the Naval Research Laboratory. The module is a modified version of a previously classified upper-stage rocket, and it could by itself provide attitude control and reboost for the station for a year or two. NASA also modified Zarya (which the U.S. owns) prior to launch to improve its station boosting and control capabilities.

The European Space Agency has agreed to provide propellant for the Service Module, according to Daniel Hedin of NASA's space development office. And NASA is now also planning to modify all its space shuttles to increase their capacity to boost the station. The fix should mean the station needs only about 30 Progress refueling boosts instead of the baseline number of 53, according to Hedin. More-over, NASA does not rule out launching the Interim Control Module sometime in 2000 even if the Service Module does launch this year, because it would provide in-surance against a future shortage of Progress rockets.

The Interim Control Module will not be the only addition to the station undertaken because of Russia's crippling budget problems. NASA is now also negotiating with Boe-ing to build a U.S. propulsion module, at an expected cost of $350 million. It would eliminate the need for about half of the currently scheduled Progress resupply flights and offer a permanent solution in the event that the Service Module never arrives.

Other aspects of the station are almost as fluid. No final decisions have yet been made on provisions for returning crew to the earth in the event of some emer-gency. In the early construction phase that role will be played by a Soyuz spacecraft attached to the station. A Soyuz, however, can transport only three astronauts, and the station's final scheduled crew numbers seven. The U.S. is planning to build a larger Crew Return Vehicle capable of bringing home all the permanent crew, but it will most likely not be ready until 2003 at the earliest, and the station will probably have a crew of more than three before then. NASA is considering buying one or more Soyuz vehicles to provide an interim emergency return capability.

In any event the U.S. crew return vehicle's final form is still undecided. The current design, based on the X-38 experimental craft, offers only nine hours of life support. NASA and the European Space Agency are discussing modifications to the design that would turn it into a transfer vehicle that could be launched on an Ariane rocket.

Even the basic design of the main American habitation module is still up for grabs. Engineers at the NASA Johnson Space Center have proposed an inflatable structure known as TransHab as a substitute for the aluminum habitation module in the present design. TransHab would have a hard composite core surrounded by Kevlar and foam layers for micrometeorite protection. Its main selling point is that it might serve to test a mode of construction that could, because of its low mass, be advantageous in future crewed moon or Mars expeditions.

But the station's value as a test bed for a future crewed mission to Mars can be questioned. The most important physical hazards facing such a crew are likely to be loss of bone mass, which seems to be a common result of prolonged weightlessness, and radiation from solar storms. Yet a vehicle designed to go to Mars could easily be furnished with artificial gravity, by separating it into two connected sections and slowly spinning them, says Ivan Bekey, a former head of advanced concepts at NASA. Furthermore, the station's orbit is too low to experience the full fury of solar storms. An earlier design would have tested five innovative space technologies, including a high-voltage power transmission system and solar-thermal power generation. They, however, were dropped from the final scheme, Bekey notes.

The International Space Station is principally a foreign-policy enterprise. And as such it may be a success. Thousands of Russian scientists and engineers who without the American bailout might have gone to well-paying jobs designing weapons for rogue states are now still at work on peaceful systems. Politicians and officials and technical experts in countries throughout the world have had the opportunity to collaborate and link their destinies in an organizationally demanding endeavor. Perhaps the value of that return cannot be measured in dollars.

KEY SPACE EXPLORATIONS OF THE NEXT DECADE

Name of Mission (Sponsor)	Main Purpose of Mission	Launch Date
THE SUN		
ACE, Advanced Composition Explorer (NASA)	Monitor solar atomic particles and the interplanetary environment	1997
TRACE, Transition Region and Coronal Explorer (NASA)	Photograph the sun's coronal plasmas in the ultraviolet range	1998
Coronas F (Russia)	Observe the sun's spectrum during a solar maximum	1999
HESSI, High Energy Solar Spectroscopic Imager (NASA)	Study solar flares through x-rays, gamma rays, and neutrons	2000
Photon (Russia)	Analyze gamma rays from the sun	2000
SST, Space Solar Telescope (China and Germany)	Study the sun's magnetic field	2001
Genesis (NASA)	Gather atomic nuclei from the solar wind and return them to the earth	2001
Solar B (Japan)	Study the sun's magnetic field around violent events	2004
Solar Probe (NASA)	Measure particles, fields, x-rays, and light in the sun's corona	2007

Continued

KEY SPACE EXPLORATIONS OF THE NEXT DECADE (*Continued*)

Name of Mission (Sponsor)	Main Purpose of Mission	Launch Date
THE MOON		
Lunar A (Japan)	Analyze the moon's subsurface soil	1999
Euromoon 2000 (ESA)	Explore the moon's south pole (two-part mission)	2000 and 2001
Selene (Japan)	Map the moon, studying fields and particles	2003
THE PLANETS		
Mars Global Surveyor (NASA)	Map Mars and relay data from other missions	1996
Planet-B (Japan)	Study interactions between the solar wind and Mars's atmosphere	1998
Mars Surveyor 1998 (NASA)	Explore a site near Mars's south pole (two-part mission)	1998 and 1999
Deep Space 2 (NASA)	Analyze Martian subsurface soil	1999
Mars Surveyor 2001 (NASA)	Land a rover on Mars (two-part mission)	2001
VESPER, Venus Sounder for Planetary Exploration (NASA)	Observe Venus's atmosphere (under study)	2002
Mars Surveyor 2003 (NASA)	Collect Martian soil samples (two-part mission, under study)	2003
Mars Express (ESA)	Analyze Martian soil, using an orbiter and two landers	2003
Europa Orbiter (NASA)	Determine if Jupiter's fourth largest moon has an ocean	2003
MESSENGER, Mercury Surface, Space Environment, Geochemistry and Ranging (NASA)	Map Mercury and its magnetic field (under study)	2004
Pluto-Kuiper Express (NASA)	Explore the solar system's only unvisited planet and the Kuiper belt (under study)	2004
Mars Surveyor 2005 (NASA)	Return Martian rock and soil samples to Earth (under study)	2005
COMETS		
CONTOUR, Comet Nucleus Tour (NASA)	Produce spectral maps of three comet nuclei	2002

Continued

KEY SPACE EXPLORATIONS OF THE NEXT DECADE (*Continued*)

Name of Mission (Sponsor)	Main Purpose of Mission	Launch Date
COMETS (Continued)		
Deep Space 4 (NASA)	Land a probe on comet Tempel 1's nucleus	2003
Rosetta (ESA and France)	Land a probe on comet Wirtanen's nucleus	2003
ASTEROID BELT		
Deep Space 1 (NASA)	Test spacecraft technologies en route to asteroid 1992 KD	1998
MUSES-C (Japan)	Return a sample of material from an asteroid	2002
DEEP SPACE		
RXTE, Rossi X-ray Timing Explorer (NASA)	Watch x-ray sources change over time	1995
Beppo-SAX (Italy and the Netherlands)	Observe x-ray sources over a wide energy range	1996
HALCA (Japan)	Study galactic nuclei and quasars via radio interferometry	1997
SWAS, Submillimeter Wave Astronomy Satellite (NASA)	Search for oxygen, water, and carbon in interstellar clouds	1998
Odin (Sweden)	Detect millimeter-wavelength emissions from oxygen and water in interstellar gas	1999
FUSE, Far Ultraviolet Spectroscopic Explorer (NASA)	Detect deuterium in interstellar space	1999
WIRE, Wide-Field Infrared Explorer (NASA)	Observe galaxy formation with a cryogenic telescope	1999
ABRIXAS, A Broad-band Imaging X-ray All-sky Survey (Germany)	Make a hard x-ray, all-sky survey	1999
SXG, Spectrum X-Gamma (Russia)	Measure x-ray emissions from pulsars, black holes, supernova remnants, and active galactic nuclei	1999
HETE II, High Energy Transient Experiment (NASA)	Study gamma-ray bursters	1999
XMM, X-ray Multi-Mirror (ESA)	Observe spectra of cosmic x-ray sources	2000

Continued

KEY SPACE EXPLORATIONS OF THE NEXT DECADE (*Continued*)

Name of Mission (Sponsor)	Main Purpose of Mission	Launch Date
DEEP SPACE (Continued)		
Astro-E (Japan)	Make high-resolution x-ray observations	2000
MAP, Microwave Anisotropy Probe (NASA)	Study the universe's origin and evolution through the cosmic-microwave background	2000
Radioastron (Russia)	Observe high-energy phenomena via radio interferometry	2000
SIRTF, Space Infrared Telescope Facility (NASA)	Make infrared observations of stars and galaxies	2001
INTEGRAL, International Gamma-Ray Astrophysics Lab (ESA)	Obtain spectra of neutron stars, black holes, gamma-ray bursters, and active galactic nuclei	2001
GALEX, Galaxy Evolution Explorer (NASA)	Observe stars, galaxies, and heavy elements at ultraviolet wavelengths (under study)	2001
Spectrum UV (Russia)	Study astrophysical objects at ultraviolet wavelengths	2001
Deep Space 3 (NASA)	Test techniques for flying spacecraft in formation	2002
Corot (France)	Search for evidence of planets around distant stars	2002
SIM, Space Interferometry Mission (NASA)	Image stars that may host earth-like planets (under study)	2005
Constellation X-Ray Mission (NASA)	Perform high-resolution x-ray spectroscopy (under study)	After 2005
OWL, Orbiting Wide-Angle Light Collectors (NASA)	Study cosmic-ray effects on earth's atmosphere (under study)	After 2005
FIRST, Far Infrared Sub-millimeter Telescope, and Planck (ESA)	Discern the fine structure of the cosmic-microwave background (combined mission)	2007
NGST, Next Generation Space Telescope (NASA)	View space at infrared wavelengths (under study)	2008
TPF, Terrestrial Plant Finder (NASA)	Find planets and protoplanets orbiting nearby stars (under study)	2010

Six Months on Mir

As the Shuttle-Mir program draws to a close, a veteran
NASA astronaut reflects on her mission on board
the Russian spacecraft and the implications for
the International Space Station

Shannon W. Lucid

For six months, at least once a day, and many times more often, I floated above the large observation window in the Kvant 2 module of Mir and gazed at the earth below or into the depths of the universe. Invariably I was struck by the majesty of the unfolding scene. But to be honest, the most amazing thing of all was that here I was, a child of the pre-Sputnik, cold war 1950s, living on a Russian space station. During my early childhood in the Texas Panhandle, I had spent a significant amount of time chasing windblown tumbleweeds across the prairie. Now I was in a vehicle that resembled a cosmic tumbleweed, working and socializing with a Russian air force officer and a Russian engineer. Just 10 years ago such a plotline would have been deemed too implausible for anything but a science-fiction novel.

In the early 1970s both the American and Russian space agencies began exploring the possibility of long-term habitation in space. After the end of the third Skylab mission in 1974, the American program focused on short-duration space shuttle flights. But the Russians continued to expand the time their cosmonauts spent in orbit, first on the Salyut space stations and later on Mir, which means "peace" in Russian. By the early 1990s, with the end of the cold war, it seemed only natural that the U.S. and Russia should cooperate in the next major step of space exploration, the construction of the International Space Station. The Russians formally joined the partnership—which also includes the European, Japanese, Canadian, and Brazilian space agencies—in 1993.

The first phase of this partnership was the Shuttle-Mir program. The National Aeronautics and Space Administration planned a series of shuttle missions to send American astronauts to the Russian space station. Each astronaut would stay on Mir for about four months, performing a wide range of peer-reviewed science experiments. The space shuttle would periodically dock with Mir to exchange crew members and deliver supplies. In addition to the science NASA's goals were to learn how to work with the Russians, to gain experience in long-duration spaceflight, and to reduce the risks involved in building the International Space Station. Astronaut Norm Thagard was the first American to live on Mir. My own arrival at the space station—eight months after the end of Thagard's mission—was the beginning of a continuous American presence in space, which has lasted for more than two years.

My involvement with the program began in 1994. At that point I had been a NASA astronaut for 15 years and had flown on four shuttle missions. Late one Friday afternoon I received a phone call from my boss, Robert "Hoot" Gibson, then the head of NASA's astronaut office. He asked if I was interested in starting full-time Russian-language instruction with the possibility of going to Russia to train for a Mir mission. My immediate answer was yes. Hoot tempered my enthusiasm by saying I was only being assigned to study Russian. This did not necessarily mean I would be going to Russia, much less flying on Mir. But because there was a possibility that I might fly on Mir and because learning Russian requires some lead time—a major understatement if ever there was one—Hoot thought it would be prudent for me to get started.

I hung up the phone and for a few brief moments stared reality in the face. The mission on which I might fly was less than a year and a half away. In that time I would have to learn a new language, not only to communicate with my crewmates in orbit but to train in Russia for the mission. I would have to learn the systems and operations for Mir and Soyuz, the spacecraft that transports Russian crews to and from the space station. Because I would be traveling to and from Mir on the space shuttle, I needed to maintain my familiarity with the American spacecraft. As if that were not enough, I would also have to master the series of experiments I would be conducting while in orbit.

It is fair at this point to ask, "Why?" Why would I wish to live and work on Mir? And from a broader perspective, why are so many countries joining together to build a new space station? Certainly one reason is scientific research. Gravity influences all experiments done on the earth except for investigations conducted in drop towers or on airplanes in parabolic flight. But on a space station, scientists can conduct long-term investigations in an environment where gravity is almost nonexistent—the microgravity environment. And the experience gained by maintaining a continuous human presence in space may help determine what is needed to support manned flights to other planets.

From a personal standpoint I viewed the Mir mission as a perfect opportunity to combine two of my passions: flying airplanes and working in laboratories. I received my private pilot's license when I was 20 years old and have been flying ever since. And before I became an astronaut I was a biochemist, earning my Ph.D. from the University of Oklahoma in 1973. For a scientist who loves flying, what could be more exciting than working in a laboratory that hurtles around the earth at 17,000 miles (27,000 kilometers) per hour?

After three months of intensive language study, I got the go-ahead to start my training at Star City, the cosmonaut training center outside Moscow. My stay there began in January 1995, in the depths of a Russian winter. Every morning I woke at five o'-clock to begin studying. As I walked to class I was always aware that one misstep on the ice might result in a broken leg, ending my dreams of a fight on Mir. I spent most of my day in classrooms listening to Mir and Soyuz system lectures—all in Russian, of course. In the evenings I continued to study the language and struggled with workbooks written in technical Russian. At midnight I finally fell exhausted into bed.

I worked harder during that year than at any other time in my life. Going to graduate school while raising toddlers was child's play in comparison. (Fortunately, my three children were grown by this point, and my husband was able to visit me in Russia.) At last, in February 1996, after I had passed all the required medical and technical exams, the Russian spaceflight commission certified me as a Mir crew member. I traveled to Baikonur, Kazakhstan, to watch the launch of the Soyuz carrying my crewmates—Commander Yuri Onufriyenko, a Russian air force officer, and flight engineer Yuri Usachev, a Russian civilian—to Mir. Then I headed back to the U.S. for three weeks of training with the crew of shuttle mission STS-76. On March 22, 1996, we lifted off from the Kennedy Space Center on the shuttle Atlantis. Three days later the shuttle docked with Mir, and I officially joined the space station crew for what was planned to be a four-and-a-half-month stay.

■ LIVING IN MICROGRAVITY

My first days on Mir were spent getting to know Onufriyenko and Usachev—we spoke exclusively in Russian—and the layout of the space station. Mir has a modular design and was built in stages. The first part, the Base Block, was launched in February 1986. Attached to one end of the Base Block is Kvant 1, launched in 1987, and at the other end is Mir's transfer node, which serves the same function as a hallway in a house. Instead of being a long corridor with doors, though, the transfer node is a ball with six hatches. Kvant 2 (1989), Kristall (1990), and Spektr (1995) are each docked to a hatch. During my stay on Mir the Russians launched Priroda, the final module of the space station, and attached it to the transfer node. Priroda contained the laboratory where I conducted most of my experiments. I stored my personal belongings in Spektr and slept there every night. My commute to work was very short—in a matter of seconds I could float from one module to the other.

The two cosmonauts slept in cubicles in the Base Block. Most mornings the wake-up alarm went off at eight o'clock (Mir runs on Moscow time, as does the Russian mission control in Korolev). In about 20 minutes we were dressed and ready to start the day. The first thing we usually did was put on our headsets to talk to mission control. Unlike the space shuttle, which transmits messages via a pair of communications satellites, Mir is not in constant contact with the ground. The cosmonauts can talk to mission control only when the space station passes over one of the communications ground sites in Russia. These "comm passes" occurred once an orbit—about every 90 minutes—and generally lasted about 10 minutes. Commander Onufriyenko wanted each of us to be "on comm" every time it was available, in case the ground needed to talk to us. This routine worked out well because it gave us short beaks throughout the day. We gathered in the Base Block and socialized a bit before and after talking with mission control.

After the first comm pass of the day, we ate breakfast. One of the most pleasant aspects of being part of the Mir crew was that we ate all our meals together, floating around a table in the Base Block. Preflight, I had assumed that the repetitive nature of the menu would dampen my appetite, but to my surprise I was hungry for every meal. We ate both Russian and American dehydrated food that we reconstituted with hot water. We experimented with mixing various packages to create new tastes, and we each had favorite mixtures that we recommended to the others. For breakfast I liked to have a bag of Russian soup—usually borscht or vegetable—and a bag of fruit juice. For lunch or supper I liked the Russian meat-and-potato casseroles. The Russians loved the packets of American mayonnaise, which they added to nearly everything they ate.

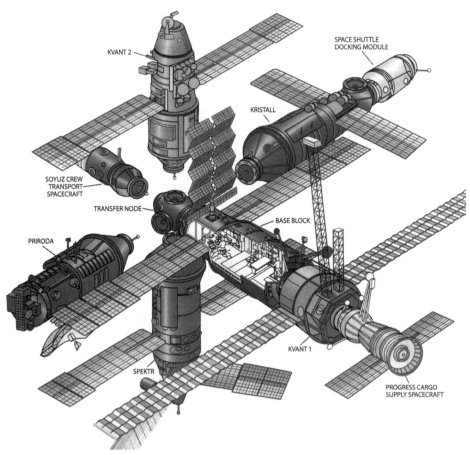

KVANT 2

SPACE SHUTTLE
DOCKING MODULE

KRISTALL

SOYUZ CREW
TRANSPORT
SPACECRAFT

TRANSFER NODE

BASE BLOCK

PRIRODA

KVANT 1

SPEKTR

PROGRESS CARGO
SUPPLY SPACECRAFT

Exploded view of Mir shows how the space station's modules fit together. Inside the Base Block is the command post, one of the treadmills, and the table where the crew members eat their meals. (Ian Worpole)

Our work schedule was detailed in a daily timeline that the Russians called the Form 24. The cosmonauts typically spent most of their day maintaining Mir's systems, while I conducted experiments for NASA . We had to exercise every day to prevent our muscles from atrophying in the weightless environment. Usually we all exercised just before lunch. There are two treadmills on Mir—one in the Base Block and the other in the Kristall module—and a bicycle ergometer is stored under a floor panel in the Base Block. We followed three exercise protocols developed by Russian physiologists; we did a different one each day, then repeated the cycle. Each protocol took about 45 minutes and alternated periods of treadmill running with exercises that involved pulling against bungee cords to simulate the gravitational forces we were no longer feeling. Toward the end of my stay on Mir I felt that I needed to be working harder, so after I finished my exercises I ran additional kilometers on the treadmill.

I'll be honest: the daily exercise was what I disliked most about living on Mir. First, it was just downright hard. I had to put on a harness and then connect it with bungee cords to the treadmill. Working against the bungees allowed me to stand flat on the device. With a little practice, I learned to run. Second, it was boring. The treadmill was so noisy you could not carry on a conversation. To keep my mind occupied, I listened to my Walkman while running, but soon I realized I'd made a huge preflight mistake. I had packed very few tapes with a fast beat. Luckily, there was a large collection of music tapes on Mir. During my six-month stay I worked through most of them.

When we had finished exercising, we usually enjoyed a long lunch, then returned to our work. Many times in the late afternoon we had a short tea break, and in the late evening we shared supper. By this point we had usually finished all the assignments on the Form 24, but there were still many housekeeping chores that needed to be done: collecting the trash, organizing the food supply, sponging up the water that had condensed on cool surfaces. Clutter was a problem on Mir. After we had unloaded new supplies from the unmanned Progress spacecraft that docked with the space station once every few months, we could put human wastes and trash into the empty vehicles, which would burn up on reentry into the atmosphere. But there was usually no room left on Progress for the many pieces of scientific equipment that were no longer in use.

After supper mission control would send us the Form 24 for the next day on the teleprinter. If there was time, we had tea and a small treat—cookies or candy— before the last comm pass of the day, which usually occurred between 10 and 11 at night. Then we said good night to one another and went to our separate sleeping areas. I floated into Spektr, unrolled my sleeping bag, and tethered it to a handrail. I usually spent some time reading and typing letters to home on my computer (we used a ham radio packet system to send the messages to the ground controllers, who sent them to my family by E-mail). At midnight I turned out the light and floated into my sleeping bag. I always slept soundly until the alarm went off the next morning.

■ QUAIL EGGS AND DWARF WHEAT

Our routine on Mir rarely changed, but the days were not monotonous. I was living every scientist's dream. I had my own lab and worked independently for much of the day. Before one experiment became dull, it was time to start another, with new equipment and in a new scientific field. I discussed my work at least once a day with Bill Gerstenmaier, the NASA flight director, or Gaylen Johnson, the NASA flight surgeon, both at Russian mission control. They coordinated my activities with the principal investigators—the American and Canadian scientists who had proposed and designed the experiments. Many times when we started a new experiment, Gerstenmaier arranged for the principal investigator to be listening to our radio conversations, so they would be ready to answer any questions I might have. And this was in the middle of the night back in the U.S.!

My role in each experiment was to do the onboard procedures. Then the data and samples were returned to the earth on the space shuttle and sent to the principal investigators for analysis and publication. I believe my experience on Mir clearly shows the value of performing research on manned space stations. During some of the experiments, I was able to observe subtle phenomena that a video or still camera would miss. Because I was familiar with the science in each experiment, I could sometimes examine the results on the spot and modify the procedures as needed. Also, if there was a malfunction in the scientific equipment, I or one of my crewmates could usually fix it. Only 1 of the 28 experiments scheduled for my mission failed to yield results because of a breakdown in the equipment.

I started my work on Mir with a biology experiment examining the development of embryos in fertilized Japanese quail eggs. The eggs were brought to Mir on the same shuttle flight that I took, then transferred to an incubator on the space station. Over the next 16 days I removed the 30 eggs one by one from the incubator and placed them in a 4 percent paraformaldehyde solution to fix the developing embryos for later analysis. Then I stored the samples at ambient temperature.

This description makes it sound like a simple experiment, but it required creative engineering to accomplish the procedure in a microgravity environment. NASA and Russian safety rules called for three layers of containment for the fixative solution; if a drop escaped, it could float into a crew member's eye and cause severe burns. Engineers at the NASA Ames Research Center designed a system of interlocking clear bags for inserting the eggs into the fixative and cracking them open. In addition, the entire experiment was enclosed in a larger bag with gloves attached to its surface, which allowed me to reach inside the bag without opening it.

Investigators at Ames and several universities analyzed the quail embryos at the end of my mission to see if they differed from embryos that had developed in an incubator on the ground. Remarkably, the abnormality rate among the Mir embryos was 13 percent—more than four times higher than the rate for the control embryos. The investigators believe two factors may have increased the abnormality rate: the slightly higher temperature in the Mir incubator and the much higher radiation levels on the space station. Other experiments determined that the average radiation exposure on Mir is the equivalent of getting eight chest x-rays a day. NASA scientists believe, however, that an astronaut would have to spend at least several years in orbit to raise appreciably his or her risk of developing cancer.

I was also involved in a long-running experiment to grow wheat in a greenhouse on the Kristall module. American and Russian scientists wanted to learn how wheat seeds would grow and mature in a microgravity environment. The experiment had an important potential application: growing plants could provide oxygen and food for long-term spaceflight. Scientists focused on the dwarf variety of wheat because of its short growing season. I planted the seeds in a bed of zeolite, an absorbent granular material. A computer program controlled the amount of light and moisture the plants received. Every day we photographed the wheat stalks and monitored their growth.

At selected times we harvested a few plants and preserved them in a fixative solution for later analysis on the ground. One evening, after the plants had been growing for about 40 days, I noticed seed heads on the tips of the stalks. I shouted excitedly to my crewmates, who floated by to take a look. John Blaha, the American astronaut who succeeded me on Mir, harvested the miniature plants a few months later and brought more than 300 seed heads back to the earth. But scientists at Utah State University discovered that all the seed heads were empty. The investigator speculated that low levels of ethylene in the space station's atmosphere may have interfered with the pollination of the wheat. In subsequent research on Mir, astronaut Michael Foale planted a variety of rapeseed that successfully pollinated.

The microgravity environment on the space station also provided an excellent platform for experiments in fluid physics and materials science. Scientists sought to further improve the environment by minimizing vibrations. Mir vibrates slightly as it orbits the earth and although the shaking is imperceptible to humans, it can have an effect on sensitive experiments. The movements of the crew and airflows on the station can also cause vibrations. To protect experiments from these disturbances, we placed them on the Microgravity Isolation Mount, a device built by the Canadian Space Agency. The top half of the isolation mount floats free, held in place solely by electromagnetic fields.

After running an extensive check of the mount, I used it to isolate a metallurgical experiment. I placed metal samples in a specially designed furnace, which heated them to a molten state. Different liquid metals were allowed to diffuse in small tubes, then slowly cooled. The principal investigators wanted to determine how molten metals would diffuse without influence of convection. (In a microgravity environment warmer liquids and gases do not rise, and colder ones do not sink.) After analyzing the results, they learned that the diffusion rate is much slower than on the earth. During the procedure one of the brackets in the furnace was bent out of alignment, threatening the completion of the experiment. But flight engineer Usachev simply removed the bracket, put it on a workbench, and pounded it straight with a hammer. Needless to say, this kind of repair would have been impossible if the experiment had taken place on an unmanned spacecraft.

Many of the experiments provided useful data for the engineers designing the International Space Station. The results from our investigations in fluid physics are helping the space station's planners build better ventilation and life-support systems. And our research on how flames propagate in microgravity may lead to improved procedures for fighting fires on the station.

■   SAFETY IN SPACE

Throughout my mission I also performed a series of earth observations. Many scientists had asked NASA to photograph parts of the planet under varying seasonal and lighting conditions. Oceanographers, geologists, and climatologists would incorporate the photographs into their research. I usually took the pictures from the Kvant 2 observation window with a handheld Hasselblad camera. I discovered that during a long spaceflight, as opposed to a quick space shuttle jaunt, I could see the flow of seasons across the face of the globe. When I arrived on Mir at the end of March, the higher latitudes of the Northern Hemisphere were covered with ice and snow. Within a few weeks, though, I could see huge cracks in the lakes as the ice started to break up. Seemingly overnight, the Northern Hemisphere glowed green with spring.

We also documented some unusual events on the earth's surface. One day as we passed over Mongolia we saw giant plumes of smoke, as though the entire country were on fire. The amount of smoke so amazed us that we told the ground controllers about it. Days later they informed us that news of huge forest fires was just starting to filter out of Mongolia.

For long-duration manned spaceflight, the most important consideration is not the technology of the spacecraft but the composition of the crew. The main reason for the success of our Mir mission was the fact that Commander Onufriyenko, flight engineer Usachev, and I were so compatible. It would have been very easy for language, gender, or culture to divide us, but this did not happen. My Russian crewmates always made sure that I was included in their conversations. Whenever practical, we worked on projects together. We did not spend time criticizing one another—if a mistake was made, it was understood, corrected, and then forgotten. Most important, we laughed together a lot.

The competence of my crewmates was one of the reasons I always felt safe on Mir. When I began my mission the space station had been in orbit for 10 years, twice as long as it had been designed to operate. Onufriyenko and Usachev had to spend most of their time maintaining the station, replacing parts as they failed and monitoring systems critical to life support. I soon discovered that my crewmates could fix just about anything. Many spare parts are stored on Mir, and more are brought up as needed on the Progress spacecraft. Unlike the space shuttle, Mir cannot return to the earth for repairs, so the rotating crews of cosmonauts are trained to keep the station functioning.

Furthermore, the crews on Mir have ample time to respond to most malfunctions. A hardware failure on the space shuttle demands immediate attention because the shuttle is the crew's only way to return to the earth. If a piece of vital equipment breaks down, the astronauts have to repair the damage quickly or end the mission early, which has happened on a few occasions. But Mir has a lifeboat: at least one Soyuz spacecraft is always attached to the space station. If a hardware failure occurs on Mir, it does not threaten the crew's safe return home. As long as the space station remains habitable, the crew members can analyze what happened, talk to mission control, and then correct the malfunction or work around the problem.

Only two situations would force the Mir crew to take immediate action: a fire inside the space station or a rapid depressurization. Both events occurred on Mir in 1997, after I left the station. In each case the crew members were able to contain the damage quickly.

My mission on the space station was supposed to end in August 1996, but my ride home—shuttle mission STS-79—was delayed for six weeks while NASA engineers studied abnormal burn patterns on the solid-fuel boosters from a previous shuttle flight. When I heard about the delay, my first thought was, "Oh no, not another month and a half of treadmill running!" Because of the delay, I was still on Mir when a new Russian crew arrived on the Soyuz spacecraft to relive Onufriyenko and Usachev. By the time I finally came back on the shuttle Atlantis on September 26, 1996, I had logged 188 days in space—an American record that still stands.

This June Astronaut Andrew Thomas—the last of the seven NASA astronauts who have lived on Mir over the past three years—is scheduled to return to the earth, ending the Shuttle-Mir program. Based on my own experience I believe there are several lessons that should be applied to the operation of the International Space Station. First, the station crew must be chosen carefully. Even if the space station has the latest in futuristic technology, if the crew members do not enjoy working together, the flight will be a miserable experience. Second, NASA must recognize that a long-duration flight is as different from a shuttle flight as a marathon is from a 100-yard dash. On a typical two-week shuttle flight, NASA ground controllers assign every moment of the crew's time to some task. But the crew on a long-duration flight must be treated more like scientists in a laboratory on the earth. They must have some control over their daily schedules.

Similarly, when a crew trains for a science mission on the space shuttle, the members practice every procedure until it can be done without even having to think about it. Training for a mission on the International Space Station needs to be different. When a crew member starts a new experiment on a long-duration flight, it might be up to six months after he or she trained for the procedure. The astronaut will need to spend some time reviewing the experiment. Therefore, their training should be skill-based. Crew members should learn the skills they will need during their mission rather than practice every specific procedure. Also, crew members on a long-duration flight need to be active partners in the scientific investigations they perform. Experiments should be designed such that the astronaut knows the science involved and can make judgment calls on how to proceed. An intellectually engaged crew member is a happy crew member.

When I reflect on my six months on Mir, I have no shortage of memories. But there is one that captures the legacy of the Shuttle-Mir program. One evening Onufriyenko, Usachev, and I were floating around the table after supper. We were drinking tea, eating cookies, and talking. The cosmonauts were very curious about my childhood in Texas and Oklahoma. Onufriyenko talked about the Ukrainian village where he grew up, and Usachev reminisced about his own Russian village. After a while we realized we had all grown up with the same fear: an atomic war between our two countries.

I had spent my grade school years living in terror of the Soviet Union. We practiced bomb drills in our classes, all of us crouching under our desks, never questioning why. Similarly Onufriyenko and Usachev had grown up with the knowledge that U.S. bombs or missiles might zero in on their villages. After talking about our childhoods some more, we marveled at what an unlikely scenario had unfolded. Here we were, from countries that were sworn enemies a few years earlier, living together on a space station in harmony and peace. And, incidentally, having a great time.

—May 1998

ABOUT THE AUTHORS*

Tim Beardsley is a *Scientific American* contributing editor.

David C. Black has done extensive research in theoretical astrophysics and planetary science, focusing on studies of star and planetary-system formation. Black received his Ph.D. in physics from the University of Minnesota. He worked for the National Aeronautics and Space Administration Ames Research Center from 1972 to 1988, when he assumed his current position as director of the Lunar and Planetary Institute in Houston. Black has studied the composition of noble gases in meteorites; he was the first to determine that meteorites contain material that originated from beyond the solar system.

Jonathan Bland-Hawthorn received his Ph.D. in astronomy and astrophysics from the University of Sussex and the Royal Greenwich Observatory. He is now a research astronomer at the Anglo-Australian Observatory in Sydney.

Hans Böhringer is a staff member of the Max Planck Institute for Extraterrestrial Physics in Garching, Germany, where he works with Ulrich G. Briel. He is a theorist who studies galaxy clusters, cosmology, and the interstellar medium.

Alan P. Boss is a professor in the Department of Terrestrial Magnetism at the Carnegie Institution of Washington.

Ulrich G. Briel is a staff member of the Max Planck Institute for Extraterrestrial Physics in Garching, Germany, where he works with Hans Böhringer. He is an observer who tested and calibrated the Roentgen Satellite (ROSAT) instrument that made the temperature maps discussed in the article included in this book. He met J. Patrick Henry in the late 1970s while working at the Smithsonian Astrophysical Observatory on one of the instruments on the Einstein X-Ray Observatory satellite.

John K. Cannizzo received his Ph.D. in theoretical astrophysics in 1984 from the University of Texas at Austin. He is a Humboldt Fellow at the Max Planck Institute for Astrophysics in Garching, Germany. He has attempted to model accretion disks in cataclysmic variables and around supermassive black holes.

Gerald Cecil, an associate professor of astronomy and physics at the University of North Carolina at Chapel Hill and director of the Morehead Observatory there, received his doctorate from the University of Hawaii.

Eric J. Chaisson is deeply involved with both the research and public-information aspects of the Hubble project. He is a senior staff scientist and director of educational programs at the Space Telescope Science Institute, located on the Johns Hopkins University campus. He is an adjunct professor of physics at Hopkins and an associate at the Harvard College Observatory. Chaisson earned a Ph.D. in astrophysics at Harvard in 1972. He joined the Space Telescope Science Institute in 1987. Chaisson has written extensively on relativity and cosmology. His interests include the thermodynamic evolution of material systems and public understanding of science and mathematics.

Lawrence Colin completed his Ph.D. in electrical engineering at Stanford University in 1964. Since then he has worked as an aerospace technologist at the Ames Research Center, where he has specialized in atmospheric and space physics.

James W. Cronin, a professor of physics at the University of Chicago since 1971, earned his master's degree from the university in 1953 and his doctorate in 1955. In 1980 he shared the Nobel Prize with Val L. Fitch for work on symmetry violations in the decay of mesons.

Mark Dickinson is an Alan C. David Postdoctoral Fellow at Johns Hopkins University and at the Space Telescope Science Institute in Baltimore, Maryland, where he previously held an Aura Fellowship.

Michael Disney is a professor of astronomy at the University of Wales in Cardiff, U.K. For 20 years he was a member of the European Space Agency's Space Telescope Faint Object Camera team. He received his Ph.D. from University College London in 1968. His other scientific interests include hidden galaxies, bird flight, and the environmental dangers posed by oil supertankers.

Gerald J. Fishman is the principal investigator for the Burst And Transient Source Experiment (BATSE) and a senior astrophysicist at the National Aeronautics and Space Administration Marshall Space Flight Center in Huntsville, Alabama. He has received the NASA Medal for Exceptional Scientific Achievement three times and in 1994 was awarded the Bruno Rossi Prize of the American Astronomical Society.

Wendy L. Freedman is a staff member at the Carnegie Institution's observatories in Pasadena, California. Born in Toronto, she received at Ph.D. in astronomy and astrophysics from the University of Toronto in 1984 and, in 1987, became the first woman to join Carnegie's scientific staff. In 1994 she received the Marc Aaronson Prize for her contributions to the study of extragalactic distance and stellar populations of galaxies. A co-leader of the Hubble Space Telescope Key Project to measure the Hubble constant, she is also a member of the National Research Council's Committee on Astronomy and Astrophysics, the executive board of the Center for Particle Astrophysics, and the National Aeronautics and Space Administration's scientific oversight committee planning the Next Generation Space Telescope.

Thomas K. Gaisser, a professor of physics at the University of Delaware, has concentrated on the interpretation of atmospheric cosmic-ray cascades; he earned his doctorate from Brown University in 1967. In 1995 he spent two months in Antarctica setting up cosmic-ray detectors.

Tom Gehrels was inspired to study celestial objects upon attending a class given by Jan Oort in the Netherlands, who surmised the existence of a distant shell of comets now called the Oort cloud. Gehrels is professor of planetary sciences at the University of Arizona at Tucson, a Sarabhai Professor at the Physical Research Laboratory in India, and principal investigator of the Spacewatch program at Kitt Peak, Arizona, where he hunts for comets and asteroids.

Everett K. Gibson, Jr., was part of the team, along with David S. McKay, Kathie Thomas-Keprta, and Christopher S. Romanek, to first report evidence of past biological activity within the ALH84001 meteorite. Gibson, a geochemist and meteorite specialist, is a senior scientist in the National Aeronautics and Space Administration Johnson Center's Earth Sciences and Solar System Exploration Division in Houston, Texas.

Alan H. Guth is the Victor Weisskopf Professor of Physics at the Massachusetts Institute of Technology. He is the originator of the inflationary-universe theory and the author of *The Inflationary Universe: The Quest for a New Theory of Cosmos Origins*.

Dieter H. Hartmann is a theoretical astrophysicist at Clemson University in South Carolina; he obtained his Ph.D. in 1989 from the University of California, Santa Cruz. Apart from gamma-ray astronomy, his primary interests are the chemical dynamics and evolution of galaxies and stars.

J. Patrick Henry is an astronomy professor at the University of Hawaii who enjoys sitting on his lanai and thinking about large-scale structure while watching the sailboats off Diamond Head. He met Ulrich G. Briel in the late 1970s while working at the Smithsonian Astrophysical Observatory on one of the instruments on the Einstein X-Ray Observatory satellite.

James E. Hesser's work has focused on the ages and compositions of globular star clusters, which are among the oldest constituents of the galaxy. He received his B.A. from the University of Kansas and his Ph.D. in atomic and molecular physics from Princeton. He works with Sidney van den Bergh at Dominion Astrophysical Observatory, National Research Council of Canada, in Victoria, British Columbia.

Craig J. Hogan studies the edge of the visible universe. He is chair of the astronomy department and professor in the physics and astronomy departments at the University of Washington. Hogan grew up in Los Angeles and received his A.B. from Harvard College in 1976 and his Ph.D. from the University of Cambridge in 1980. After postdoctoral fellowships at the University of Chicago and the California Institute of Technology, he joined the faculty of Steward Observatory at the University of Arizona for five years. He moved to Seattle in 1990.

Bruce M. Jakosky is a professor of geology and a member of the Laboratory for Atmospheric and Space Physics at the University of Colorado at Boulder. He is also an investigator on the Mars Global Surveyor mission currently in orbit around Mars. He is the author of *The Search for Life on Other Planets*.

David C. Jewitt developed his passion for astronomy as a youngster in England. As a professor at the Massachusetts Institute of Technology he began collaborative work with Jane X. Luu, who was a graduate student at MIT in 1986. In 1988 Jewitt became professor at the University of Hawaii. It was during this time that he and Luu, who was then doing her postdoctoral fellowship at the Harvard-Smithsonian Center for Astrophysics, discovered the first Kuiper belt object.

Ronald H. Kaitchuck earned a Ph.D. in astronomy in 1981 from Indiana University. He is a faculty member in the Department of Physics and Astronomy at Ball State University in Muncie, Indiana. Kaitchuck specializes in time-resolved spectroscopy of binary stars.

Jeffrey S. Kargel received his doctorate in planetary sciences from the University of Arizona in 1990. He remained at Arizona's Lunar and Planetary Science Laboratory doing postdoctoral research on the icy moons of the outer solar system and then joined the U.S. Geological Survey's astrogeology group in Flagstaff. He has worked with Robert G. Strom on various projects in planetary science for over a decade.

Richard G. Kron has served on the faculty of the Department of Astronomy and Astrophysics at the University of Chicago since 1978. He is also a member of the experimental astrophysics group at Fermi National Accelerator Laboratory. He enjoys observing near Lake Geneva, Wisconsin.

Kenneth R. Lang is professor of astronomy at Tufts University. His recent illustrated book *Sun, Earth and Sky,* describes all aspects of the sun and its interactions with the earth. Lang has also written more than 150 professional articles and four additional books, which have been translated into seven languages. Among them is the classic reference *Astrophysical Formulae.*

David H. Levy is a writer and lecturer and an avid amateur astronomer. He has written over 15 books, somewhat fewer than the number of comets he has discovered.

Andrei Linde is one of the originators of inflationary theory. After graduating from Moscow University, he received his Ph.D. at the P. N. Lebedev Physics Institute in Moscow, where he began probing the connections between particle physics and cosmology. He became a professor of physics at Stanford University in 1990. A detailed description of inflationary theory is given in his book *Particle Physics and Inflationary Cosmology.*

Shannon w. Lucid is an astronaut at the National Aeronautics and Space Administration Johnson Space Center in Houston, Texas. She has participated in five spaceflights, including her mission on Mir, logging a total of 223 days in orbit. She is currently the astronaut representative to the Shuttle-Mir program. She is still an active-duty astronaut and hopes to be assigned to another NASA spaceflight.

Janet G. Luhmann received her Ph.D. in astronomy from the University of Maryland in 1974. Since 1980 she has been a research scientist at the Institute of Geophysics and Planetary Physics at the University of California, Los Angeles. Much of her work has concerned the interaction between planets and the solar wind.

Jane X. Luu came to southern California as a child and as a refugee of Vietnam. She became enamored of astronomy almost by accident, during a summer spent at the Jet Propulsion Laboratory in Pasadena. She began her collaborative work with David C. Jewitt in 1986 as a graduate student at the Massachusetts Institute of Technology. It was during her postdoctoral fellowship at the Harvard-Smithsonian Center for Astrophysics that she and Jewitt discovered the first Kuiper belt object. In 1994 Luu jointed the faculty of Harvard University.

F. Duccio Macchetto is associate director for science programs for the Hubble Space Telescope (HST) project at the Space Telescope Science Institute in Baltimore, Maryland. As a member of the European Space Agency (ESA), he has served as the ESA Hubble project scientist and as the principal investigator for ESA's Faint Object Camera onboard the telescope.

David S. McKay was part of the team, along with Everett K. Gibson, Kathie Thomas-Keprta, and Christopher S. Romanek, to first report evidence of past biological activity within the ALH84001 meteorite. McKay, a geologist and expert on planetary regoliths, is a senior scientist in the National Aeronautics and Space Administration Johnson Center's Earth Sciences and Solar System Exploration Division in Houston, Texas.

P. James E. Peebles is a professor of physics at Princeton University, where in 1958 he began an illustrious career in gravitational physics. Most of his free time is spent with his three grandchildren.

Tsvi Piran has studied general relativity and astrophysics for over 20 years. He is a professor at the Hebrew University of Jerusalem, where he received his Ph.D. in 1976. He has also worked at the University of Oxford, Harvard University, Kyoto University, and Fermilab. Piran, with Steven Weinberg of the University of Texas, established the Jerusalem Winter School for Theoretical Physics.

James B. Pollack earned his Ph.D. in astronomy from Harvard University in 1965. He has been a research scientist at the National Aeronautics and Space Administration Ames Research Center since 1970. During those years he has participated in numerous NASA planetary missions, including the Galileo mission to Jupiter in 1994.

Christopher S. Romanek was part of the team, along with Everett K. Gibson, Kathie Thomas-Keprta, and David S. McKay, to first report evidence of past biological activity within the ALH84001 meteorite. Romanek, a former National Research Council postdoctoral fellow at the National Aeronautics and Space Administration Johnson Center, is now with the Department of Geology and the Savannah River Ecology Laboratory at the University of Georgia. His specialty is low-temperature geochemistry and stable-isotope mass spectrometry.

Vera Rubin is a staff member at the Department of Terrestrial Magnetism of the Carnegie Institution of Washington, where she has been since 1965. That same year she became the first woman permitted to observe at Palomar Observatory. The author of more than 200 papers on the structure of the Milky Way, motions within galaxies, and large-scale motions in the universe, she received Carnegie Mellon University's Dickson Prize for Science in 1994 and the Royal Astronomical Society's Gold Medal in 1996. President Bill Clinton awarded her the National Medal of Science in 1993 and appointed her to the President's Committee on the National Medal of Science in 1995.

Bradley E. Schaefer is assistant professor of Physics and Astronomy at Yale University.

David N. Schramm, who was Louis Block Distinguished Service Professor in the Physical Sciences and vice president for research at the University of Chicago, died in a tragic plane crash in 1998.

Carolyn S. Shoemaker is on the staff of the Lowell Observatory. She also works as a visiting scientist at the U.S. Geological Survey and is research professor at Northern Arizona University.

Eugene M. Shoemaker served as a geologist for the U.S. Geological Survey from 1948 to 1993, where he organized the Branch of Astrogeology. Until his death in a car accident in 1997, he was scientist emeritus with the U.S.G.S. and held a staff position at Lowell Observatory.

Paul J. Steinhardt is professor of physics at Princeton University. His research has spanned problems in particle physics, astrophysics, cosmology, and condensed matter physics, from theoretical work on the properties of elementary particles and superstrings related to cosmology to the study of astrophysical observations that test cosmological models. Particularly, he has focused on the inflationary model of the universe, which theorizes that the universe underwent faster-than-light expansion during the first instants after the big bang. His work has also included study of glassy solids and quasicrystals in condensed matter physics. He was elected to the National Academy of Sciences in 1998.

Robert G. Strom began his career working as a petroleum geologist, but he became involved in lunar exploration efforts during the 1960s and joined the faculty of the University of Arizona, where he continues to teach and conduct research. He has participated on National Aeronautics and Space Administration science teams assembled for the Apollo program, for the Mariner mission to Venus and Mercury, and for the Voyager missions to the outer solar system. He has worked together with Jeffrey S. Kargel on various projects in planetary science for over a decade.

Leonard Susskind is one of the early inventors of string theory. He holds a Ph.D. from Cornell University and has been a professor at Stanford University since 1978. He has made many contributions to elementary-particle physics, quantum field theory, cosmology, and, most recently, to the theory of black holes. His current studies in gravitation have led him to suggest that information can be compressed into one lower dimension, a concept he calls the holographic universe.

Simon P. Swordy, an associate professor at the University of Chicago, has been active in cosmic-ray measurement since 1976. He earned his Ph.D. from the University of Bristol in 1979.

Kathie Thomas-Keprta was part of the team, along with Everett K. Gibson, Christopher Romanek, and David S. McKay, to first report evidence of past biological activity within the ALH84001 meteorite. Thomas-Keprta, a senior scientist at Lockheed Martin, is a biologist who applies electron microscopy to the study of meteorites, interplanetary dust particles, and lunar samples.

Edwin L. Turner is chair of astrophysical sciences at Princeton and director of the 3.5-meter ARC telescope in New Mexico. He has a personal, cultural, and religious interest in Japan.

Sidney van den Bergh has a longtime interest in the classification and evolution of galaxies and in problems related to the age and size of the universe. He received his undergraduate degree from Princeton University and a doctorate in astronomy from the University of Gottingen. He now works with James E. Hesser at Dominion Astrophysical Observatory, National Research Council of Canada, in Victoria, British Columbia.

Sylvain Veilleux, now an assistant professor of astronomy at the University of Maryland, received his Ph.D. from the University of California, Santa Cruz.

Paul R. Weissman is a senior research scientist at the Jet Propulsion Laboratory in Pasadena, California, where he specializes in studies of the physics and dynamics of comets. He is also the project scientist for NASA's Deep Space 4/Champollion mission, which is scheduled to land on short-period comet Tempel 1 in 2005. Weissman has written more than 80 refereed scientific papers and is one of the three editors of the *Encyclopedia of the Solar System* published by Academic Press.

*This information was compiled at the time the articles were originally published in *Scientific American;* some biographies may not be completely up-to-date.

Index